T0341198

South Asian Mammals

South Asian Mammals

An Updated Checklist and Their Scientific Names

Chelmala Srinivasulu

CRC Press is an imprint of the
Taylor & Francis Group, an **informa** business

CRC Press
Taylor & Francis Group
6000 Broken Sound Parkway NW, Suite 300
Boca Raton, FL 33487-2742

Printed on acid-free paper

International Standard Book Number-13: 978-1-1386-0926-6 (Hardback)

Library of Congress Cataloging-in-Publication Data

Names: Srinivasulu, C., author.
Title: South Asian mammals : an updated checklist and their scientific names / Chelmala Srinivasulu
Description: Boca Raton : Taylor & Francis, 2018. | Includes bibliographical references and indexes
Identifiers: LCCN 2018028403| ISBN 9781138609266 (hardback : alk. paper) | ISBN 9780429466199 (ebook)
Subjects: LCSH: Mammals--South Asia. | Mammals--South Asia--Nomenclature
Classification: LCC QL729.S64 S66 2018 | DDC 599.0954--dc23LC record available at https://lccn.loc.gov/2018028403

Visit the Taylor & Francis Web site at
http://www.taylorandfrancis.com

and the CRC Press Web site at
http://www.crcpress.com

Dedication

Adi and Bhari
in mundi mundi

Contents

Preface

Ashes blow away in the wind, but names do not,
for I may be ashes someday, but my name shan't be forgotten.

A 'name' is perhaps the singular means by which the identity of a person, a living organism, or an item is established. The importance of names of the species in biology becomes even more important due to their sheer number. More than 1.9 million named species are known and almost four times as many are yet to be named.

Quod in nomine? What's in a name? This question has bothered each one of us who is interested in biology or understanding species diversity present on Earth, especially so you who are presently reading this book. With approximately 10 percent of the world's mammals present in South Asia, the scientific names used for identifying each of the 524 species is indeed a long list for those who are not aware of what these names mean. Embarking on this journey, one will travel through time and understand the meanings of the scientific names by which our mammals are known and appreciate the wisdom of authors who named the organisms for the first time.

As a student in high school biology class, I used to raptly listen to my biology teacher explaining scientific names. He used to explain roots, derivations, and meanings of some common plant and animal names occurring in the environs of a small town, Jamalpur, in Bihar, India. I used to eagerly wait for the weekly field excursions outside the school campus wherein Mr. Choudhury, my biology teacher, taught us about scientific names. Perhaps that's when the passion of understanding scientific names began in me and became stronger with the passage of time.

As a university teacher, the scientific names still fascinate me as much as they used to when I was a schoolboy. This book provides an updated checklist of mammals known to occur in South Asia, and explains the scientific names (including valid species and subspecies, and synonyms) of 524 species of mammals known to be present in South Asia. I hope that readers will enjoy the book as much as I enjoyed writing it.

Chelmala Srinivasulu
Osmania University

Acknowledgements

This work is based on research contributions by numerous scientists who have described the mammals in South Asia and adjoining areas from 1758 to March 2018. Scouring through numerous types of descriptions, we find that many more individuals also contributed to mammalogy of the region by actively participating in mammal collections, preserving them and donating them to museums. I thank all these wonderful souls for their unflinching efforts in documenting our wonderful mammals.

Special thanks to my friends who have helped in enriching my knowledge through engaging and stimulating discussions over the past two decades, and to my teachers and my colleagues at Osmania University for believing in me and encouraging me. The librarians of the various libraries both in India and elsewhere, the Biodiversity Heritage Library, and Archive.org made available numerous resources wherein the type descriptions appeared from time to time.

I love my family for being there for me and steadfastly trusting me on this project. This journey of five years would not have been completed without them cheering me.

About the Book

This book explains the scientific names (all given and available names) of 524 species of mammals occurring, or known to occur, in South Asia from the countries of Afghanistan, Pakistan, India, Nepal, Bhutan, Bangladesh, Sri Lanka and Maldives. The species listing is following Srinivasulu and Srinivasulu (2012) with inclusion of species described after December 2011. The checklist provided in this book is based on the best of my knowledge on mammalian diversity as on 31 March 2018. Any omissions are purely unintentional.

This book consists of 4 chapters – (1) Chapter 1 deals with the updated checklist of mammals of South Asia, (2) Chapter 2 deals with the scientific nomenclature and its understanding, (3) Chapter 3 provides the explanation of generic and specific names along with the synonyms, and (4) Chapter 4 is a dictionary of names in alphabetical order for easy tracking by the reader. Additionally, the list of languages used in autochtony and vernacular names are provided as an appendix. A detailed list of 545 published references dealing with published species descriptions is also provided as a bibliography.

Each binomial name of mammal species dealt with in Chapter 3 of this book is followed by the name of the authority and year of description, accepted English name, and explanation of generic name, followed by explanation of the specific name, general distribution, explanation of subspecific names, other English name/s for nominate form and/or synonyms, and some common vernacular names. Chapter 4 also includes, besides the Latin generic and specific names dealt with in Chapter 3, explanations of synonyms listed under recognized species.

Author

Chelmala Srinivasulu PhD, FLS, FZS, is Assistant Professor of Zoology at Osmania University, Hyderabad, India, with a research focus on biodiversity conservation, systematics, and faunal taxonomy in South Asia. He is passionate about all things small, such as volant and non-volant small mammals, reptiles, and spiders, groups in which he has discovered and named new taxa.

List of Abbreviations and Symbols

>	derived from or from
adj.	adjective
Arab.	Arabic
Assam.	Assamese or Asamiya
b.	born
Balt.	Balti
Beng.	Bengali or Bangla
Bhut.	Bhutia
Bihar.	Bihari
Bundel.	Bundeli
cf.	compare (Latin *confer*)
comp.	comparision
d.	died
Dakh.	Dakhini (also Dakhani)
Dima.	Dimasa
Fr.	French
Ger.	German
Gond.	Gondi
Gr.	Greek
Guj.	Gujarati
Hind.	Hindi
Ital.	Italian
J. Syn.	junior synonym
Jap.	Japanese
Kan.	Kannada or Kanarese
Kash.	Kashmiri
Kathi.	Kathiawari dialect of Gujarati
Khám.	Khámti
Khan.	Khandeshi
Khas.	Khasi
Kin.	Kinnauri
Kiran.	Kiranti
Kon.	Konda
Kum.	Kumaoni
Kurg.	Kurgi
Kut.	Kutchi
L.	Latin
Lad.	Ladakhi
Lah.	Lahauli

Lepch.	Lepcha
Mal.	Malayalam
Mani.	Manipuri
Mar.	Marathi
Marw.	Marwari
Med. L.	Medieval Latin
Mis.	Mishmi
Mod. L.	Modern Latin
Mon.	Monpa
Mongol.	Mongolian
Myth.	Mythology
Nep.	Nepali
Nic.	Nicobarese
nom.	nominate
Odi.	Odiya
OEN	other English name/s
Old Fr.	Old French
Old L.	Old Latin
Pahari W.	Western Pahari
Pard.	Pardhi
Pash.	Pashto
Pers.	Persian
Punj.	Punjabi
Russ.	Russian
S. Syn.	senior synonym
Sans.	Sanskrit
Sant.	Santali (also Santhali)
Sind.	Sindhi
Sinh.	Sinhala
Subsp.	Subspecies
super.	Superlative
Sylh.	Sylheti
Tam.	Tamil
Tam. (SL)	Sri Lankan Tamil
Tel.	Telugu
Tib.	Tibeti
v.	verb
VN	vernacular name/s
Wad.	Waddari
Yan.	Yan

1 South Asian Mammals
An Updated Checklist

The mammalian diversity of South Asia (from Afghanistan, Bangladesh, Bhutan, India, Maldives, Nepal, Pakistan, and Sri Lanka [Figure 1.1]) is represented by 524 species belonging to 215 genera in 14 orders, which represent approximately 9.4 percent of the world's mammalian diversity. The present checklist of mammals of South Asia is an updated list of currently known species (cutoff date: March 31, 2018). Each species entry includes the scientific name, authority, common name, and details of subspecies recognized in South Asia with country-level distribution. Recent taxonomic updates and nomenclatural changes have been incorporated.

The first mammals (33 species) of the region were named by Carolus Linnaeus in the 10th edition of the *Systema Naturae*. Between 1758 and 1799, as many as 91 mammal names were published. In the nineteenth century, 307 species were described; in the twentieth century, 111 species; and in the twenty-first century, 14 species (Figure 1.2). An earlier work on South Asian mammals, published in 2012, listed 506 species (Srinivasulu and Srinivasulu, 2012).

While providing an updated checklist (with details on subspecies, names of countries where they occur, and endemicity), this work also aims to create interest in and explain the scientific names of mammals in South Asia.

CHECKLIST OF MAMMALS OF SOUTH ASIA

Order Proboscidea Illiger, 1811
Family Elephantidae Gray, 1821

1. *Elephas maximus* Linnaeus, 1758 Asian Elephant
   Subspecies *maximus* (Endemic to Sri Lanka), *indicus* (Bangladesh, Bhutan, India, and Nepal)

Order Sirenia Illiger, 1811
Family Dugongidae Gray, 1821

2. *Dugong dugon* (Müller, 1776) Dugong (Arabian Sea, Bay of Bengal and Indian Ocean: India, Maldives, Pakistan, and Sri Lanka)

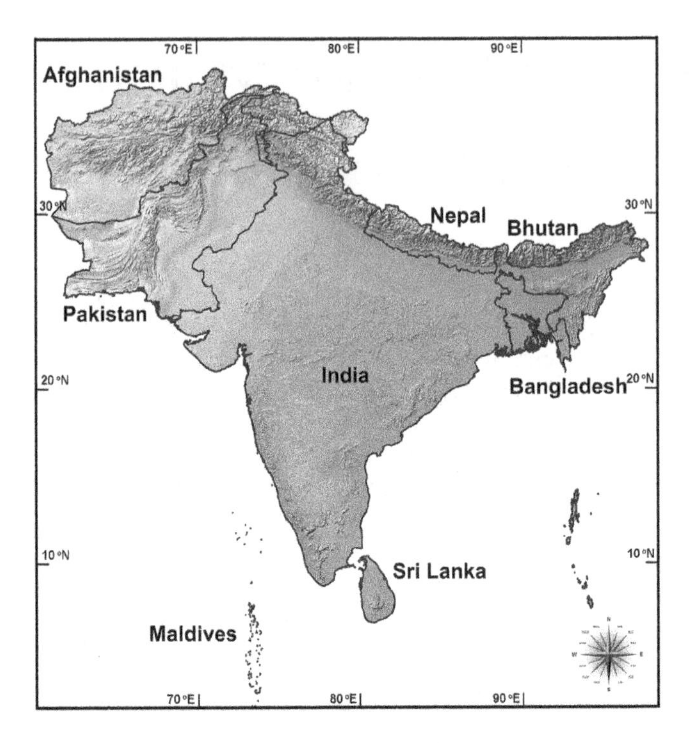

FIGURE 1.1 Map of South Asia depicting major physiographic features and political boundaries of countries.

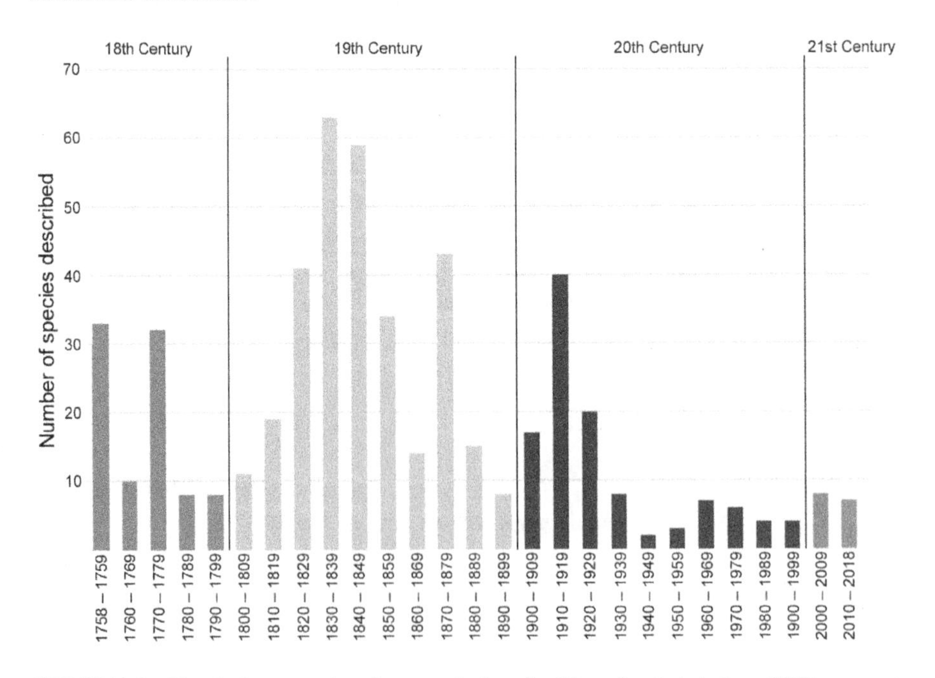

FIGURE 1.2 Trend of new species of mammals described from South Asia from 1758 to present.

Order Scandentia Wagner, 1855
Family Tupaiidae Gray, 1825

3. *Anathana ellioti* (Waterhouse, 1850) Madras Treeshrew (Endemic to India)
4. *Tupaia belangeri* (Wagner, 1841) Northern Treeshrew (Bangladesh, Bhutan, India, and Nepal)
5. *Tupaia nicobarica* (Zelebor, 1869) Nicobar Treeshrew
   Subspecies *nicobarica* (Endemic India), *surda* (Endemic India)

Order Primates Linnaeus, 1758
Family Lorisidae Gray, 1821

6. *Loris tardigradus* (Linnaeus, 1758) Red Slender Loris
   Subspecies *tardigradus* (Endemic to Sri Lanka), *nycticeboides* (Endemic to Sri Lanka)
7. *Loris lydekkerianus* Cabrera, 1908 Grey Slender Loris
   Subspecies *lydekkerianus* (Endemic to India), *malabaricus* (Endemic to India), *grandis* (Endemic to Sri Lanka), *nordicus* (Endemic to Sri Lanka)
8. *Nycticebus bengalensis* (Lacépède, 1800) Bengal Slow Loris (Bangladesh, Bhutan, India)

Family Cercopithecidae Gray, 1821
Subfamily Cercopithecinae Gray, 1821

9. *Macaca silenus* (Linnaeus, 1758) Lion-tailed Macaque (Endemic to India)
10. *Macaca sinica* (Linnaeus, 1771) Toque Macaque
    Subspecies *sinica* (Endemic to Sri Lanka), *aurifrons* (Endemic to Sri Lanka), *opisthomelas* (Endemic to Sri Lanka)
11. *Macaca mulatta* (Zimmermann, 1780) Rhesus Macaque (Afghanistan, Bangladesh, Bhutan, India, Nepal, and Pakistan)
12. *Macaca radiata* (E. Geoffroy, 1812) Bonnet Macaque
    Subspecies *radiata* (Endemic to India), *diluta* (Endemic to India)
13. *Macaca fascicularis* (Raffles, 1821) Crab-eating Macaque
    Subspecies *aureus* (Bangladesh), *umbrosus* (Endemic to India)
14. *Macaca arctoides* (I. Geoffroy, 1831) Stump-tailed Macaque (Bangladesh and India)
15. *Macaca assamensis* (McClelland, 1839) Assam Macaque
    Subspecies *assamenis* (Bhutan and India), *pelops* (Bangladesh, Bhutan, India, and Nepal)
16. *Macaca leonina* (Blyth, 1863) Northern Pig-tailed Macaque (Bangladesh and India)
17. *Macaca munzala* Sinha et al., 2005 Arunachal Macaque (Bhutan and India)
18. *Macaca leucogenys* Li et al., 2015 White-cheeked Macaque (India)

Subfamily Colobinae Jerdon, 1867

19. *Semnopithecus vetulus* (Erxleben, 1777) Purple-faced Langur
    Subspecies *vetulus* (Endemic to Sri Lanka), *nestor* (Endemic to Sri Lanka), *monticola* (Endemic to Sri Lanka), *philbricki* (Endemic to Sri Lanka)
20. *Semnopithecus entellus* (Dufresne, 1797) Northern Plains Gray Langur (Endemic to Bangladesh and India)
21. *Semnopithecus johnii* (Fischer, 1829) Nilgiri Langur (Endemic to India)
22. *Semnopithecus schistaceus* Hodgson, 1840 Central Himalayan Langur (Bhutan, India, Nepal, and Pakistan)
23. *Semnopithecus hypoleucos* Blyth, 1841 Dark-legged Malabar Langur (Endemic to India)
24. *Semnopithecus dussumieri* I. Geoffroy, 1843 Southern Plains Gray Langur (Endemic to India)
25. *Semnopithecus priam* Blyth, 1844 Coromandel Gray Langur
    Subspecies *priam* (Endemic to India), *thersites* (Endemic to India and Sri Lanka)
26. *Semnopithecus ajax* (Pocock, 1928) Himalayan Gray Langur (Endemic to India)
27. *Semnopithecus hector* (Pocock, 1928) Lesser Hill Langur (Endemic to Bhutan, India, and Nepal)
28. *Trachypithecus pileatus* (Blyth, 1843) Capped Langur
    Subspecies *pileatus* (India), *brahma* (Endemic to India), *durga* (Bangladesh and India), *tenebricus* (Endemic to Bhutan and India)
29. *Trachypithecus phayrei* (Blyth, 1847) Phayre's Leaf Monkey (Bangladesh and India)
30. *Trachypithecus geei* (Khajuria, 1956) Gee's Golden Langur (Endemic to Bhutan and India)

Family Hylobatidae Gray, 1871

31. *Hoolock hoolock* (Harlan, 1834) Hoolock Gibbon
    Subspecies *hoolock* (Bangladesh and India), *mishmiensis* (Endemic to India)
32. *Hoolock leuconedys* Groves, 1967 Eastern Hoolock Gibbon (India)

Order Rodentia
Suborder Sciuromorpha Brandt, 1855
Family Sciuridae Hemprich, 1820
Subfamily Ratufinae Moore, 1959

33. *Ratufa macroura* (Pennant, 1769) Grizzled Giant Squirrel
    Subspecies *macroura* (Endemic to Sri Lanka), *melanochra* (Endemic to Sri Lanka), *dandolena* (Endemic to India and Sri Lanka)
34. *Ratufa indica* (Erxleben, 1777) Indian Giant Squirrel (Endemic to India)
    Subspecies *indica*, *maxima*, *centralis*
35. *Ratufa bicolor* (Sparrman, 1778) Black Giant Squirrel
    Subspecies *gigantea* (Bangladesh, Bhutan, India, and Nepal)

Subfamily Sciurinae Fischer de Waldheim, 1817

36. *Belomys pearsonii* (Gray, 1842) Hairy-footed Flying Squirrel (Bhutan, India, and Nepal)
37. *Biswamoyopterus biswasi* Saha, 1981 Namdapha Flying Squirrel (Endemic to India)
38. *Eoglaucomys fimbriatus* (Gray, 1837) Small Kashmir Flying Squirrel
    Subspecies *fimbriatus* (India and Pakistan), *baberi* (Afghanistan, India, and Pakistan)
39. *Eupetaurus cinereus* Thomas, 1888 Woolly Flying Squirrel (Bhutan, India, and Pakistan)
40. *Hylopetes alboniger* (Hodgson, 1836) Parti-coloured Flying Squirrel (Bhutan, India, and Nepal)
41. *Petaurista petaurista* (Pallas, 1766) Red Giant Flying Squirrel
    Subspecies *albiventer* (Afghanistan, Bangladesh, Bhutan, India, Nepal, and Pakistan)
42. *Petaurista magnificus* (Hodgson, 1836) Himalayan Giant Flying Squirrel
    Subspecies *magnificus* (Bhutan, India, and Nepal), *hodgsoni* (India and Nepal)
43. *Petaurista philippensis* (Elliot, 1839) South Indian Giant Flying Squirrel (India and Sri Lanka)
44. *Petaurista elegans* (Müller, 1840) Spotted Giant Flying Squirrel
    Subspecies *caniceps* (Bhutan, India, and Nepal)
45. *Petaurista nobilis* (Gray, 1842) Noble Giant Flying Squirrel
    Subspecies *nobilis* (Endemic to India and Nepal), *singhei* (Endemic to Bhutan and India)
46. *Petaurista mechukaensis* Choudhury, 2007 Mechuka Giant Flying Squirrel (Endemic to India)
47. *Petaurista mishmiensis* Choudhury, 2009 Mishmi Hills Giant Flying Squirrel (Endemic to India)
48. *Petaurista siangensis* Choudhury, 2013 Mebo Giant Flying Squirrel (Endemic to India)
49. *Petinomys fuscocapillus* (Jerdon, 1847) Travancore Flying Squirrel (Endemic to India and Sri Lanka)

Subfamily Callosciurinae Pocock, 1923

50. *Callosciurus erythraeus* (Pallas, 1799) Pallas's Squirrel
    Subspecies *erythraeus* (Bangladesh, Bhutan, and India), *erythrogaster* (India), *intermedia* (India)
51. *Callosciurus pygerythrus* (I. Geoffroy Saint-Hilaire, 1832) Irrawady Squirrel
    Subspecies *lokroides* (Bhutan, India, and Nepal), *blythii* (Bangladesh and India), *stevensi* (Assam)
52. *Dremomys lokriah* (Hodgson, 1836) Orange-bellied Himalayan Squirrel
    Subspecies *lokriah* (Bangladesh, Bhutan, India, and Nepal), *macmillani* (India), *garonum* (Endemic to India)

53. *Dremomys pernyi* (Milne-Edwards, 1867) Pernyi's Long-nosed Squirrel
    Subspecies *pernyi* (India), *howelli* (India)
54. *Dremomys rufigenis* (Blanford, 1878) Red-cheeked Squirrel (India)
55. *Funambulus palmarum* (Linnaeus, 1766) Three-striped Palm Squirrel (Endemic to South Asia)
    Subspecies *palmarum* (India), *brodiei* (Sri Lanka), *robertsoni* (India), *bellaricus* (India), *matugamaensis* (Sri Lanka)
56. *Funambulus tristriatus* (Waterhouse, 1837) Jungle Striped Squirrel (Endemic to India)
    Subspecies *tristriatus*, *wroughtoni*, *numarius*
57. *Funambulus sublineatus* (Waterhouse, 1838) Dusky-striped Squirrel
    Subspecies *sublineatus* (Endemic to India), *obscurus* (Endemic to Sri Lanka)
58. *Funambulus layardi* (Blyth, 1849) Layard's Striped Squirrel (Endemic to Sri Lanka)
59. *Funambulus pennantii* Wroughton, 1905 Northern Palm Squirrel
    Subspecies *pennantii* (Endemic to India), *lutescens* (Endemic to India and Pakistan), *argentescens* (India and Pakistan), *chhattisgarhi* (Endemic to Bangladesh and India), *gangutrianus* (Endemic to Bhutan, India, and Nepal)
60. *Tamiops macclellandi* (Horsfield, 1839) Himalayan Striped Squirrel (Bhutan, India, and Nepal)

Subfamily Xerinae Osborn, 1910

61. *Spermophilopsis leptodactylus* (Lichtenstein, 1823) Long-clawed Ground Squirrel
    Subspecies *bactrianus* (Afghanistan)
62. *Marmota himalayana* (Hodgson, 1841) Himalayan Marmot (India and Nepal)
63. *Marmota caudata* (Geoffroy, 1844) Long-tailed Marmot
    Subspecies *caudata* (Afghanistan and India), *aurea* (Pakistan)
64. *Spermophilus fulvus* (Lichtenstein, 1823) Yellow Ground Squirrel (Afghanistan)

Family Gliridae Muirhead, 1819
Subfamily Leithiinae Lydekker, 1896

65. *Dryomys nitedula* (Pallas, 1778) Forest Dormouse
    Subspecies *pictus* (Afghanistan and Pakistan)
66. *Dryomys niethammeri* Holden, 1996 Niethhammer's Forest Dormouse (Endemic to Pakistan)

Family Dipodidae Fischer, 1817
Subfamily Allactaginae Vinogradov, 1925

67. *Allactaga elater* (Lichtenstein, 1828) Small Five-toed Jerboa (Afghanistan and Pakistan)
68. *Allactaga williamsi* Thomas, 1897 William's Jerboa (Afghanistan)

69. *Allactaga hotsoni* Thomas, 1920 Hotson's Five-toed Jerboa (Afghanistan and Pakistan)

Subfamily Cardiocraniinae Vinogradov, 1925

70. *Salpingotulus michaelis* (Fitzgibbon, 1966) Balochistan Pygmy Jerboa (Endemic to Pakistan)

Subfamily Dipodinae G. Fischer, 1817

71. *Jaculus blanfordi* (Murray, 1884) Blanford's Jerboa (Afghanistan and Pakistan)

Subfamily Sicistinae Allen, 1901

72. *Sicista concolor* (Büchner, 1892) Chinese Birch Mouse (India and Pakistan)

Superfamily Muroidea Illiger, 1811
Family Platacanthomyidae Alston, 1876

73. *Platacanthomys lasiurus* Blyth, 1859 Malabar Spiny Dormouse (Endemic to India)

Family Spalacidae Gray, 1821
Subfamily Rhizomyinae Winge, 1887

74. *Cannomys badius* (Hodgson, 1841) Bay Bamboo Rat (Bangladesh, India, and Nepal)
75. *Rhizomys pruinosus* Blyth, 1851 Hoary Bamboo Rat (India)

Family Calomyscidae Vorontsov and Potapova, 1979

76. *Calomyscus baluchi* Thomas, 1920 Baluchi Mouse-like Hamster (Afghanistan and Pakistan)
77. *Calomyscus hotsoni* Thomas, 1920 Hotson's Mouse-like Hamster (Pakistan)
78. *Calomyscus elburzensis* Goodwin, 1938 Goodwin's Mouse-like Hamster (Afghanistan)

Family Cricetidae Fischer, 1817
Subfamily Cricetinae Fischer, 1817

79. *Cricetulus migratorius* (Pallas, 1773) Little Grey Hamster
    Subspecies *cinerascens* (Afghanistan and Pakistan), *fulvus* (Afghanistan and India)
80. *Cricetulus alticola* Thomas, 1917 Ladakh Hamster (India and Nepal)

Subfamily Arvicolinae

81. *Alticola roylei* (Gray, 1842) Royle's Vole (Endemic to India)
82. *Alticola stoliczkanus* (Blanford, 1875) Stoliczka's Vole (India and Nepal)
83. *Alticola argentatus* (Severtzov, 1879) Silver Mountain Vole (Afghanistan, India, and Pakistan)
84. *Alticola albicauda* (True, 1894) White-tailed Mountain Vole (Endemic to India)
85. *Alticola montosa* (True, 1894) Kashmir Mountain Vole (India and Pakistan)
86. *Blanfordimys afghanus* (Thomas, 1912) Afghan Vole (Afghanistan)
87. *Blanfordimys bucharensis* (Vinogradov, 1930) Bucharian Vole (Afghanistan)
88. *Ellobius talpinus* (Pallas, 1770) Northern Mole Vole (Afghanistan and Pakistan)
89. *Ellobius fuscocapillus* (Blyth, 1842) Afghan Mole Vole (Afghanistan and Pakistan)
90. *Eothenomys melanogaster* (Milne-Edwards, 1871) Pere David's Vole (India)
91. *Hyperacrius wynnei* (Blanford, 1881) Murree Vole (India and Pakistan)
92. *Hyperacrius fertilis* (True, 1894) Subalpine Kashmir Vole (Endemic to India and Pakistan)
93. *Microtus ilaeus* Thomas, 1912 Kazkhstan Vole (Afghanistan)
94. *Neodon sikimensis* (Hodgson, 1849) Sikkim Vole (Bhutan, India, and Nepal)
95. *Neodon juldaschi* (Severtzov, 1879) Juniper Vole (Afghanistan, India, and Pakistan)
96. *Phaiomys leucurus* Blyth, 1863 Blyth's Vole (India and Nepal)

Family Muridae Illiger, 1811
Subfamily Deomyinae Thomas, 1888

97. *Acomys dimidiatus* (Cretzschmar, 1826) Arabian Spiny Mouse (Pakistan)

Subfamily Gerbillinae Gray, 1825

98. *Gerbillus nanus* Blanford, 1875 Balochistan Gerbil (Afghanistan, India, and Pakistan)
99. *Gerbillus gleadowi* Murray, 1886 Little Hairy-footed Gerbil (India and Pakistan)
100. *Gerbillus aquilus* Schlitter and Stezer, 1972 Swarthy Gerbil (Afghanistan and Pakistan)
101. *Meriones meridianus* (Pallas, 1773) Mid-day Jird (Afghanistan)
102. *Meriones libycus* Lichtenstein, 1823 Libyan Jird (Afghanistan and Pakistan)
103. *Meriones crassus* Sundevall, 1842 Sundevall's Jird (Afghanistan and Pakistan)

104. *Meriones hurrianae* (Jerdon, 1867) Indian Desert Gerbil (India and Pakistan)
105. *Meriones persicus* (Blanford, 1875) Persian Jird
    Subspecies *persicus* (Afghanistan and Pakistan), *baptistae* (Pakistan)
106. *Meriones zarudnyi* Heptner, 1937 Zarudny's Jird (Afghanistan)
107. *Rhombomys opimus* (Lichtenstein, 1823) Great Gerbil (Afghanistan and Pakistan)
108. *Tatera indica* (Hardwicke, 1807) Indian Gerbil
    Subspecies *indica* (Afghanistan, Bangladesh, India, Nepal, and Pakistan), *cuvieri* (Endemic to India and Sri Lanka)

Subfamily Murinae Illiger, 1811

109. *Apodemus draco* (Barrett-Hamilton, 1900) South China Field Mouse (India)
110. *Apodemus pallipes* Barrett-Hamilton, 1900 Himalayan Field Mouse (Afghanistan, India, Nepal, and Pakistan)
111. *Apodemus latronum* Thomas, 1911 Sichuan Field Mouse (India)
112. *Apodemus rusiges* Miller, 1913 Kashmir Field Mouse (Endemic to India)
113. *Apodemus gurkha* Thomas, 1924 Himalayan Wood Mouse (Endemic to Nepal)
114. *Bandicota indica* (Bechstein, 1800) Greater Bandicoot Rat
    Subspecies *indica* (India and Sri Lanka), *nemorivaga* (Bangladesh, India, and Nepal), *malabarica* (Endemic to India)
115. *Bandicota bengalensis* (Gray and Hardwicke, 1833) Lesser Bandicoot Rat
    Subspecies *bengalensis* (Bangladesh, Bhutan, India, Nepal, Pakistan, and Sri Lanka), *wardi* (India)
116. *Berylmys bowersi* (Anderson, 1879) Bower's White-toothed Rat (India)
117. *Berylmys mackenziei* (Thomas, 1916) Mackenzie's White-toothed Rat (India)
118. *Berylmys manipulus* (Thomas, 1916) Manipur White-toothed Rat (India)
119. *Chiropodomys gliroides* (Blyth, 1856) Pencil-tailed Tree Mouse (India)
120. *Cremnomys cutchicus* Wroughton, 1912 Cutch Rock-rat (Endemic to India)
121. *Cremnomys elvira* (Ellerman, 1947) Large Rock-rat (Endemic to India)
122. *Dacnomys millardi* Thomas, 1916 Millard's Rat (India and Nepal)
123. *Diomys crumpi* Thomas, 1917 Crump's Mouse (India and Nepal)
124. *Golunda ellioti* Gray, 1837 Indian Bush-rat
    Subspecies *ellioti* (Bangladesh, Bhutan, India, Nepal, and Pakistan), *newara* (Endemic to Sri Lanka)
125. *Hadromys humei* (Thomas, 1886) Hume's Rat (Endemic to India)
126. *Leopoldamys edwardsi* (Thomas, 1882) Edward's Long-tailed Giant Rat (India)
127. *Leopoldamys sabanus* (Thomas, 1887) Long-tailed Giant Rat (Bangladesh)
128. *Madromys blanfordi* (Thomas, 1881) White-tailed Wood Rat (Endemic to Bangladesh, India, and Sri Lanka)

129. *Micromys minutus* (Pallas, 1771) Eurasian Harvest Mouse
    Subspecies *erythrotis* (India)
130. *Millardia meltada* (Gray, 1837) Soft-furred Metad (Endemic to Bangladesh, India, Nepal, Pakistan, and Sri Lanka)
131. *Millardia gleadowi* (Murray, 1885) Sand-coloured Metad (Endemic to India and Pakistan)
132. *Millardia kondana* Mishra and Dhanda, 1975 Kondana Rat (Endemic to India)
133. *Mus musculus* Linnaeus, 1758 House Mouse
    Subspecies *bactrianus* (Afghanistan, Pakistan, and India), *castaneus* (Afghanistan, Bangladesh, Bhutan, India, Nepal, Pakistan, and Sri Lanka), *domesticus* (India, Pakistan, and Nepal), *musculus* (India and Pakistan)
134. *Mus platythrix* Bennett, 1832 Brown Spiny Mouse (Endemic to India)
135. *Mus booduga* (Gray, 1837) Common Indian Field Mouse (Bangladesh, India, Nepal, Pakistan, and Sri Lanka)
136. *Mus saxicola* Elliot, 1839 Brown Spiny Mouse
    Subspecies *saxicola* (Endemic to India), *sadhu* (Endemic to India and Pakistan), *gurkha* (Endemic to India and Nepal)
137. *Mus cervicolor* Hodgson, 1845 Fawn-coloured Mouse (Bhutan, India, Nepal, and Pakistan)
138. *Mus terricolor* Blyth, 1851 Earth-coloured Mouse (Bangladesh, India, Nepal, and Pakistan)
139. *Mus famulus* Bonhote, 1898 Bonhote's Mouse (Endemic to India)
140. *Mus phillipsi* Wroughton, 1912 Wroughton's Small Spiny Mouse (Endemic to India and Nepal)
141. *Mus cookii* Ryley, 1914 Ryley's Spiny Mouse (Bangladesh, Bhutan, India, and Nepal)
142. *Mus mayori* (Thomas, 1915) Mayor's Mouse (Endemic to Sri Lanka)
143. *Mus pahari* Thomas, 1916 Sikkim Mouse (Bhutan and India)
144. *Mus fernandoni* (Phillips, 1932) Ceylon Spiny Mouse (Endemic to Sri Lanka)
145. *Nesokia indica* (Gray and Hardwicke, 1832) Short-tailed Bandicoot-rat (Afghanistan, Bangladesh, India, Nepal, and Pakistan)
146. *Niviventer niviventer* (Hodgson, 1836) Himalayan Niviventer (Bhutan, India, and Nepal)
147. *Niviventer fulvescens* (Gray, 1847) Indomalayan Niviventer (India, Pakistan, and Nepal)
148. *Niviventer brahma* (Thomas, 1914) Brahma Niviventer (India)
149. *Niviventer eha* (Wroughton, 1916) Smoke-bellied Niviventer (India and Nepal)
150. *Niviventer langbianis* (Robinson and Kloss, 1922) Lang Bian Niviventer (India)
151. *Rattus rattus* (Linnaeus, 1758) House Rat (Afghanistan, Bangladesh, Bhutan, India, Nepal, Pakistan, and Sri Lanka)
152. *Rattus norvegicus* (Berkenhout, 1769) Brown Rat (India, Pakistan, and Sri Lanka)

153. *Rattus tanezumi* (Temminck, 1844) Oriental House Rat (Afghanistan, Bangladesh, Bhutan, India, Nepal, and Pakistan)
154. *Rattus nitidus* (Hodgson, 1845) Himalayan Rat (Bhutan, India, and Nepal)
155. *Rattus pyctoris* (Hodgson, 1845) Turkestan Rat (Afghanistan, Bangladesh, Bhutan, India, Nepal, and Pakistan)
156. *Rattus exulans* (Peale, 1848) Polynesian Rat (Bangladesh)
157. *Rattus andamanensis* (Blyth, 1860) Indochinese Forest Rat (Bhutan, India, and Nepal)
158. *Rattus palmarum* (Zelebor, 1869) Zelebor's Nicobar Rat (Endemic to India)
159. *Rattus burrus* (Miller, 1902) Miller's Nicobar Rat (Endemic to India)
160. *Rattus stoicus* (Miller, 1902) Andaman Rat (Endemic to India)
161. *Rattus satarae* Hinton, 1918 Sahyadris Forest Rat (Endemic to India)
162. *Rattus montanus* Phillips, 1932 Sri Lankan Mountain Rat (Endemic to Sri Lanka)
163. *Rattus ranjiniae* Agrawal and Ghosh, 1969 Ranjini's Field Rat (Endemic to India)
164. *Srilankamys ohiensis* (Phillips, 1929) Ohiya Rat (Endemic to Sri Lanka)
165. *Vandeleuria oleracea* (Bennett, 1832) Asiatic Long-tailed Climbing Mouse (Bangladesh, Bhutan, India, Nepal, and Sri Lanka)
166. *Vandeleuria nilagirica* (Jerdon, 1867) Nilgiri Long-tailed Tree Mouse (Endemic to India)
167. *Vandeleuria nolthenii* Phillips, 1929 Sri Lankan Highland Tree Mouse (Endemic to Sri Lanka)

Infraorder Hystricognathi Brandt, 1855
Family Hystricidae G. Fischer, 1817

168. *Atherurus macrourus* (Linnaeus, 1758) Asiatic Brush-tailed Porcupine (India)
169. *Hystrix brachyura* Linnaeus, 1758 Malayan Porcupine
    Subspecies *bengalensis* (India), *hodgsoni* (India and Nepal), *subcristata* (Bangladesh and India)
170. *Hystrix indica* Kerr, 1792 Indian Crested Porcupine (Afghanistan, India, Pakistan, Nepal, and Sri Lanka)

Order Lagomorpha Brandt, 1855
Family Ochotonidae Thomas, 1897

171. *Ochotona roylei* (Ogilby, 1839) Royle's Pika
    Subspecies *roylei* (India and Pakistan), *nepalensis* (Nepal)
172. *Ochotona rufescens* (Gray, 1842) Afghan Pika (Afghanistan and Pakistan)
173. *Ochotona curzoniae* (Hodgson, 1858) Black-lipped Pika (India)
174. *Ochotona thibetana* (Milne-Edwards, 1871) Moupin Pika
    Subspecies *sikimaria* (Bhutan, India, and Nepal)
175. *Ochotona ladacensis* (Günther, 1875) Ladakh Pika (India and Pakistan)

176. *Ochotona macrotis* (Günther, 1875) Large-eared Pika
Subspecies *auritus* (India), *macrotis* (Afghanistan, Bhutan, India, Nepal, and Pakistan), *wollastoni* (Nepal)
177. *Ochotona nubrica* Thomas, 1922 Nubra Pika
Subspecies *nubrica* (India), *lhasaensis* (Nepal)
178. *Ochotona forresti* Thomas, 1923 Forrest's Pika (Bhutan, India, and Nepal)
179. *Ochotona himalayana* Feng, 1973 Himalayan Pika (Nepal)

Family Leporidae Fischer, 1817

180. *Caprolagus hispidus* (Pearson, 1839) Hispid Hare (Endemic to India and Nepal)
181. *Lepus tolai* Pallas, 1778 Tolai Hare (Afghanistan)
182. *Lepus nigricollis* Cuvier, 1823 Black-napped Hare
Subspecies *nigricollis* (Endemic to India), *ruficaudatus* (Endemic to Bangladesh, Bhutan, India, Nepal, and Pakistan), *aryabertensis* (Endemic to Nepal), *dayanus* (Endemic to Afghanistan, India, and Pakistan), *simcoxi* (Endemic to India), *singhala* (Endemic to Sri Lanka), *sadiya* (Endemic to India)
183. *Lepus oiostolus* Hodgson, 1840 Woolly Hare
Subspecies *pallipes* (India and Nepal), *hypsibius* (India)
184. *Lepus tibetanus* Waterhouse, 1841 Desert Hare
Subspecies *tibetanus* (Afghanistan, India, and Pakistan), *craspedotis* (Endemic to Pakistan)

Order Erinaceomorpha Gregory, 1910
Family Erinaceidae Fischer, 1817

185. *Hemiechinus auritus* (Gmelin, 1770) Long-eared Hedgehog
Subspecies *megalotis* (Afghanistan and Pakistan)
186. *Hemiechinus collaris* (Gray, 1830) Collared Hedgehog (Endemic to India and Pakistan)
187. *Paraechinus hypomelas* (Brandt, 1836) Brandt's Hedgehog
Subspecies *hypomelas* (Afghanistan and Pakistan), *blanfordi* (Endemic to Pakistan)
188. *Paraechinus micropus* (Blyth, 1846) Indian Hedgehog (Endemic to India and Pakistan)
189. *Paraechinus nudiventris* (Horsfield, 1851) Madras Hedgehog (Endemic to India)

Order Soricomorpha Gregory, 1910
Family Soricidae Fischer, 1817

190. *Crocidura leucodon* (Hermann, 1780) Bicoloured White-toothed Shrew (India)

191. *Crocidura gmelini* (Pallas, 1811) Gmelin's White-toothed Shrew (Afghanistan, India, and Pakistan)
192. *Crocidura fuliginosa* (Blyth, 1855) Southeast Asian Shrew (India)
193. *Crocidura horsfieldii* (Tomes, 1856) Horsfield's Shrew (Endemic to India, Nepal, and Sri Lanka)
194. *Crocidura attenuata* Milne-Edwards, 1872 Grey Shrew (Bhutan, India, and Nepal)
195. *Crocidura andamanensis* Miller, 1902 Andaman White-toothed Shrew (Endemic to India)
196. *Crocidura nicobarica* Miller, 1902 Nicobar White-toothed Shrew (Endemic to India)
197. *Crocidura pullata* Miller, 1911 Kashmir White-toothed Shrew (Endemic to India and Pakistan)
198. *Crocidura hispida* Thomas, 1913 Andaman Shrew (Endemic to India)
199. *Crocidura pergrisea* Miller, 1913 Pale Grey Shrew (Endemic to India)
200. *Crocidura rapax* G. Allen, 1923 Chinese White-toothed Shrew (India)
201. *Crocidura vorax* G. Allen, 1923 Voracious Shrew (India)
202. *Crocidura zarudnyi* Ognev, 1928 Zarudny's Rock Shrew (Afghanistan and Pakistan)
203. *Crocidura miya* Phillips, 1929 Sri Lankan Long-tailed Shrew (Endemic to Sri Lanka)
204. *Crocidura jenkinsi* Chakraborty, 1978 Jenkin's Andaman Spiny Shrew (Endemic to India)
205. *Crocidura hikmiya* Meegaskumbura et al., 2007 Sinharaja Shrew (Endemic to Sri Lanka)
206. *Crocidura gathornei* Jenkins, 2013 Gathorne's Rock Shrew (Endemic to India)
207. *Feroculus feroculus* (Kelaart, 1850) Kelaart's Long-clawed Shrew (Endemic to India and Sri Lanka)
208. *Solisorex pearsoni* Thomas, 1924 Pearson's Long-clawed Shrew (Endemic to Sri Lanka)
209. *Suncus murinus* (Linnaeus, 1766) House Shrew (Afghanistan, Bangladesh, Bhutan, India, Pakistan, Nepal, and Sri Lanka)
210. *Suncus etruscus* (Savi, 1822) Savi's Pygmy Shrew (Bhutan, India, Nepal, Pakistan, and Sri Lanka)
211. *Suncus montanus* (Kelaart, 1850) Sri Lankan Highland Shrew (Endemic to Sri Lanka)
212. *Suncus niger* (Horsfield, 1851) Indian Highland Shrew (Endemic to India)
213. *Suncus stoliczkanus* (Anderson, 1877) Anderson's Shrew (Endemic to Bangladesh, India, Nepal, and Pakistan)
214. *Suncus dayi* (Dobson, 1888) Day's Shrew (Endemic to India)
215. *Suncus zeylanicus* Phillips, 1928 Sri Lankan Shrew (Endemic to Sri Lanka)
216. *Suncus fellowesgordoni* Phillips, 1932 Ceylon Pygmy Shrew (Endemic to Sri Lanka)
217. *Anourosorex squamipes* Milne-Edwards, 1872 Chinese Mole-Shrew (India)

218. *Anourosorex assamensis* Anderson, 1875 Assam Mole-Shrew (Endemic to India)
219. *Anourosorex schmidi* Petter, 1963 Giant Mole-Shrew (Endemic to Bhutan and India)
220. *Chimarrogale himalayica* (Gray, 1842) Himalayan Water Shrew (Bhutan, India, and Nepal)
221. *Episoriculus caudatus* (Horsfield, 1851) Hodgson's Brown-toothed Shrew (India and Nepal)
222. *Episoriculus leucops* (Horsfield, 1855) Long-tailed Brown-toothed Shrew (Bhutan, India, and Nepal)
223. *Episoriculus macrurus* (Blanford, 1888) Arboreal Brown-toothed Shrew (India and Nepal)
224. *Episoriculus sacratus* (Thomas, 1911) Sichuan Brown-toothed Shrew
    Subspecies *umbrinus* (India)
225. *Episoriculus baileyi* Thomas, 1914 Long-tailed Brown-toothed Shrew (India and Nepal)
226. *Episoriculus soluensis* Gruber, 1969 Solu Brown-toothed Shrew (Endemic to India and Nepal)
227. *Nectogale elegans* Milne-Edwards, 1870 Web-footed Shrew (Bhutan, India, and Nepal)
228. *Sorex minutus* Linnaeus, 1766 Eurasian Pygmy Shrew (India, Nepal, and Pakistan)
229. *Sorex bedfordiae* Thomas, 1911 Lesser Striped Shrew (Nepal)
230. *Sorex planiceps* Miller, 1911 Kashmir Pygmy Shrew (India and Pakistan)
231. *Sorex excelsus* Allen, 1923 Chinese Highland Shrew (Nepal)
232. *Soriculus nigrescens* (Gray, 1842) Sikkim Large-clawed Shrew
    Subspecies *nigrescens* (India and Nepal), *minor* (Endemic to Bhutan and India)

Family Talpidae Fischer, 1817

233. *Euroscaptor micrura* (Hodgson, 1841) Himalayan Mole (Bhutan, India, and Nepal)
234. *Parascaptor leucura* (Blyth, 1850) Indian Mole (Bangladesh and India)

Order Chiroptera Blumenbach, 1779
Suborder Megachiroptera Dobson, 1875
Family Pteropodidae Gray, 1821

235. *Cynopterus sphinx* (Vahl, 1797) Greater Short-nosed Fruit Bat
    Subspecies *sphinx* (Bangladesh, India, Nepal, Pakistan, and Sri Lanka), *scherzeri* (Endemic to India), *angulatus* (Bhutan)
236. *Cynopterus brachyotis* (Müller, 1838) Lesser Short-nosed Fruit Bat
    Subspecies *brachyotis* (India), *brachysoma* (Endemic to India), *ceylonensis* (Endemic to India and Sri Lanka)
237. *Eonycteris spelaea* (Dobson, 1871) Lesser Dawn Bat (India and Nepal)

238. *Latidens salimalii* Thonglongya, 1972 Salim Ali's Fruit Bat (Endemic to India)
239. *Macroglossus sobrinus* Andersen, 1911 Greater Long-nosed Fruit Bat (India)
240. *Megaerops niphanae* Yenbutra and Fenten, 1983 Ratanaworabhan's Fruit Bat (Bangladesh and India)
241. *Pteropus giganteus* (Brünnich, 1782) Indian Flying Fox
    Subspecies *giganteus* (Bangladesh, India, Nepal, Pakistan, and Sri Lanka), *ariel* (Endemic to Maldives), *leucocephalus* (Bhutan, India, and Nepal)
242. *Pteropus hypomelanus* Temminck, 1853 Variable Flying Fox
    Subspecies *geminorum* (India), *satyrus* (Endemic to India), *maris* (Endemic to Maldives)
243. *Pteropus melanotus* Blyth, 1863 Black-eared Flying Fox
    Subspecies *melanotus* (Endemic to India), *tytleri* (Endemic to India)
244. *Pteropus faunulus* Miller, 1902 Nicobar Flying Fox (Endemic to India)
245. *Rousettus aegyptiacus* (E. Geoffroy, 1810) Egyptian Rousette
    Subspecies *arabicus* (Pakistan)
246. *Rousettus leschenaultii* (Desmarest, 1820) Leschenault's Rousette
    Subspecies *leschenaultii* (Bangladesh, Bhutan, India, Nepal, and Pakistan), *seminudus* (Endemic to Sri Lanka)
247. *Sphaerias blanfordi* (Thomas, 1891) Blanford's Fruit Bat (Bhutan, India, and Nepal)

Suborder Microchiroptera Dobson, 1875
Family Rhinolophidae Bell, 1836

248. *Rhinolophus ferrumequinum* (Schreber, 1774) Greater Horseshoe Bat
    Subspecies *tragatus* (India and Nepal), *proximus* (Afghanistan, India, and Pakistan)
249. *Rhinolophus hipposideros* (Bechstein, 1800) Lesser Horseshoe Bat
    Subspecies *midas* (Afghanistan, India, and Pakistan)
250. *Rhinolophus affinis* Horsfield, 1823 Intermediate Horseshoe Bat
    Subspecies *himalayanus* (Bangladesh, Bhutan, India, and Nepal)
251. *Rhinolophus pusillus* Temminck, 1834 Least Horseshoe Bat
    Subspecies *gracilis* (Endemic to India), *blythii* (India and Nepal)
252. *Rhinolophus trifoliatus* Temminck, 1834 Trefoil Horseshoe Bat (India)
253. *Rhinolophus luctus* Temminck, 1835 Woolly Horseshoe Bat
    Subspecies *perniger* (Bangladesh, India, and Nepal)
254. *Rhinolophus rouxii* Temminck, 1835 Lesser Rufous Horseshoe Bat
    Subspecies *rouxii* (India), *rubidus* (Endemic to Sri Lanka)
255. *Rhinolophus lepidus* Blyth, 1844 Blyth's Horseshoe Bat
    Subspecies *lepidus* Bangladesh and India, *monticola* (Afghanistan, India, Nepal, and Pakistan)
256. *Rhinolophus macrotis* Blyth, 1844 Big-eared Horseshoe Bat
    Subspecies *macrotis* (Bangladesh, India, and Nepal), *topali* (Endemic to Pakistan)

257. *Rhinolophus mitratus* Blyth, 1844 Mitred Horseshoe Bat (Endemic to India)
258. *Rhinolophus subbadius* Blyth, 1844 Little Nepalese Horseshoe Bat (Bangladesh, India, and Nepal)
259. *Rhinolophus pearsonii* Horsfield, 1851 Pearson's Horseshoe Bat (Bangladesh, Bhutan, India, and Nepal)
260. *Rhinolophus blasii* Peters, 1866 Blasius's Horseshoe Bat
    Subspecies *meyeroehmi* (Afghanistan and Pakistan)
261. *Rhinolophus andamanensis* Dobson, 1872 Homfray's Horseshoe Bat (Endemic to India)
262. *Rhinolophus yunanensis* Dobson, 1872 Dobson's Horseshoe Bat (India)
263. *Rhinolophus beddomei* Andersen, 1905 Beddome's Horseshoe Bat
    Subspecies *beddomei* (Endemic to India), *sobrinus* (Endemic to Sri Lanka)
264. *Rhinolophus sinicus* Andersen, 1905 Chinese Horseshoe Bat (India and Nepal)
265. *Rhinolophus cognatus* Andersen, 1906 Andaman Horseshoe Bat
    Subspecies *cognatus* (Endemic to India), *famulus* (Endemic to India)
266. *Rhinolophus bocharicus* Kastchenko and Akimov, 1917 Central Asian Horseshoe Bat (Afghanistan)
267. *Rhinolophus shortridgei* K. Andersen, 1918 Shortridge's Horseshoe Bat (India)
268. *Rhinolophus indorouxii* Chattopadhyay et al., 2012 Greater Rufous Horseshoe Bat (Endemic to India)

Family Hipposideridae Lydekker, 1891

269. *Asellia tridens* (E. Geoffroy, 1813) Geoffroy's Trident Roundleaf Bat
    Subspecies *murraiana* (Afghanistan, India, and Pakistan)
270. *Coelops frithii* Blyth, 1848 East Asian Tailless Roundleaf Bat (Bangladesh and India)
271. *Hipposideros speoris* (Schneider, 1800) Schneider's Roundleaf Bat (Endemic to India and Sri Lanka)
272. *Hipposideros diadema* (E. Geoffroy, 1813) Diadem Roundleaf Bat
    Subspecies *nicobarensis* (Endemic to India), *masoni* (India)
273. *Hipposideros armiger* (Hodgson, 1835) Great Roundleaf Bat (India and Nepal)
274. *Hipposideros fulvus* Gray, 1838 Fulvus Roundleaf Bat
    Subspecies *fulvus* (India and Sri Lanka), *pallidus* (Endemic to Afghanistan, India, Nepal, and Pakistan)
275. *Hipposideros galeritus* Cantor, 1846 Cantor's Roundleaf Bat
    Subspecies *brachyotus* (Endemic to Bangladesh, India, and Sri Lanka)

276. *Hipposideros ater* Templeton, 1848 Dusky Roundleaf Bat
Subspecies *ater* (Endemic to India and Sri Lanka), *nallamalaensis* (Endemic to India)
277. *Hipposideros lankadiva* Kelaart, 1850 Kelaart's Roundleaf Bat
Subspecies *lankadiva* (Endemic to Sri Lanka), *indus* (Endemic to Bangladesh and India), *gyi* (India and Nepal)
278. *Hipposideros cineraceus* Blyth, 1853 Ashy Roundleaf Bat (India, Nepal, and Pakistan)
279. *Hipposideros nicobarulae* Miller, 1902 Nicobar Roundleaf Bat (Endemic to India)
280. *Hipposideros gentilis* Andersen, 1918 Andersen's Roundleaf Bat (Bangladesh, India, and Nepal)
281. *Hipposideros pomona* Andersen, 1918 Pomona Roundleaf Bat (Endemic to India)
282. *Hipposideros grandis* G.M. Allen, 1936 Grand Roundleaf Bat
Subspecies *leptophyllus* (Bangladesh and India)
283. *Hipposideros durgadasi* Khajuria, 1970 Durga Das's Roundleaf Bat (Endemic to India)
284. *Hipposideros hypophyllus* Kock and Bhat, 1994 Leafletted Roundleaf Bat (Endemic to India)
285. *Hipposideros khasiana* Thabah et al., 2006 Khasian Roundleaf Bat (India)
286. *Triaenops persicus* Dobson, 1871 Persian Trident Bat (Pakistan)

Family Megadermatidae H. Allen, 1864

287. *Megaderma spasma* (Linnaeus, 1758) Lesser False Vampire Bat
Subspecies *horsfieldii* (Endemic to India), *ceylonense* (Endemic to Sri Lanka), *majus* (Bangladesh and India)
288. *Lyroderma lyra* (E. Geoffroy, 1810) Greater False Vampire Bat (Afghanistan, Bangladesh, India, Nepal, Pakistan, and Sri Lanka)

Family Rhinopomatidae Bonaparte, 1838

289. *Rhinopoma hardwickii* Gray, 1831 Lesser Mouse-tailed Bat (Afghanistan, Bangladesh, India, and Pakistan)
290. *Rhinopoma microphyllum* (Brünnich, 1872) Greater Mouse-tailed Bat
Subspecies *microphyllum* (Afghanistan and Pakistan), *kinneari* (Endemic to Bangladesh and India)
291. *Rhinopoma muscatellum* Thomas, 1903 Small Mouse-tailed Bat
Subspecies *seianum* (Afghanistan and Pakistan)

Family Emballonuridae Gervais, 1855

292. *Saccolaimus saccolaimus* (Temminck, 1838) Pouch-bearing Tomb Bat
Subspecies *crassus* (Bangladesh, India, and Sri Lanka)

293. *Taphozous perforatus* E. Geoffroy, 1818 Egyptian Tomb Bat (India and Pakistan)
294. *Taphozous longimanus* Hardwicke, 1825 Long-winged Tomb Bat (Bangladesh, India, Nepal, and Sri Lanka)
295. *Taphozous nudiventris* Cretzschmar, 1830 Naked-rumped Tomb Bat
Subspecies *kachhensis* (Afghanistan, Bangladesh, India, and Pakistan)
296. *Taphozous melanopogon* Temminck, 1841 Black-bearded Tomb Bat (Bangladesh, India, and Sri Lanka)
297. *Taphozous theobaldi* Dobson, 1872 Theobald's Tomb Bat
Subspecies *secatus* (Endemic to India)

Family Molossidae Gill, 1872

298. *Chaerephon plicatus* (Buchanan, 1800) Wrinkle-lipped Free-tailed Bat
Subspecies *plicatus* (Afghanistan and India), *insularis* (Endemic to Sri Lanka)
299. *Otomops wroughtoni* (Thomas, 1913) Wroughton's Giant Mastiff Bat (India)
300. *Tadarida teniotis* (Rafinesque, 1814) European Free-tailed Bat (Afghanistan and India)
301. *Tadarida aegyptiaca* (E. Geoffroy, 1818) Egyptian Free-tailed Bat
Subspecies *tragatus* (Endemic to Bangladesh and India), *thomasi* (India, Pakistan, and Sri Lanka), *sindhica* (Afghanistan and Pakistan)

Family Vespertilionidae Gray, 1821
Subfamily Vespertilioninae Miller, 1897
Tribe Eptisicini Volleth and Heller, 1994

302. *Arielulus circumdatus* (Temminck, 1840) Bronze Sprite (India and Nepal)
303. *Cassistrellus dimissus* (Thomas, 1916) Surat Serotine (Nepal)
304. *Eptesicus serotinus* (Schreber, 1774) Serotine
Subspecies *pashtonus* (Afghanistan and Pakistan)
305. *Eptesicus pachyomus* Tomes, 1857 Tomes' Serotine (Afghanistan, India, and Pakistan)
306. *Eptesicus pachyotis* (Dobson, 1871) Thick-eared Bat (Bangladesh and India)
307. *Eptesicus ognevi* Bobrinski, 1918 Ognev's Serotine (Afghanistan and India)
308. *Eptesicus gobiensis* Bobrinski, 1926 Gobi Big Brown Bat
Subspecies *kashgaricus* (Afghanistan and India), *centrasiaticus* (Nepal)
309. *Eptesicus tatei* Ellerman and Morrison-Scott, 1951 Sombre Bat (Endemic to India)
310. *Hesperoptenus tickelli* (Blyth, 1851) Tickell's Bat (Bangladesh, Bhutan, India, Nepal, and Sri Lanka)
311. *Rhyneptesicus nasutus* (Dobson, 1877) Sindh Bat (Afghanistan and Pakistan)

Tribe Nycticeiini Gervais, 1855

312. *Scotoecus pallidus* (Dobson, 1876) Desert Yellow Lesser House Bat (India and Pakistan)
313. *Scotomanes ornatus* (Blyth, 1851) Harlequin Bat
    Subspecies *ornatus* (Bangladesh, Bhutan, India, and Nepal), *imbrensis* (India)
314. *Scotophilus kuhlii* Leach, 1821 Lesser Asiatic Yellow House Bat (Bangladesh, India, Pakistan, and Sri Lanka)
315. *Scotophilus heathii* (Horsfield, 1831) Greater Asiatic Yellow House Bat (Afghanistan, Bangladesh, India, Nepal, Pakistan, and Sri Lanka)

Tribe Pipistrellini Tate, 1942

316. *Nyctalus noctula* (Schreber, 1774) Noctule
    Subspecies *labiata* (Afghanistan, India, Nepal, and Pakistan)
317. *Nyctalus leisleri* (Kuhl, 1817) Leisler's Noctule (Afghanistan, India, and Pakistan)
318. *Nyctalus montanus* (Barret-Hamilton, 1906) Mountain Noctule (Afghanistan, India, and Nepal)
319. *Pipistrellus pipistrellus* (Schreber, 1774) Common Pipistrelle
    Subspecies *aladdin* (Afghanistan, India, and Pakistan)
320. *Pipistrellus kuhlii* (Kuhl, 1817) Kuhl's Pipistrelle
    Subspecies *lepidus* (Afghanistan, India, and Pakistan)
321. *Pipistrellus coromandra* (Gray, 1838) Indian Pipistrelle (Afghanistan, Bangladesh, Bhutan, India, Nepal, Pakistan, and Sri Lanka)
322. *Pipistrellus javanicus* (Gray, 1838) Javan Pipistrelle
    Subspecies *camortae* (Endemic to India), *babu* (Afghanistan, Bangladesh, India, Nepal, and Pakistan), *peguensis* (India)
323. *Pipistrellus abramus* (Temminck, 1840) Japanese Pipistrelle (India)
324. *Pipistrellus tenuis* (Temminck, 1840) Least Pipistrelle
    Subspecies *mimus* (Afghanistan, Bangladesh, India, Nepal, Pakistan, and Sri Lanka)
325. *Pipistrellus ceylonicus* (Kelaart, 1852) Kelaart's Pipistrelle
    Subspecies *ceylonicus* (Endemic to Sri Lanka), *indicus* (Bangladesh, India, and Pakistan)
326. *Pipistrellus paterculus* Thomas, 1915 Mount Popa Pipistrelle (India and Nepal)
327. *Scotozous dormeri* Dobson, 1875 Dormer's Pipistrelle (Bangladesh, India, and Pakistan)

Tribe Plecotinini Gray, 1866

328. *Barbastella darjelingensis* (Hodgson, 1855) Large Barbastelle (Afghanistan, Bhutan, India, Nepal, and Pakistan)
329. *Otonycteris hemprichii* Peters, 1859 Hemprich's Desert Bat (Afghanistan, India, and Pakistan)

330. *Otonycteris leucophaea* (Severcov, 1873) Turkenstani Long-eared Bat (Afghanistan)
331. *Plecotus homochrous* Hodgson, 1847 Hodgson's Long-eared Bat (India, Nepal, and Pakistan)
332. *Plecotus wardi* Thomas, 1911 Ward's Long-eared Bat (Afghanistan, India, Nepal, and Pakistan)
333. *Plecotus strelkovi* Spitzenberger, 2006 Strelkov's Long-eared Bat (Afghanistan)

Tribe Vespertilionini Gray, 1821

334. *Falsistrellus affinis* (Dobson, 1871) Chocolate Pipistrelle (India, Nepal, and Sri Lanka)
335. *Hypsugo savii* (Bonaparte, 1837) Savi's Pipistrelle
    Subspecies *austenianus* (Bangladesh and India), *caucasicus* (Afghanistan and India)
336. *Hypsugo cadornae* (Thomas, 1916) Cadorna's Pipistrelle (India)
337. *Ia io* Thomas, 1902 Great Evening Bat (India and Nepal)
338. *Philetor brachypterus* (Temminck, 1840) Rohu's Bat (India and Nepal)
339. *Tylonycteris fulvida* (Blyth, 1850) Blyth's Lesser Bamboo Bat (Bangladesh and India)
340. *Tylonycteris aurex* (Thomas, 1915) Thomas' Lesser Bamboo Bat (Endemic to India)
341. *Tylonycteris malayana* Chasen, 1940 Malayan Greater Bamboo Bat
    Subspecies *malayana* (India), *eremtaga* (Endemic to India)
342. *Vespertilio murinus* Linnaeus, 1758 Particoloured Bat (Afghanistan and India)

Subfamily Myotinae Tate, 1942

343. *Myotis emarginatus* (E. Geoffroy, 1806) Geoffroy's Myotis
    Subspecies *desertorum* (Afghanistan)
344. *Myotis formosus* (Hodgson, 1835) Hodgson's Myotis (Afghanistan, Bangladesh, India, and Nepal)
345. *Myotis hasseltii* (Temminck, 1840) Lesser Large-footed Myotis (India and Sri Lanka)
346. *Myotis horsfieldii* (Temminck, 1840) Horsfield's Myotis
    Subspecies *dryas* (Endemic to India), *peshwa* (Endemic to India)
347. *Myotis muricola* (Gray, 1846) Hairy-faced Myotis (Bhutan, India, Pakistan, and Nepal)
348. *Myotis siligorensis* (Horsfield, 1855) Himalayan Whiskered Myotis (India and Nepal)
349. *Myotis blythii* (Tomes, 1857) Lesser Myotis (Afghanistan, India, Nepal, and Pakistan)
350. *Myotis davidii* (Peters, 1869) David's Myotis (Afghanistan)

351. *Myotis annectans* (Dobson, 1871) Hairy-faced Myotis (India)
352. *Myotis laniger* (Peters, 1871) Chinese Water Myotis (India)
353. *Myotis nipalensis* (Dobson, 1871) Nepal Whiskered Myotis (Afghanistan, India, Nepal, and Pakistan)
354. *Myotis longipes* (Dobson, 1873) Kashmir Cave Myotis (Afghanistan, India, and Nepal)
355. *Myotis montivagus* (Dobson, 1874) Burmese Whiskered Myotis (India)
356. *Myotis peytoni* Wroughton and Ryley, 1913 Peyton's Whiskered Myotis (Endemic to India)
357. *Myotis sicarius* Thomas, 1915 Mendelli's Mouse-eared Myotis (India and Nepal)
358. *Myotis csorbai* Topál, 1997 Csorba's Myotis (Endemic to Nepal)
359. *Submyotodon caliginosus* (Tomes, 1859) Tomes' Hairy-faced Myotis (Afghanistan, India, and Pakistan)

Subfamily Murininae Miller, 1907

360. *Harpiocephalus harpia* (Temminck, 1840) Lesser Hairy-winged Bat
    Subspecies *lasyurus* (India), *madrassius* (Endemic to India)
361. *Murina aurata* Milne-Edwards, 1872 Little Tube-nosed Bat (India and Nepal)
362. *Murina cyclotis* Dobson, 1872 Round-eared Tube-nosed Bat
    Subspecies *cyclotis* (India and Nepal), *eileenae* (Endemic to Sri Lanka)
363. *Murina huttoni* (Peters, 1872) Hutton's Tube-nosed Bat (India, Nepal, and Pakistan)
364. *Murina leucogaster* Milne-Edwards, 1872 Greater Tube-nosed Bat
    Subspecies *rubex* (India and Nepal)
365. *Murina tubinaris* (Scully, 1881) Scully's Tube-nosed Bat (India and Pakistan)
366. *Murina feae* (Thomas, 1891) Fea's Tube-nosed Bat (India)
367. *Murina pluvialis* Ruedi et al., 2012 Rainforest Tube-nosed Bat (Endemic to India)
368. *Murina jaintiana* Ruedi et al., 2012 Jaintia Tube-nosed Bat (Endemic to India)
369. *Murina guilleni* Soisook et al., 2013 Guillen's Tube-nosed Bat
    Subspecies *nicobarensis* (Endemic to India)
370. *Harpiola grisea* (Peters, 1872) Peter's Tube-nosed Bat (Endemic to India)

Subfamily Kerivoulinae Miller, 1907

371. *Kerivoula picta* (Pallas, 1767) Painted Woolly Bat (Bangladesh, India, Nepal, and Sri Lanka)
372. *Kerivoula hardwickii* (Horsfield, 1824) Hardwicke's Woolly Bat (Bangladesh, India, Sri Lanka, and Pakistan)
373. *Kerivoula lenis* Thomas, 1916 Lenis Woolly Bat (India)

Family Miniopteridae Miller, 1907

374. *Miniopterus fuliginosus* (Hodgson, 1835) Eastern Long-fingered Bat (Afghanistan, India, Sri Lanka, and Nepal)
375. *Miniopterus pusillus* Dobson, 1876 Small Long-fingered Bat (India and Nepal)
376. *Miniopterus pallidus* (Thomas, 1905) Pallid Long-fingered Bat (Afghanistan)
377. *Miniopterus magnater* Sanborn, 1931 Western Long-fingered Bat
 Subspecies *macrodens* (India)

Order Pholidota Weber, 1904
Family Manidae Gray, 1821

378. *Manis crassicaudata* Gray, 1827 Indian Pangolin (Bangladesh, Bhutan, India, Nepal, Pakistan, and Sri Lanka)
379. *Manis pentadactyla* Linnaeus, 1758 Chinese Pangolin
 Subspecies *auritus* (Bangladesh, Bhutan, India, and Nepal)

Order Carnivora Bowdich, 1821
Suborder Caniformia Kretzoi, 1938
Family Canidae Fischer, 1817

380. *Canis aureus* Linnaeus, 1758 Golden Jackal
 Subspecies *indicus* (Afghanistan, Bangladesh, Bhutan, India, Nepal, and Pakistan), *naria* (Endemic to India and Sri Lanka)
381. *Canis lupus* Linnaeus, 1758 Wolf
 Subspecies *pallipes* (Afghanistan, Bangladesh, Bhutan, India, Nepal, and Pakistan), *chanco* (Afghanistan, India, and Pakistan)
382. *Cuon alpinus* (Pallas, 1811) Dhole
 Subspecies *dukhunensis* (Endemic to India), *primaevus* (Bhutan, India, and Nepal), *laniger* (India), *adustus* (Bangladesh and India)
383. *Vulpes vulpes* (Linnaeus, 1758) Red Fox
 Subspecies *griffithi* (Afghanistan and Pakistan), *montana* (Bhutan, India, and Nepal), *pusilla* (India and Pakistan)
384. *Vulpes corsac* (Linnaeus, 1768) Corsac Fox (Afghanistan)
385. *Vulpes bengalensis* (Shaw, 1800) Bengal Fox (Endemic to Bangladesh, Bhutan, India, Nepal, and Pakistan)
386. *Vulpes rueppelli* (Schinz, 1825) Ruppell's Fox
 Subspecies *zarudnyi* (Afghanistan)
387. *Vulpes ferrilata* Hodgson, 1842 Tibetan Fox (India and Nepal)
388. *Vulpes cana* Blanford, 1877 Blanford's Fox (Afghanistan and Pakistan)

Family Mustelidae Fischer de Waldheim, 1817

389. *Aonyx cinerea* (Illiger, 1815) Oriental Small-clawed Otter
 Subspecies *concolor* (Bangladesh, Bhutan, India, and Nepal), *nirnai* (Endemic to India)

390. *Lutra lutra* (Linnaeus, 1758) European Otter
 Subspecies *nair* (Endemic to India and Sri Lanka), *aurobrunneus* (India and Nepal), *monticola* (Bangladesh, Bhutan, India, and Nepal), *kutab* (Afghanistan, India, and Pakistan)
391. *Lutrogale perspicillata* (I. Geoffroy Saint-Hilaire, 1826) Smooth-coated Otter
 Subspecies *perspicillata* (Bangladesh, Bhutan, India, and Nepal), *sindica* (India and Pakistan)

Subfamily Mustelinae Fischer, 1817

392. *Arctonyx collaris* F. Cuvier, 1825 Hog-Badger
 Subspecies *collaris* (Bhutan, India, and Nepal), *consul* (Bangladesh and India)
393. *Martes foina* (Erxleben, 1777) Beech Marten
 Subspecies *toufoeus* (India), *intermedia* (Afghanistan, India, Nepal, and Pakistan)
394. *Martes flavigula* (Boddaert, 1785) Yellow-throated Marten (Afghanistan, Bhutan, India, Pakistan, and Nepal)
395. *Martes gwatkinsii* Horsfield, 1851 Nilgiri Marten (Endemic to India)
396. *Meles meles* (Linnaeus, 1758) European Badger (Afghanistan)
397. *Mellivora capensis* (Schreber, 1776) Honey Badger
 Subspecies *indicus* (India and Pakistan), *inauritus* (Afghanistan, India, and Nepal)
398. *Melogale moschata* (Gray, 1831) Chinese Ferret-Badger
 Subspecies *millsi* (India)
399. *Melogale personata* I. Geoffroy Saint-Hilaire, 1831 Burmese Ferret-Badger
 Subspecies *nipalensis* (Bhutan, India, and Nepal)
400. *Mustela erminea* Linnaeus, 1758 Ermine
 Subspecies *ferghanae* (Afghanistan, India, Nepal, and Pakistan)
401. *Mustela nivalis* Linnaeus, 1758 Least Weasel
 Subspecies *stoliczkana* (Extinct in South Asia; Afghanistan)
402. *Mustela sibirica* Pallas, 1773 Siberian Weasel
 Subspecies *subhemachalana* (Bhutan, India, and Nepal), *hodgsoni* (India), *canigula* (India and Pakistan)
403. *Mustela altaica* Pallas, 1811 Mountain Weasel
 Subspecies *temon* (Bhutan, India, Nepal, and Pakistan)
404. *Mustela kathiah* Hodgson, 1835 Yellow-bellied Weasel
 Subspecies *kathiah* (Bhutan, India, and Nepal), *caporiaccoi* (India and Pakistan)
405. *Mustela strigidorsa* Gray, 1853 Back-striped Weasel (India and Nepal)
406. *Vormela peregusna* (Guldenstaedt, 1770) Marbled Polecat (Afghanistan and Pakistan)

Family Ailuridae Gray, 1843

407. *Ailurus fulgens* F. G. Cuvier, 1825 Red Panda (Bhutan, India, and Nepal)

Family Ursidae Fischer de Waldheim, 1817

408. *Helarctos malayanus* (Raffles, 1822) Sun Bear (Bangladesh and India)
409. *Melursus ursinus* (Shaw, 1791) Sloth Bear
Subspecies *ursinus* (Endemic to Bangladesh, Bhutan, India, and Nepal), *inornatus* (Endemic to Sri Lanka)
410. *Ursus arctos* Linnaeus, 1758 Brown Bear
Subspecies *isabellinus* (Afghanistan, India, Nepal, and Pakistan)
411. *Ursus thibetanus* (G. Cuvier, 1823) Asian Black Bear
Subspecies *thibetanus* (Bangladesh, Bhutan, India, and Nepal), *gedrosianus* (Endemic to Pakistan), *laniger* (Endemic to Afghanistan, India, and Pakistan)

Family Felidae Fischer de Waldheim, 1817

412. *Acinonyx jubatus* (Griffith, 1821) Cheetah
Subspecies *venaticus* (Extinct in South Asia; Afghanistan, India, and Pakistan)
413. *Caracal* (Schreber, 1776) Caracal
Subspecies *schmitzi* (Afghanistan, India, and Pakistan)
414. *Felis chaus* Schreber, 1777 Jungle Cat
Subspecies *affinis* (Bhutan, India, and Nepal), *kutas* (Bangladesh and India), *maimanah* (Afghanistan), *prateri* (Endemic to India and Pakistan), *kelaarti* (Endemic to India and Sri Lanka)
415. *Felis silvestris* Schreber, 1777 Wild Cat
Subspecies *ornata* (Afghanistan, Pakistan, and India)
416. *Felis margarita* Loche, 1858 Sand Cat
Subspecies *scheffeli* (Endemic to Pakistan)
417. *Lynx lynx* (Linnaeus, 1758) European Lynx
Subspecies *isabellinus* (Afghanistan, India, and Nepal)
418. *Otocolobus manul* (Pallas, 1776) Pallas's Cat
Subspecies *ferruginea* (Afghanistan and Pakistan), *nigripecta* (India)
419. *Pardofelis temminckii* (Vigors and Horsfield, 1827) Asiatic Golden Cat (Bangladesh, Bhutan, India, and Nepal)
420. *Pardofelis marmorata* (Martin, 1837) Marbled Cat
Subspecies *charltonii* (Bangladesh, Bhutan, India, and Nepal)
421. *Prionailurus bengalensis* (Kerr, 1792) Leopard Cat
Subspecies *bengalensis* (Bangladesh and India), *horsfieldii* (Bhutan, India, and Nepal), *trevelyani* (Afghanistan, India, and Pakistan)
422. *Prionailurus rubiginosus* (I. Geoffroy, 1831) Rusty-spotted Cat
Subspecies *rubiginosus* (Endemic to India), *phillipsi* (Endemic to Sri Lanka)
423. *Prionailurus viverrinus* (Bennett, 1833) Fishing Cat (Bangladesh, Bhutan, India, Nepal, Pakistan, and Sri Lanka)

424. *Neofelis nebulosa* (Griffith, 1821) Clouded Leopard
Subspecies *macrosceloides* (Bangladesh, Bhutan, India, and Nepal)
425. *Panthera leo* (Linnaeus, 1758) Lion
Subspecies *persica* (Endemic to India)
426. *Panthera pardus* (Linnaeus, 1758) Leopard
Subspecies *saxicolor* (Afghanistan), *fusca* (Afghanistan, Bangladesh, Bhutan, India, Nepal, and Pakistan), *kotiya* (Endemic to Sri Lanka)
427. *Panthera tigris* (Linnaeus, 1758) Tiger
Subspecies *tigris* (Bangladesh, Bhutan, India, and Nepal), *virgata* (Extinct; Afghanistan)
428. *Panthera uncia* (Schreber, 1775) Snow Leopard (Afghanistan, Bhutan, India, Nepal, and Pakistan)

Family Prionodontidae Pocock, 1933

429. *Prionodon pardicolor* Hodgson, 1842 Spotted Linsang (Bhutan, India, and Nepal)

Family Hyaenidae Gray, 1821

430. *Hyaena hyaena* (Linnaeus, 1758) Striped Hyaena (Afghanistan, India, Nepal, and Pakistan)

Family Herpestidae Bonaparte, 1845

431. *Urva edwardsii* (E. Geoffroy Saint-Hilaire, 1818) Indian Grey Mongoose
Subspecies *edwardsii* (Endemic to India), *nyula* (Endemic to Bangladesh, Bhutan, India, and Nepal), *ferrugineus* (Endemic to Afghanistan, India, and Pakistan), *montanus* (Endemic to Pakistan), *lanka* (Endemic to Sri Lanka)
432. *Urva vitticollis* Bennett, 1835 Striped-necked Mongoose
Subspecies *vitticollis* (Endemic to India and Sri Lanka), *inornatus* (Endemic to India)
433. *Urva auropunctatus* (Hodgson, 1836) Small Indian Mongoose
Subspecies *auropunctatus* (Endemic to Bangladesh, Bhutan, India, and Nepal), *pallipes* (Afghanistan, India, and Pakistan)
434. *Urva urva* (Hodgson, 1836) Crab-eating Mongoose (Bangladesh, Bhutan, India, and Nepal)
435. *Urva smithii* Gray, 1837 Ruddy Mongoose
Subspecies *smithii* (Endemic to India), *thysanurus* (Endemic to India), *zeylanius* (Endemic to Sri Lanka)
436. *Urva fuscus* Waterhouse, 1838 Indian Brown Mongoose
Subspecies *fuscus* (Endemic to India), *flavidens* (Endemic to Sri Lanka), *maccarthiae* (Endemic to Sri Lanka), *siccatus* (Endemic to Sri Lanka), *rubidior* (Endemic to Sri Lanka)
437. *Urva palustris* Ghose, 1965 Indian Marsh Mongoose (Endemic to India)

Family Viverridae Gray, 1821

438. *Arctictis binturong* (Raffles, 1821) Binturong
Subspecies *albifrons* (Bangladesh, Bhutan, India, and Nepal)
439. *Arctogalidia trivirgata* (Gray, 1832) Small-toothed Palm Civet
Subspecies *millsi* (Bangladesh and India)
440. *Paguma larvata* (Hamilton-Smith, 1827) Masked Palm Civet
Subspecies *grayi* (India and Nepal), *tytleri* (Endemic to India), *wroughtoni* (Endemic to India and Pakistan), *neglecta* (Bangladesh, Bhutan, India, and Nepal)
441. *Paradoxurus hermaphroditus* (Pallas, 1777) Common Palm Civet
Subspecies *hermaphroditus* (Bangladesh, India, and Sri Lanka), *bondar* (Endemic to India and Nepal), *pallasi* (India and Nepal), *nictitatans* (Endemic to India), *scindiae* (Endemic to India), *vellerosus* (Endemic to India and Pakistan)
442. *Paradoxurus aureus* F. Cuvier, 1822 Golden Wet-zone Palm Civet (Endemic to Sri Lanka)
443. *Paradoxurus montanus* Kelaart, 1852 Sri Lankan Brown Palm Civet (Endemic to Sri Lanka)
444. *Paradoxurus jerdoni* Blanford, 1885 Jerdon's Palm Civet
Subspecies *jerdoni* (Endemic to India), *caniscus* (Endemic to India)
445. *Paradoxurus stenocephalus* Groves et al., 2009 Golden Dry-zone Palm Civet (Endemic to Sri Lanka)
446. *Viverra zibetha* (Linnaeus, 1758) Large Indian Civet (Bangladesh, India, and Nepal)
447. *Viverra civettina* Blyth, 1862 Malabar Large Spotted Civet (Endemic to India)
448. *Viverricula indica* (E. Geoffroy Saint-Hilaire, 1803) Small Indian Civet
Subspecies *indica* (Bangladesh, India, and Pakistan), *deserti* (Endemic to India and Pakistan), *baptistae* (Endemic to Bhutan, India, and Nepal), *mayori* (Endemic to Sri Lanka), *wellsi* (Endemic to India)

Order Perissodactyla Owen, 1848
Family Equidae Gray, 1821

449. *Equus hemionus* Pallas, 1775 Asian Wild Ass
Subspecies *khur* (Endemic to India and Pakistan), *blanfordi* (Extinct; Afghanistan and Pakistan)
450. *Equus kiang* Moorcroft, 1841 Tibetan Wild Ass
Subspecies *kiang* (India), *polyodon* (India and Nepal)

Family Rhinocerotidae Gray, 1821

451. *Dicerorhinus sumatrensis* (Fischer, 1814) Sumatran Rhinoceros
Subspecies *lasiotis* (Extinct; Bangladesh, Bhutan, and India)

452. *Rhinoceros unicornis* Linnaeus, 1758 Great One-horned Rhinoceros (Afghanistan [extinct], Bangladesh [extinct], Bhutan, India, Nepal, and Pakistan [extinct])
453. *Rhinoceros sondaicus* Desmarest, 1822 Lesser One-horned Rhinoceros
    Subspecies *inermis* (Extinct; Bangladesh, Bhutan, and India)

Order Artiodactyla Owen, 1848
Family Suidae Gray, 1821

454. *Porcula salvania* (Hodgson, 1847) Pygmy Hog (Endemic to Bhutan and India)
455. *Sus scrofa* (Linnaeus, 1758) Wild Boar
    Subspecies *vittatus* (India), *cristatus* (Afghanistan, Bangladesh, Bhutan, India, Sri Lanka, and Nepal), *davidi* (Endemic to India and Pakistan)

Family Tragulidae Milne-Edwards, 1864

456. *Moschiola meminna* (Erxleben, 1777) White Spotted Chevrotain (Endemic to Sri Lanka)
457. *Moschiola indica* (Gray, 1843) Indian Spotted Chevrotain (Endemic to India and Nepal)
458. *Moschiola kathygre* Groves and Meijaard, 2005 Yellow Striped Chevrotain (Endemic to Sri Lanka)

Family Moschidae Gray, 1821

459. *Moschus chrysogaster* Hodgson, 1839 Golden-bellied Musk Deer
    Subspecies *chrysogaster* (Bhutan, India, and Nepal), *sifanicus* (India)
460. *Moschus leucogaster* Hodgson, 1839 White-bellied Musk Deer (Bhutan, India, and Nepal)
461. *Moschus fuscus* Li, 1981 Dwarf Musk Deer (Bhutan, India, and Nepal)
462. *Moschus cupreus* Grubb, 1982 Kashmir Musk Deer (Endemic to Afghanistan, India, and Pakistan)

Family Cervidae Goldfuss, 1820

463. *Axis axis* (Erxleben, 1777) Spotted Deer (Endemic to Bangladesh, Bhutan, India, Nepal, and Sri Lanka)
464. *Axis porcinus* (Zimmermann, 1780) Hog-Deer (Bangladesh, Bhutan, India, Nepal, Pakistan, and Sri Lanka)
465. *Cervus elaphus* Linnaeus, 1758 Red Deer
    Subspecies *wallichi* (Extinct; Bhutan and India), *hanglu* (Endemic to India), *yarkandensis* (Afghanistan)

466. *Muntiacus vaginalis* (Boddaert, 1785) Northern Red Muntjak
Subspecies *vaginalis* (Bangladesh, Bhutan, India, Nepal, and Pakistan), *aureus* (Endemic to India), *malabaricus* (Endemic to India and Sri Lanka)
467. *Muntiacus putaoensis* Amato et al., 1999 Leaf Muntjak (India)
468. *Rucervus duvaucelii* (Cuvier, 1823) Swamp Deer
Subspecies *duvaucelii* (Endemic to India, Nepal, and Pakistan [extinct]), *branderi* (Endemic to India), *ranjitsinhi* (Endemic to Bangladesh [extinct] and India)
469. *Rucervus eldii* (McClelland, 1842) Brow-antlered Deer (Endemic to India)
470. *Rusa unicolor* (Kerr, 1792) Sambar (Bangladesh, Bhutan, India, Nepal, and Sri Lanka)

Family Bovidae Gray, 1821
Subfamily Antilopinae Gray, 1821

471. *Antilope cervicapra* (Linnaeus, 1758) Blackbuck
Subspecies *cervicapra* (Endemic to India), *rajputanae* (Endemic to India)
472. *Gazella subgutturosa* (Guldenstaedt, 1780) Goitered Gazelle (Afghanistan and Pakistan)
473. *Gazella bennettii* (Sykes, 1831) Indian Gazelle
Subspecies *bennettii* (Endemic to India), *christyi* (Endemic to India), *fuscifrons* (Endemic to Afghanistan, India, and Pakistan), *salinarum* (Endemic to India and Pakistan)
474. *Procapra picticaudata* Hodgson, 1846 Tibetan Gazelle (India)

Subfamily Bovinae Gray, 1821

475. *Bos gaurus* H. Smith, 1827 Indian Bison
Subspecies *gaurus* (India and Nepal), *laosiensis* (Bangladesh, Bhutan, and India)
476. *Bos mutus* (Przewalski, 1883) Wild Yak (India)
477. *Boselaphus tragocamelus* (Pallas, 1766) Nilgai (Endemic to Bangladesh [extinct], India, Pakistan, and Nepal)
478. *Bubalus arnee* (Kerr, 1792) Wild Buffalo
Subspecies *arnee* (Endemic to India, Nepal, and Sri Lanka [extinct]), *fulvus* (Endemic to Bangladesh [extinct], Bhutan, and India)
479. *Tetracerus quadricornis* (de Blainville, 1816) Four-horned Antelope
Subspecies *quadricornis* (Endemic to India), *iodes* (Endemic to India and Nepal), *subquadricornis* (Endemic to India)

Subfamily Caprinae Gray, 1821

480. *Budorcas taxicolor* Hodgson, 1850 Takin
Subspecies *taxicolor* (India), *whitei* (Bhutan and India)

481. *Capra sibirica* (Pallas, 1776) Siberian Ibex (Afghanistan, India, and Pakistan)
482. *Capra aegagrus* Erxleben, 1777 Wild Goat
    Subspecies *blythii* (Pakistan), *chialtanensis* (Endemic to Pakistan)
483. *Capra falconeri* (Wagner, 1839) Markhor
    Subspecies *falconeri* (Endemic to Afghanistan, India, and Pakistan), *megaceros* (Endemic to Afghanistan and Pakistan), *heptneri* (Afghanistan)
484. *Capricornis thar* (Hodgson, 1831) Himalayan Serow (Bhutan, India, and Nepal)
485. *Hemitragus jemlahicus* (H. Smith, 1826) Himalayan Tahr (Bhutan, India, and Nepal)
486. *Naemorhedus goral* (Hardwicke, 1825) Himalayan Goral
    Subspecies *goral* (Bhutan and India), *bedfordi* (India, Nepal, and Pakistan)
487. *Naemorhedus griseus* Milne-Edwards, 1872 Chinese Goral (India)
488. *Naemorhedus baileyi* Pocock, 1914 Red Goral (India)
489. *Nilgiritragus hylocrius* (Ogilby, 1838) Nilgiri Tahr (Endemic to India)
490. *Ovis ammon* (Linnaeus, 1758) Argali
    Subspecies *hodgsoni* (India and Nepal), *polii* (Afghanistan and India)
491. *Ovis orientalis* Gmelin, 1774 Urial
    Subspecies *vignei* (Endemic to India and Pakistan), *cycloceros* (Afghanistan, India, and Pakistan), *punjabiensis* (Endemic to Pakistan)
492. *Pantholops hodgsoni* (Abel, 1826) Tibetan Antelope (India and Nepal [extinct])
493. *Pseudois nayaur* (Hodgson, 1833) Blue Sheep (Bhutan, India, and Nepal)

Order Cetacea Brisson, 1762
Suborder Mysticeti Flower, 1864
Family Balaenidae Gray, 1821

494. *Eubalaena australis* (Desmoulins, 1822) Southern Right Whale (Arabian Sea: India)

Family Balaenopteridae Gray, 1864

495. *Balaenoptera musculus* (Linnaeus, 1758) Blue Whale
    Subspecies *brevicauda* (Arabian Sea, Bay of Bengal, and Indian Ocean: India, Pakistan, and Sri Lanka), *indica* (Arabian Sea, Bay of Bengal, and Indian Ocean: India)
496. *Balaenoptera physalus* (Linnaeus, 1758) Fin Whale (Arabian Sea, Bay of Bengal, and Indian Ocean: India, Pakistan, and Sri Lanka)
497. *Balaenoptera acutorostrata* Lacepede, 1804 Common Minke Whale (Arabian Sea, Bay of Bengal, and Indian Ocean: India and Sri Lanka)
498. *Balaenoptera edeni* Anderson, 1879 Bryde's Whale (Arabian Sea, Bay of Bengal, and Indian Ocean: India, Pakistan, and Sri Lanka)

499. *Megaptera novaeangliae* (Borowski, 1781) Humpback Whale (Arabian Sea, Bay of Bengal, and Indian Ocean: India, Pakistan, and Sri Lanka)

Suborder Odontoceti Flower, 1867
Family Delphinidae Gray, 1821

500. *Delphinus capensis* Gray, 1828 Long-beaked Common Dolphin
    Subspecies *tropicalis* (Arabian Sea, Bay of Bengal, and Indian Ocean: Bangladesh, India, Maldives, Pakistan, and Sri Lanka)
501. *Feresa attenuata* Gray, 1875 Pygmy Killer Whale (Arabian Sea, Bay of Bengal, and Indian Ocean: Bangladesh, India, Maldives, Pakistan, and Sri Lanka)
502. *Globicephala macrorhynchus* Gray, 1846 Short-finned Pilot Whale (Arabian Sea, Bay of Bengal, and Indian Ocean: Bangladesh, India, Maldives, Pakistan, and Sri Lanka)
503. *Grampus griseus* (G. Cuvier, 1812) Risso's Dolphin (Arabian Sea, Bay of Bengal, and Indian Ocean: Bangladesh, India, Maldives, Pakistan, and Sri Lanka)
504. *Lagenodelphis hosei* Fraser, 1956 Fraser's Dolphin (Arabian Sea, Bay of Bengal, and Indian Ocean: India, Maldives, Pakistan, and Sri Lanka)
505. *Orcaella brevirostris* (Owen, 1866) Irrawady Dolphin (Bay of Bengal: Bangladesh and India)
506. *Orcinus orca* (Linnaeus, 1758) Killer Whale (Arabian Sea, Bay of Bengal, and Indian Ocean: Bangladesh, India, Maldives, Pakistan, and Sri Lanka)
507. *Peponocephala electra* (Gray, 1846) Melon-headed Dolphin (Arabian Sea, Bay of Bengal, and Indian Ocean: Bangladesh, India, Maldives, Pakistan, and Sri Lanka)
508. *Pseudorca crassidens* (Owen, 1846) False Killer Whale (Arabian Sea, Bay of Bengal, and Indian Ocean: Bangladesh, India, Maldives, Pakistan, and Sri Lanka)
509. *Sousa chinensis* (Osbeck, 1765) Indopacific Humpback Dolphin (Arabian Sea, Bay of Bengal, and Indian Ocean: Bangladesh, India, Pakistan, and Sri Lanka)
510. *Stenella longirostris* (Gray, 1828) Spinner Dolphin (Arabian Sea, Bay of Bengal, and Indian Ocean: Bangladesh, India, Maldives, Pakistan, and Sri Lanka)
511. *Stenella coeruleoalba* (Mayen, 1833) Striped Dolphin (Arabian Sea, Bay of Bengal, and Indian Ocean: Bangladesh, India, Maldives, Pakistan, and Sri Lanka)
512. *Stenella attenuata* (Gray, 1846) Pantropical Spotted Dolphin (Arabian Sea, Bay of Bengal, and Indian Ocean: Bangladesh, India, Maldives, Pakistan, and Sri Lanka)
513. *Steno bredanensis* (Lesson, 1828) Rough-toothed Dolphin (Arabian Sea, Bay of Bengal, and Indian Ocean: Bangladesh, India, Maldives, Pakistan, and Sri Lanka)

514. *Tursiops truncatus* (Montagu, 1821) Bottle-nosed Dolphin (Arabian Sea, Bay of Bengal, and Indian Ocean: Bangladesh, India, Maldives, Pakistan, and Sri Lanka)
515. *Tursiops aduncus* (Ehrenberg, 1833) Indo-Pacific Bottlenose Dolphin (Arabian Sea, Bay of Bengal, and Indian Ocean: Bangladesh, India, Pakistan, and Sri Lanka)

Family Phocoenidae Gray, 1825

516. *Neophocaena phocaenoides* (G. Cuvier, 1829) Finless Porpoise (Arabian Sea, Bay of Bengal, and Indian Ocean: Bangladesh, India, Pakistan, and Sri Lanka)

Family Physeteridae Gray, 1821

517. *Physeter macrocephalus* Linnaeus, 1758 Sperm Whale (Arabian Sea, Bay of Bengal, and Indian Ocean: Bangladesh, India, Maldives, Pakistan, and Sri Lanka)

Family Kogiidae Gill, 1871

518. *Kogia breviceps* (Blainville, 1838) Pygmy Sperm Whale (Arabian Sea, Bay of Bengal, and Indian Ocean: Bangladesh, India, Maldives, Pakistan, and Sri Lanka)
519. *Kogia sima* (Owen, 1866) Dwarf Sperm Whale (Arabian Sea, Bay of Bengal, and Indian Ocean: Bangladesh, India, Maldives, Pakistan, and Sri Lanka)

Family Platanistidae Gray, 1846

520. *Platanista gangetica* (Roxburgh, 1801) Gangetic Dolphin
    Subspecies *gangetica* (Endemic to Bangladesh, Bhutan, India, and Nepal), *minor* (Endemic to Pakistan)

Family Ziphiidae Gray, 1865

521. *Indopacetus pacificus* (Longman, 1926) Tropical Bottlenose Whale (Indian Ocean: Maldives and Sri Lanka)
522. *Mesoplodon densirostris* (Blainville, 1817) Blainville's Beaked Whale (Arabian Sea and Indian Ocean: India, Pakistan, and Sri Lanka)
523. *Mesoplodon hotaula* Deraniyagala, 1963 Deraniyagala's Beaked Whale (Indian Ocean: India and Sri Lanka)
524. *Ziphius cavirostris* G. Cuvier, 1823 Goose-beaked Whale (Arabian Sea, Bay of Bengal, and Indian Ocean: Bangladesh, India, Maldives, Pakistan, and Sri Lanka)

2 Understanding Scientific Names

Scientific nomenclature is an integral part of taxonomy and classification. Correct and stable names of taxa create order in systematics. Zoologists have been striving to achieve this for centuries. In many parts of the world, scientific nomenclature is a dreaded subject, and many show disinterest in it, disregarding it as 'Greek and Latin'. Undoubtedly, scientific nomenclature is mostly in 'Latin', but once we understand the basics of nomenclature, and the governing rules and principles, it transforms into an interesting science. The origin and meaning of generic and specific names, along with their English names, provide an insight into the behaviour, morphology, habit, and also the people involved with species collection and identification, as well as the geographical area where the species was first detected.

PARTS OF SCIENTIFIC NAMES

The scientific name of a mammal consists of four parts. For example, *Elephas maximus* (Linnaeus, 1758), is the full scientific name of the Asian Elephant. The first two parts, the binomen *Elephas maximus*, are written in a Latin or neo-Latin form and are traditionally printed in italics or, when handwritten, underlined. The third and fourth parts of the Asian Elephant's scientific name – (Linnaeus, 1758) – reveal the author of the specific name and the year in which that name was first validly published as a binomen.

The first part of the name denotes the genus, distinguishing a group of related species or an isolated, distinctive species. It is in the form of a noun or a substantivised adjective treated as a noun, is unique in the zoological world, and should always begin with an upper-case letter. If the same genus is subsequently referred to, but in different specific combinations, the convention is to use the initial abbreviation for the generic term, provided it does not cause confusion (e.g. *Macaca silenus*, *M. sinica*, *M. mulatta*, *M. radiata*, *M. fascicularis*, *M. arctoides*, *M. assamensis*, *M. leonina*, and *M. munzala*, which are all in the genus *Macaca*; but not *M. crassicaudata*, since *M. crassicaudata* refers to *Manis crassicaudata*). The second part of the binomen, beginning with a lower-case letter, is the specific or trivial name, distinguishing the species within the genus, and although taking many forms, it is commonly an adjective or a noun in the genitive case. Only in combination with a generic name does it have any validity or make any sense, but it can be used in more than one genus; thus, *Ratufa indica* (Erxleben, 1777), *Tatera indica* (Hardwicke, 1807), *Bandicota indica* (Bechstein, 1800), *Nesokia indica* (Gray and Hardwicke, 1832), *Hystrix indica* (Kerr, 1792), *Viverricula indica* (E. Geoffroy Saint-Hilaire, 1803), and *Moschiola indica* (Gray, 1843) are all valid names in zoology. Within the genus, however, no two species, subspecies, or forms may bear the same specific name.

The scientific nomen of the Tiger *Panthera tigris* (Linnaeus, 1758) depicts the placement of the author's name and year in parenthesis after a specific name. This is to indicate that the current generic classification differs from that assigned by the original author. Linnaeus originally described the Tiger *Panthera tigris* (Linnaeus, 1758) in the genus *Felis*, which was later reassigned to the genus *Panthera* by Reginald Innes Pocock in 1929.

Some species of mammals are represented by many distinct races (populations of species occupying distinct geographic range, distinguished by recognizable morphological characters). Such mammals are polytypic, and the distinct populations are known as subspecies. The monotypic species does not have any subspecies. In nomenclature, subspecies are designated by adding a third name to the binomen, creating a trinomen. The two subspecies of Tiger *Panthera tigris* (Linnaeus, 1758) that are recognised from South Asia include the nominate subspecies, the Bengal Tiger – *Panthera tigris tigris* (Linnaeus, 1758) – and the now extinct Caspian Tiger *Panthera tigris virgata* (Illiger, 1815). The nominate subspecies' trinomen is created by repeating the specific epithet. Later names for distinct races of tigers – *Panthera tigris virgata* (Illiger, 1815), *Panthera tigris altaica* (Temminck, 1844), *Panthera tigris sondaica* (Temminck, 1844), *Panthera tigris amoyensis* (Hilzheimer, 1905), *Panthera tigris balica* (Schwarz, 1912), *Panthera tigris sumatrae* (Pocock, 1929), and *Panthera tigris corbetti* (Mazak, 1968) – were based on morphologically distinguished populations of the tiger in its original range, and are based on mostly on geographical range [*P. t. altaica* Temminck, 1844, *P. t. sondaica* Temminck, 1844, *P. t. amoyensis* Hilzheimer, 1905, *P. t. balica* Schwarz, 1912, and *P. t. sumatrae* Pocock, 1929], characteristics [*P. t. virgata* (Illiger, 1815)], or after a scientist [*P. t. corbetti* Mazak, 1968].

Extreme cases of polytypism, as in House Rat *Rattus rattus* (Linnaeus, 1758), have led to naming of numerous forms (with as many as 87 names) throughout its range, which on subsequent observations and studies have led to the conclusion that most of these forms can interbreed with each other, and thus are currently treated as synonyms of *Rattus rattus* (Linnaeus, 1758).

Systematic publications, such as checklists, handbooks, and synopses, generally give a full citation of both generic and specific names, including original publication details, identification of the type species of the genus, type locality of species, and synonyms (Srinivasulu and Srinivasulu, 2012, p. 334). Non-systematic scientific works, popular books, field guides, and magazines usually give only the binomen (Johnsingh and Manjrekar, 2013, p. 596).

CODES OF NOMENCLATURE

Although credited as the founder of the binomial nomenclature (or binominal organismal nomenclature) and hailed as the 'Father of Modern Taxonomy', Carolus Linnaeus (b. 1707–d. 1778) was not the first to have attempted to name animals. For centuries, animals have been named and classified based on their forms and habits by Greek and Roman naturalists and philosophers. The seeds of scientific taxonomy were first sown in the works of an Italian botanist Andrea Cesalpino (b. 1519–d. 1603), who first provided the classification of plants in his work *De Plantis*. In the seventeenth century, John Ray (b. 1627–d. 1705) furthered the art and science of taxonomy, and his work (followed by Linnaeus's), unlike earlier works, categorised animals based on their morphologies rather than their place of occurrence. Scientific nomenclature can be broadly classified into two broad categories – (1) Pre-Linnaean and (2) Linnaean. In the former, the scientific names of animals were Latin *nomina specifica*, that is, they were binomial, trinomial, or polynomial names; often inconsistent; and paragraph-long, serving as diagnosis, description, and key to identification. Furthermore, the names of the species kept constantly changing. In Linnaean nomenclature, the scientific names of animals are Latin *nomina trivialia*; that is, they are always binomial in structure, divorced from diagnosis and description, and constant. Unlike popularly believed, Carolus Linnaeus was not the inventor of modern binomial nomenclature, and the honour for developing such a system belongs to Gaspard 'Gaspar' Bauhin (b. 1560–d. 1624) and Johann 'Jean' Bauhin (b. 1541–d. 1613), the Swiss brothers who formalised a method of naming plants and animals using two words, the first representing the genus and the second representing the species. What favoured Carolus Linnaeus's work to be considered over the Bauhin brothers' work is the brothers' inconsistency in applying the binomial nomenclature in their works. Linnaeus employed a more scientific and consistent approach, applying numerous rules and principles to ensure taxonomic orderliness. He was the first to consistently group species under various categories or genera, and was the first to conceptualise the kingdoms of nature in his *Systema Naturae* (a series of 12 editions of the book; the 13th edition was authored by J.F. Gmelin after Linnaeus's death), categorising animate and inanimate things into an animal/vegetable/mineral scheme. The first edition of this work was published in 1735. Volume 1 of the 10th edition of *Systema Naturae* (*Systema naturæ per regna tria naturæ, secundum classes, ordines, genera, species, cum characteribus, differentiis, synonymis, locis* – translated, '*System of nature through the three kingdoms of nature, according to classes, orders, genera and species, with characters, differences, synonyms, places*'), along with another monumental work, *Svenska Spindlar uti sina hufvud-slägter indelte samt under några och sextio särskildte arter beskrefne och med illuminerade figurer uplyste* in Swedish or *Aranei Svecici, descriptionibus et figuris æneis illustrati, ad genera subalterna redacti, speciebus ultra LX determinati* (– translated 'Swedish Spiders divided into their principal genera as well as in few and sixty distinct species described with illuminated figures enlightened') by Carl Alexander Clerck (b. 1709–d. 1765) are considered as the starting point of zoological nomenclature. Although the later work by Clerck was published in 1757, to bring in consistency with the year of publication and acceptance of names included therein as valid zoological names, it was agreed to use 1758 as the publication date of this work, as Carolus Linnaeus's work was published in 1758. Linnaeus's work divided the Animal Kingdom into six classes – Mammalia (all mammals, including bats; in the first edition, however, whales and the West Indian Manatee were listed alongside the fishes), Aves (birds), Amphibia (included amphibians, reptiles, and non-bony fishes), Pisces (bony fishes), Insecta (all arthropods, crustaceans, arachnids, and myriapods), and Vermes (invertebrates, including 'worms', molluscs, and echinoderms).

Written nomenclatural rules in zoology were compiled since the late 1830s, ranging from *Strickland's Codes* (Strickland, 1841, 1842), *Kiesenwetter Code of*

Entomological Nomenclature (von Kiesenwetter, 1858), *Report on Nomenclature in Zoology and Botany* (Dall, 1877), and *American Ornithologists Union Code* (Coues et al., 1886) to *Règles Applicables a la Nomenclature des Êtres Organisés* (Blanchard, 1881, 1889), to name a few. In order to establish a set of commonly accepted international rules for all disciplines and countries, and to replace the varied conventions and unwritten rules, zoologists made efforts during the first and second International Zoological Congresses held at Paris (1889) and Moscow (1892). In the third International Congress for Zoology held at Leiden (1895), the compiling of '*International Rules on Zoological Nomenclature*' was first proposed, and it was officially published in French in 1905 (with English and German translated copies) (ICZN, 1905). With this, the International Commission on Zoological Nomenclature (ICZN) came into existence. Many amendments and modifications were subsequently passed by various zoological congresses (Boston 1907, Graz 1910, Monaco 1913, Budapest 1927, Padova 1930, Paris 1948, Copenhagen 1953, London 1958). During the International Zoological Congress of 1958 held at London, the Editorial Committee elaborated a completely new version of the nomenclatural rules, which were finally published as the first edition of the ICZN Code (popularly known as International Code on Zoological Nomenclature) on 9 November 1961. The fourth edition is in effect from 1 January 2000 and supercedes the previous editions (ICZN, 1999). The Code proper comprises the Preamble, 90 Articles (grouped in 18 Chapters) and the Glossary. Each Article consists of one or more mandatory provisions, which are sometimes accompanied by Recommendations and/or illustrative examples. In interpreting the Code, the meaning of a word or expression is to be taken as that given in the Glossary.

As it is not the scope of this book to detail the Code, interested readers are advised to log on to *http://www.iczn.org/iczn/index.jsp* for the online version. However, a brief outline of the principles that govern the zoological nomenclature is included here for clarity.

Zoological nomenclature is defined in the Code as 'the system of scientific names applied to taxonomic units (taxa; singular: taxon) of extant or extinct animals', and the term 'animals' refers to the Metazoa and also to protistan taxa when workers treat them as animals for the purposes of nomenclature. The date 1 January 1758, is arbitrarily fixed in this Code as the date of the starting point of zoological nomenclature. No name or nomenclatural act published before 1 January 1758, enters zoological nomenclature, but information (such as descriptions or illustrations) published before that date may be used. New names should be in Latin form, and should be euphonious and easily memorable. No scientific name should hurt anybody's sentiment on any grounds, and any document or debate conducted involving any zoological nomenclature should be courteous and friendly. The scientific name of a taxon of higher rank than the species group consists of one word (i.e. the name is ***uninominal***); it must begin with an upper-case letter.

Principle of Binomial Nomenclature

The scientific name of a species, and not of a taxon of any other rank, is a combination of two names (a ***binomen***), the first being the generic name and the second being the specific name. The generic name must begin with an upper-case letter and the specific name must begin with a lower-case letter.

Principle of Typification

Each nominal taxon in the family, genus, or species groups actually (or potentially) has a name-bearing type. The fixation of the name-bearing type of a nominal taxon provides an objective standard of reference for the application of the name it bears. No matter how the boundaries of a taxon may vary in the opinion of zoologists, the valid name of such a taxon is determined from the name-bearing type(s) considered to belong within those boundaries. Objectivity provided by typification is continuous throughout the hierarchy of names. It extends in ascending order from the species group to the family group. Thus the name-bearing type of a nominal species-group taxon is a specimen or a set of specimens (a holotype, lectotype, neotype, or syntype); that of a nominal genus-group taxon is a nominal species defined objectively by its type; and that of a nominal family-group taxon is the nominal genus on which its name is based. Once fixed, name-bearing types are stable and provide objective continuity in the application of names. Thus, the name-bearing type of any nominal taxon, once fixed in conformity with the provisions of the Code, is not subject to change except in the case of nominal genus-group taxa as provided in Article 70.3.2 (the taxonomic species actually involved in the misidentification). If choosing this exception, the author must refer to this Article and cite together both the name previously cited as type species and the name of the species selected, of nominal species-group taxa as provided in Articles 74 (name-bearing types fixed subsequently from the type series (lectotypes from syntypes)) and 75 (neotypes – a neotype is the name-bearing type of a nominal species-group taxon designated under conditions specified in this Article when no name-bearing type specimen (i.e. holotype, lectotype, syntype, or prior neotype) is believed to be extant and an author considers that a name-bearing type is necessary to define the nominal taxon objectively; the continued existence of paratypes or paralectotypes does not in itself preclude the designation of a neotype), and by use of the plenary power of the Commission (Art. 81 – Use of the Plenary Power).

Principle of Priority

The valid name of a taxon is the oldest available name applied to it unless that name has been invalidated or another name is given precedence by any provision of the Code or by any ruling of the Commission. For this reason priority applies to the validity of synonyms (the Principle of Priority requires that a taxon formed by bringing together into a single taxon at one rank two or more previously established nominal taxa within the family group, genus group, or species group takes as its valid name the name determined in accordance with the Principle of Priority and its Purpose, with change of suffix if required in the case of a family-group name), to the relative precedence of homonyms, the correctness (or otherwise) of spellings, and to the validity of nomenclatural acts (such as acts taken under the Principle of the First Reviser and the fixation of name-bearing types).

Principle of Homonymy

When two or more taxa are distinguished from each other they must not be denoted by the same name within its group. When two or more names are homonyms, only the senior, as determined by the Principle of Priority, may be used as a valid name, with few exceptions. The relative precedence of homonyms (including primary and secondary homonyms in the case of species-group names) is determined by applying the relevant provisions of the Principles of Priority and the First Reviser.

CRITERIA OF AVAILABILITY

According to the International Code for Zoological Nomenclature, a name is available only when it is published and uses a Latin or Latinised word for forming the binomen (generic and specific) name using the principle of binomial nomenclature. A scientific name may be derived from any language or even arbitrary combination of letters if this is formed to be used as a word (e.g. *eha*, acronym of E.H. Aitken, after whom the species was named). When publishing a new nomen, explicit indication of it as intentionally new by using abbreviations such as *n. sp.*, *sp. nov.*, *gen. nov.*, *nom. nov.*, and so on should be present, and fixation of name-bearing types, their correct designations, and details of deposition should be provided, failing which, the nomen will not be considered valid.

SYNONYMS

The International Code for Zoological Nomenclature strictly maintains that no taxon of the same rank can have more than one name; simply put, there will be no two names for a species. Excepting the valid name (earliest correctly published and available name), all names shall be treated as synonyms. The earliest name is called the senior synonym and the later name the junior synonym. Synonyms are also categorised as objective synonyms or subjective synonyms. Objective (nomenclatural or homotypic) synonyms refer to taxa with the same type and the same rank. According to the ICZN Code, (a) if two or more objectively synonymous generic names have been used as the basis for names in the family group, the family-group names are objective synonyms; (b) if two or more nominal genus-group taxa have the same type species, or type species with different names but based on the same name-bearing type, their names are objective synonyms; and (c) if two or more nominal species-group taxa have the same name-bearing type, their names are objective synonyms. However, in the case of subjective (taxonomic or heterotypic) synonymy, no such shared type is present, allowing for debate and taxonomic adjustments by later workers. Objective synonyms are common at the level of genera, because for various reasons two genera may contain the same type species, while subjective synonyms are common at the level of species, due to natural variation in a species or ignorance about an earlier description. Some valid names ignore senior synonym/s, and this is due to the fact that the senior name has not been used since 1899 and the junior name was in common use. The older name may be declared to be a *nomen oblitum*, and the junior name declared a *nomen protectum*. This rule exists primarily to prevent the confusion that would result if a well-known name, with a large accompanying body of literature, were to be replaced by a completely unfamiliar name.

TYPES

The principle of typification of the ICZN Code mandates that each nominal taxon in the family, genus, or species groups has actually, or potentially, a name-bearing type. The fixation of the name-bearing type of a nominal taxon provides the objective standard of reference for the application of the name it bears. The valid name from a taxon is determined only from the name-bearing type(s). Objectivity through typification is continuous through the hierarchy of names, from species to family group. Once fixed, name-bearing types are stable and provide objective continuity in the application of names. Types are international standards for scientific names. The identity of a name relies only on its type, not on its description or diagnosis.

Basing on the designation, types are either original designation types or subsequent designation types.

The original designation types are fixed in the original publication, and could be one of the following three:

Holotype: the single specimen upon which a new species-group taxon is based in the original publication
Paratypes: the remaining specimens of the original type series (see also allotype)
Syntypes: the specimens of a type series that collectively constitute the name-bearing type

The subsequent designation types are fixed in subsequent publication and could be one of the following three:

Lectotype: a syntype designated as the single name-bearing type specimen, after the establishment of a nominal species or subspecies
Paralectotypes: each specimen of the former syntype series remaining after lectotype designation
Neotype: the single specimen designated as the name-bearing type when no name-bearing type specimen is believed to be extant

Terms not regulated or recognised by the ICZN Code, but commonly used in taxonomic publications, are as follows:

Allotype: a designated specimen of opposite sex of the holotype
Cotype: a term formerly used for either syntype or paratype
Genotype: a term formerly used to designate the holotype

Topotype: a term formerly utilised for a specimen originating from the type locality (the geographical place of capture, collection, or observation of the name-bearing type of a nominal species or subspecies) of the species or subspecies to which it is thought to belong, whether or not the specimen is part of the type series.

FORMATION OF ZOOLOGICAL NAMES

As mentioned earlier, all zoological names are treated as Latin, thus invariably should also follow the rules of Latin linguistics. Latin nouns are declined and verbs are conjugated, that is, their terminations change according to their case, tense, person, and number, or, more simply, the manner in which they are used.

The commonest Latin terminations are (1) *-us* (masculine), *-a* (feminine), *-um* (neuter); and (2) *-is* (masculine), *-is* (feminine), *-e* (neuter); and (3) *-er* (masculine), *-era* (feminine), *-erum* (neuter).

General Recommendation 5 of the ICZN Code states that 'an author establishing a new genus- or species-group name should state its derivation (etymology), and in the case of a genus-group name its gender'.

Nouns are indicated in the nominative singular (*folium*, leaf) and, where the derivation is from the stem of the noun, in the genitive or possessive case also (*foliatus*, leaved). Latin verbs are shown in the present infinitive, rather than the present indicative. When the Latin names are derived from Greek, the Greek words are transliterated in accordance with generally accepted rules. Adjectival epithets or trivial names have to agree in gender with that of the genus to which they are assigned. If a species is transferred from a masculine genus to a feminine one or vice versa, the specific termination must be changed accordingly. Some species names that may look like adjectives are in fact nouns in apposition given an adjectival function, and their terminations do not change to agree with the gender of the generic name. When a species is named after a feature, Latin adjectives are used when after other species, a noun in apposition or adjective is used; when after people, a noun in genitive case is used; and when after a place, an adjectival toponym is used.

Hyphens, diaereses, diacritic marks, and diphthongs are not to be used while creating new names. The following suffixes and grammatical standards are usually used in scientific nomenclature to refer to dedication, location, comparison, relation, inception, and possession:

-aceus, *-acea*, *-aceum*, pertaining to, having the nature of (L.)
-acus, *-aca*, *-acum*, belonging to, pertaining to (L./Gr.)
-age, *-arum*, commemorating (dedication); geographic (location, toponym) (L.)
-alis, *-ale*, pertaining to, having the nature of (L.)
-anus, *-ana*, *-anum*, geographic (location, toponym); belonging to, pertaining to (L.)
-aris, *-arius*, *-aria*, *-arium*, pertaining to, having the nature of; one who (L.)
-atus, *-ata*, *-atum*, provided with, pertaining to (L.)

-ellus, *-ella*, *-ellum*, diminutive (comparison); somewhat (adj.) (L.)
-ensis, *-ense*, geographic, occurrence in (location, toponym) (L.)
-enus, *-ena*, *-enum*, relating to, formation from (L.)
-escens, becoming, somewhat (L.)
-eus, *-ea*, *-eum*, made of, having the quality of (L.)
-i, commemorating (dedication, eponym) (L.)
-icus, *-ica*, *-icum* belonging to, pertaining to (L./Gr.)
-idion, diminutive (comparison) (Gr.)
-idius, *-idia*, *-idium*, diminutive (comparison) (L. (from Gr.))
-ii, commemorating (dedication, eponym) (L.)
-illus, *-illa*, *-illum*, diminutive (comparison); somewhat (adj.) (L.)
-inus, *-ina*, *-inum*, belonging to, pertaining to; one who (L.)
-iscus, *-isca*, *-iscum*, diminutive (comparison) (L./Gr.)
-ister, *-istes*, *-istis*, *-istor*, *-istria*, agent, one who (Gr.)
-ites, agent, one who (Gr.)
-ius, *-ia*, *-ium*, diminutive (comparison); having the nature of; commemorating (dedication, eponym) (L./Gr.)
-olus, *-ola*, *-olum*, diminutive (comparison); somewhat (adj.) (L.)
-oma, formation from, relating to (Gr.)
-orius, *-oria*, *-orium*, pertaining to, having the nature of (L.)
-osus, abundance, fullness, quality of (L.)
-otes, agent, one who (Gr.)
-ter, *-tes*, *-tis*, *-tor*, *-tria*, agent, one who (Gr.)
-ulus, *-ula*, *-ulum*, diminutive (comparison); somewhat (adj.) (L.)
-unus, *-una*, *-unum*, belonging to, pertaining to (L.)

ANALYSIS OF NAMES

The scientific names of mammals, basing on their meaning and derivation, can be divided into following categories:

1. *Morphonym* (Gr. *morphē* form, *onuma* name): Morphonyms of mammals in South Asia are based on the physical characteristics, fur colour, patterns, size, and structure. As much as 31 percent of the names belong to this category, with 20 percent of all generic names, 34 percent of all specific names, and 35 percent of all subspecies names.

   The genera *Acomys*, *Anourosorex*, *Aonyx*, *Asellia*, *Atherurus*, *Belomys*, *Chiropodomys*, *Coelops*, *Crocidura*, *Cynopterus*, *Dacnomys*, *Dicerorhinos*, *Ellobius*, *Hadromys*, *Hipposideros*, *Latidens*, *Lyroderma*, *Macroglossus*, *Megaderma*, *Megaerops*, *Meriones*, *Micromys*, *Microtus*, *Miniopterus*, *Murina*, *Myotis*, *Neodon*, *Niviventer*, *Otocolobus*, *Otomops*, *Otonycteris*, *Paradoxurus*, *Phaiomys*, *Platacanthomys*, *Plecotus*, *Prionodon*, *Pteropus*, *Rhinolophus*, *Rhinopoma*, *Rhombomys*, *Rousettus*, *Saccolaimus*, *Triaenops*, and *Tylonycteris* belong to this category.

   The specific epithets *abramus*, *acutorostrata*, *albicauda*, *alboniger*, *aquilus*, *argentatus*, *ater*, *attenuata*, *aurata*, *aureus*, *aurex*, *auritus*, *auropunctatus*, *badius*, *bicolor*, *brachyotis*, *brachypterus*, *brachyura*, *breviceps*, *brevirostris*, *caliginosus*, *cana*, *caudata*, *caudatus*, *cavirostris*, *cervicolor*, *chrysogaster*, *cineraceus*, *cinerea*, *cinereus*, *circumdatus*, *coeruleoalba*, *collaris*, *concolor*, *crassicaudata*, *crassidens*, *crassus*, *cupreus*, *cyclotis*, *densirostris*, *diadema*, *electra*, *emarginatus*, *fascicularis*, *ferrilata*, *ferrumequinum*, *fimbriatus*, *flavigula*, *formosus*, *fulgens*, *fuliginosa*, *fuliginosus*, *fulvescens*, *fulvus*, *fuscocapillus*, *fuscus*, *galeritus*, *giganteus*, *grisea*, *griseus*, *hipposideros*, *hispida*, *hispidus*, *homochrous*, *hypoleucos*, *hypomelanus*, *hypomelas*, *hypophyllus*, *jubatus*, *laniger*, *larvata*, *lasyurus*, *lenis*, *leptodactylus*, *leucodon*, *leucogaster*, *leucogenys*, *leuconedys*, *leucophaea*, *leucops*, *leucura*, *leucurus*, *longimanus*, *longipes*, *macrocephalus*, *macrorhynchus*, *macrotis*, *macroura*, *macrourus*, *macrurus*, *magnater*, *magnificus*, *marmorata*, *maximus*, *megalotis*, *melanogaster*, *melanopogon*, *melanotus*, *microphyllum*, *micropus*, *micrura*, *minutus*, *mitratus*, *nanus*, *nasutus*, *niger*, *nigricollis*, *nigriscens*, *nitidus*, *nivalis*, *niviventer*, *nudiventris*, *oiostolus*, *opimus*, *ornatus*, *pachyomus*, *pachyotis*, *pallidus*, *pallipes*, *pardicolor*, *pentadactyla*, *perforatus*, *pergrisea*, *personata*, *perspicillata*, *picta*, *picticaudata*, *pileatus*, *planiceps*, *platythrix*, *plicatus*, *pruionosus*, *pullata*, *pusillus*, *pyctoris*, *pygerythrus*, *radiata*, *rubidus*, *rubiginosus*, *rufescens*, *rufigenis*, *rusiges*, *saccolaimus*, *schistaceus*, *sima*, *squamipes*, *stenocephalus*, *strigidorsa*, *subbadius*, *subgutturosa*, *sublineatus*, *taxicolor*, *teniotis*, *tenuis*, *terricolor*, *tridens*, *trifoliatus*, *tristriatus*, *trivirgata*, *truncatus*, *tubinaris*, *unicolor*, *vetulus*, and *vitticollis* belong to this category.

The subspecific nomen *adustus*, *albifrons*, *albiventer*, *angulatus*, *argentescens*, *aurea*, *aureus*, *aurifrons*, *auritus*, *aurobrunneus*, *brachyotus*, *brachysoma*, *brevicauda*, *caniceps*, *canigula*, *caniscus*, *castaneus*, *cinerascens*, *concolor*, *craspedotis*, *crassus*, *cristatus*, *cycloceros*, *diluta*, *erythrogaster*, *erythrotis*, *ferruginea*, *ferrugineus*, *flavidens*, *fulvidus*, *fulvus*, *fusca*, *fuscifrons*, *gracilis*, *grandis*, *inauritus*, *indicus*, *inermis*, *inornatus*, *intermedia*, *iodes*, *isabellinus*, *labiata*, *laniger*, *lasiotis*, *lasyurus*, *leptophyllus*, *leucocephalus*, *lutescens*, *macrodens*, *macrosceloides*, *majus*, *malabaricus*, *maxima*, *megaceros*, *melanochra*, *minor*, *nictitatans*, *nigripecta*, *obscurus*, *ophistomelas*, *ornata*, *pallidus*, *pallipes*, *pelops*, *perniger*, *pictus*, *polyodon*, *pusilla*, *rubex*, *rubidior*, *rubidus*, *ruficaudatus*, *sadhu*, *saxicolor*, *secatus*, *seminudus*, *subcristata*, *tenebricus*, *thysanurus*, *tragatus*, *umbrinus*, *umbrosus*, *vellerosus*, and *vittatus* belong to this category.

2. *Eponym* (Gr. *epōnumos*, named after): An eponym commemorates a real person or a mythical or fictional character. Most eponyms are commemorative names after the collector or discoverer of the species or a fellow scientist who either helped directly or indirectly in species collection or has contributed to the appropriate group or area. Of nearly 140 mammalogists, collectors, and scientists whose names were commemorated in the names of South Asian mammals (both specific and subspecific names), only 6 names

are based on female names. It is one of the most popular forms of nomenclature, which contributes for 19 percent of names of mammals in South Asia, including just over one-fifth of all specific and subspecific names. As many as 10 names are based on mythical or classical names, including two names based on the female gender. Among species eponyms, a maximum of three species was named for William Thomas Blanford and Brian Houghton Hodgson each, followed by nine individuals who had two species each named after them. As many as four subspecies have been named after Brian Houghton Hodgson, and three subspecies after Edward Blyth, followed by five individuals who had two subspecies each named after them. Brian Houghton Hodgson and Reginald Innes Pocock have based their eponyms on classical and mythical characters.

As many as five genera have also been named after persons, and these include *Biswamoyopterus*, *Blanfordimys*, *Leopoldamys*, *Millardia*, and *Steno*. Although the use of personal names in the formation of compound genus names has long been objectionable, names such as *Biswamoyopterus*, *Blanfordimys*, and *Leopoldamys* are examples of such exception for South Asian mammals.

The species *ajax*, *baileyi*, *beddomei*, *bedfordiae*, *belangeri*, *bennettii*, *biswasi*, *blanfordi*, *blasii*, *blythii*, *bowersi*, *brahma*, *cadornae*, *cookii*, *crumpi*, *csorbai*, *curzoniae*, *dayi*, *dormeri*, *durgadasi*, *dussumieri*, *duvaucelii*, *edeni*, *edwardsi*, *edwardsii*, *eha*, *eldii*, *ellioti*, *davidii*, *falconeri*, *fellowesgordoni*, *fernandoni*, *forresti*, *frithii*, *gathornei*, *geei*, *gleadowi*, *gmelini*, *guilleni*, *gwatkinsii*, *hardwickii*, *harpia*, *heathii*, *hector*, *hemprichii*, *hodgsoni*, *horsfieldii*, *hosei*, *humei*, *huttoni*, *io*, *jenkinsi*, *jerdoni*, *johnii*, *juldaschi*, *kuhlii*, *layardi*, *leisleri*, *leschenaultii*, *lydekkerianus*, *macclellandi*, *mackenziei*, *margarita*, *mayori*, *michaelis*, *millardi*, *niethammeri*, *niphanae*, *nolthenii*, *ognevi*, *pearsoni*, *pearsonii*, *pennantii*, *pernyi*, *phayrei*, *phillipsi*, *priam*, *ranjiniae*, *reuppelli*, *rouxii*, *roylei*, *salimalii*, *savii*, *schmidi*, *shortridgei*, *silenus*, *smithii*, *stoliczkanus*, *strelkovi*, *tatei*, *temminckii*, *theobaldi*, *tickelli*, *wardi*, *williamsi*, *wroughtoni*, *wynnei*, and *zarudnyi* belong to this category.

The subspecies *austenianus*, *baberi*, *baptistae*, *bedfordi*, *blanfordi*, *blythii*, *brahma*, *branderi*, *brodiei*, *caporiaccoi*, *charltonii*, *christyi*, *cuvieri*, *davidi*, *dayanus*, *durga*, *eileenae*, *grayi*, *griffithi*, *gyi*, *heptneri*, *hodgsoni*, *horsfieldii*, *howelli*, *kelaarti*, *kinneari*, *maccarthiae*, *macmillani*, *mayori*, *meyeroehmi*, *millsi*, *minor*, *murraiana*, *nestor*, *pallasi*, *peytoni*, *philbricki*, *phillipsi*, *polii*, *prateri*, *ranjitsinhi*, *robertsoni*, *scheffeli*, *scherzeri*, *schmitzi*, *scindiae*, *simcoxi*, *singhei*, *stevensi*, *thersites*, *thomasi*, *trevelyani*, *tytleri*, *vignei*, *wallachi*, *wardi*, *wellsi*, *whitei*, *wollastoni*, *wroughtoni*, and *zarudnyi* belong to this category.

The Latin genitives are formed from personal names by adding *-i* to a man's name if it ends in a vowel or *-er* (the vowel itself often being changed to *i*, e.g. *millardi* for Walter Samuel Millard); *-ii* if it ends in a consonant, has been Latinised, or has a Latin form (e.g. *blasii* for Johann

Heinrich Blasius); and by adding *-ae* to a woman's name (e.g. *ranjiniae* for P.V. Ranjini). Although no South Asian mammal examples have plural eponyms, the plural forms are given the terminations *-orum* (masculine) or *-arum* (feminine). Personal names also take the form of adjectives agreeing in gender with the generic name (e.g. *lydekkerianus* for Richard Lydekker), but this is no longer recommended. Rules for the formation of eponyms have changed and been flouted over the years, hence the occurrence of the genitive terminations *-i* and *-ii* commemorating the same person, which can be found in the literature; this becomes evident when the synonyms are considered.

3. *Taxonym* (Gr. *taxis* arrangement; *onuma* name): A variety of names is included here, all suggestive of relationship or resemblance, diminutives, generic combinations, combinations of generic and substantive names, comparison, affiliation, or questionable affinities. As many as 16 percent of the names of South Asian mammals are taxonyms. These are more frequent in generic names (39 percent) than specific names (12 percent) and subspecific names (7 percent). The use of Greek combining forms *-oidēs* (resembling), *pseudo-* (false), and *-opsis* (appearance), and Latin *sub-* (near to), are especially common. Epithets such as *affinis* and *similis* imply a comparative or relative degree (i.e. the species so designated are considered similar to another species or group of species).

The genera *Acinonyx*, *Ailurus*, *Apodemus*, *Arctitis*, *Arctogalidia*, *Arctonyx*, *Arielulus*, *Balaenoptera*, *Berylmys*, *Boselaphus*, *Budorcas*, *Callomyscus*, *Callosciurus*, *Cannomys*, *Capricornis*, *Caprolagus*, *Chaerephon*, *Cricetulus*, *Diomys*, *Dremomys*, *Dryomys*, *Eoglaucomys*, *Eonycteris*, *Eothenomys*, *Episoriculus*, *Eubalaena*, *Eupetaurus*, *Euroscaptor*, *Falsistrellus*, *Feroculus*, *Globicephalus*, *Grampus*, *Harpiocephalus*, *Harpiola*, *Helarctos*, *Hemiechinus*, *Hemitragus*, *Hylopetes*, *Ia*, *Indopacetus*, *Kogia*, *Lagenodelphis*, *Lutrogale*, *Megaptera*, *Melogale*, *Mesoplodon*, *Moschiola*, *Neofelis*, *Neophocaena*, *Nilgiritragus*, *Nycticebus*, *Orcaella*, *Orcinus*, *Paguma*, *Pantholops*, *Paraechinus*, *Parascaptor*, *Pardofelis*, *Peponocephala*, *Petaurista*, *Petinomys*, *Philetor*, *Porcula*, *Prionailurus*, *Procapra*, *Pseudois*, *Pseudorca*, *Rhinoceros*, *Rhizomys*, *Rhyneptesicus*, *Rucervus*, *Salpingotulus*, *Semnopithecus*, *Solisorex*, *Spermophilopsis*, *Sphaerias*, *Stenella*, *Tadarida*, *Tamiops*, *Tetracerus*, *Trachypithecus*, *Tursiops*, *Vandeleuria*, *Viverricula*, and *Vormela* belong to this category.

The species *aduncus*, *affinis*, *alpinus*, *annectans*, *arctoides*, *armiger*, *burrus*, *cervicapra*, *chaus*, *civettina*, *cognatus*, *dimidiatus*, *dimissus*, *draco*, *elegans*, *elvira*, *entellus*, *erythreus*, *exulans*, *famulus*, *faunulus*, *feroculus*, *fertilis*, *gliroides*, *hemionus*, *hermaphroditus*, *ilaeus*, *leonina*, *lepidus*, *luctus*, *lyra*, *meridianus*, *mulatta*, *murinus*, *musculus*, *nebulosa*, *nitedula*, *nobilis*, *palmarum*, *paterculus*, *petaurista*, *phocaenoides*, *pomona*, *porcinus*, *quadricornis*, *sacratus*, *sicarius*, *sobrinus*, *speoris*, *sphinx*, *talpinus*, *tragocamelus*, *unicornis*, *ursinus*, *vaginalis*, and *zibetha* belong to this category.

The subspecies *affinis*, *aladdin*, *ariel*, *babu*, *consul*, *domesticus*, *dryas*, *famulus*, *gentilis*, *lepidus*, *lokrioides*, *midas*, *neglecta*, *numarius*, *nycticeboides*, *peshwa*, *primaevus*, *proximus*, *satyrus*, *sobrinus*, and *viverrinus* belong to this category.

4. *Toponym* (Gr. *topos* place; *onuma* name): As many as 15 percent of the names of South Asian mammals are based on type locality or geographic range of the species. These are more frequently used specifically than generically, and 17 percent of the specific names and less than 1 percent of the generic names are of this category. As many as 23 percent of subspecific names are toponyms. Some of the names simply commemorate a larger geographic area from where the species has been reported (e.g. *aegyptiaca*, *assamensis*, *bengalensis*, *ceylonicus*, *indicus*), while some names indicate smaller area from where the types have been collected (e.g. *darjeelingensis*, *elburzensis*, *putaoensis*, *sabanus*, *siangensis*). The most numerous and accurate toponyms refer to India (seven specific and four subspecific names), Sri Lanka (three specific and four subspecific names), and Nepal (one specific and two subspecific names). Some toponyms allude to the type locality without direct reference to the actual place or habitat (*australis*, *centralis*, *desertorum*, *dukhunensis*, *nordicus*, *pahari*, etc.). One country, Sri Lanka (*Srilankamys*), and one region, Madras (Presidency) (*Madromys*), are commemorated in generic names.

   The species *aegyptiaca*, *aegyptiacus*, *afghanus*, *altiaca*, *andamanensis*, *assamensis*, *australis*, *baluchi*, *bengalensis*, *bocharicus*, *bredanensis*, *bucharensis*, *capensis*, *ceylonicus*, *chinensis*, *coromandra*, *cutchicus*, *darjelingensis*, *elburzensis*, *etruscus*, *gangetica*, *gobiensis*, *gurkha*, *himalayana*, *himalayica*, *hurrianae*, *indica*, *jaintiana*, *javanicus*, *jemlahicus*, *khasiana*, *kondana*, *ladacensis*, *langbianis*, *lankadiva*, *libycus*, *malayana*, *malayanus*, *manipulus*, *mechkuaensis*, *mishmiensis*, *muscatellum*, *nicobarica*, *nicobarulae*, *nilagirica*, *nipalensis*, *norvegicus*, *novaeangliae*, *ohiensis*, *orientalis*, *pacificus*, *pahari*, *persicus*, *philippensis*, *putaoensis*, *sabanus*, *satarae*, *siangensis*, *sibirica*, *sikimensis*, *siligorensis*, *sinica*, *sinicus*, *soluensis*, *sondaicus*, *sumatrensis*, *thibetana*, *thibetanus*, *tibetanus*, *yunanensis*, and *zeylanicus* belong to this category.

   The subspecies *andamanensis*, *arabicus*, *aryabertensis*, *bactrianus*, *bellaricus*, *bengalensis*, *camortae*, *caucasicus*, *centralis*, *centrasiaticus*, *ceylonense*, *ceylonensis*, *chhattisgarhi*, *chialtanensis*, *desertorum*, *dukhunensis*, *ferghanae*, *gangutrianus*, *garonum*, *gedrosianus*, *geminorum*, *gurkha*, *himalayanus*, *imbrensis*, *indica*, *indus*, *kachhensis*, *kashgaricus*, *lanka*, *laosiensis*, *lhasaensis*, *madrassius*, *maimanah*, *malabarica*, *malabaricus*, *matugamaensis*, *mishmiensis*, *nallamalaensis*, *nepalensis*, *newara*, *nicobarensis*, *nipalensis*, *nordicus*, *pashtonus*, *peguensis*, *persica*, *punjabiensis*, *rajputanae*, *sadiya*, *salinarum*, *seianum*, *sifanicus*, *sikimaria*, *sindhica*, *sindica*, *singhala*, *subhemachalana*, *yarkandensis*, and *zelanius* belong to this category.

5. *Autochthonym* (Gr. *autokhthōn*, indigenous, native; *onuma*, name): Autochthonyms are scientific names based on modern native language

names and adopted from over twenty modern languages. They also include classical epithets taken or modified directly from Latin and Ancient Greek. As many as 12 percent of the names of South Asian mammals are autochthonyms. These are more frequently used generically than specifically, and 28 percent of the generic names (and less than 8 percent of the specific names) are of this category. The origin of some names, lost or unrecorded, is undoubtedly indigenous, because authors like Thomas Horsfield and Brian Houghton Hodgson frequently made use of such names without supplying an etymology. Some autochthonyms are a challenge, as many authors have not provided proper explanations while first describing the taxon.

The genera *Allactaga*, *Anathana*, *Antilope*, *Axis*, *Bandicota*, *Barbastella*, *Bos*, *Bubalus*, *Canis*, *Capra*, *Caracal*, *Cervus*, *Cuon*, *Delphinus*, *Dugong*, *Elephas*, *Equus*, *Felis*, *Feresa*, *Gazella*, *Gerbillus*, *Golunda*, *Hesperoptenus*, *Hoolock*, *Hyaena*, *Hystrix*, *Kerivoula*, *Lepus*, *Loris*, *Lutra*, *Lynx*, *Macaca*, *Marmota*, *Martes*, *Meles*, *Mus*, *Mustela*, *Nesokia*, *Ochotona*, *Ovis*, *Panthera*, *Physeter*, *Pipistrellus*, *Platanista*, *Rattus*, *Ratufa*, *Sicista*, *Sorex*, *Sousa*, *Suncus*, *Sus*, *Tatera*, *Tupaia*, *Ursus*, *Urva*, *Vespertilio*, *Viverra*, *Vulpes*, and *Ziphius* belong to this category.

The species *ammon*, *arctos*, *arnee*, *axis*, *binturong*, *booduga*, *caracal*, *corsac*, *dugon*, *elaphus*, *erminea*, *foina*, *gaurus*, *goral*, *hikmiya*, *hoolock*, *hotaula*, *hyaena*, *kathiah*, *khur*, *kiang*, *leo*, *lokriah*, *lupus*, *lutra*, *lynx*, *manul*, *meles*, *meltada*, *meminna*, *miya*, *munzala*, *nayaur*, *noctula*, *orca*, *pardus*, *peregusna*, *physalus*, *pipistrellus*, *rattus*, *scrofa*, *serotinus*, *tanezumi*, *thar*, *tigris*, *tolai*, *uncia*, *urva*, and *vulpes* belong to this category.

The subspecies *bondar*, *chanco*, *dandolena*, *hanglu*, *kotiya*, *kutab*, *kutas*, *nair*, *naria*, *nirnai*, *nyula*, *temon*, and *toufoeus* belong to this category.

6. *Bionym* (Gr. *bios* life; *onuma* name): These names are based on the habitat and behaviour of the mammals towards the habitat or environment. A mere 4 percent of the names of South Asian mammals are bionyms. These are more frequently used generically (7 percent) than specifically (3 percent) and subspecifically (4 percent).

   The genera *Alticola*, *Chimarrogale*, *Cremnomys*, *Eptesicus*, *Hyperacrius*, *Mellivora*, *Melursus*, *Naemorhedus*, *Nectogale*, *Scotoecus*, *Scotomanes*, *Scotophilus*, *Scotozous*, and *Taphozous* belong to this category.

   The species *aegargus*, *alticola*, *excelsus*, *hylocrius*, *kathygre*, *montanus*, *montivagus*, *montosa*, *muricola*, *palustris*, *pluvialis*, *saxicola*, *silvestris*, and *spelaea* belong to this category.

   The subspecies *deserti*, *eremtaga*, *hypsibius*, *maris*, *mimus*, *montana*, *montanus*, *monticola*, *nemorivaga*, *siccatus*, and *tropicalis* belong to this category.

7. *Ergonym* (Gr. *ergon*, work, occupation; *onuma*, name): Ergonyms refer to display, typical habits, or temperament. A mere 2 percent of the names of South Asian mammals are ergonyms. These include the genera *Funambulus*,

Hypsugo, *Jaculus*, *Manis*, and *Nyctalus*; species *elator*, *latronum*, *migratorius*, *moschata*, *mutus*, *oleracea*, *rapax*, *spasma*, *stoicus*, *tardigradus*, and *vorax*; and subspecies *surda*.

8. *Phagonym* (Gr. *phageinto* eat; *onuma* name): Phagonym is based on dietary habits. Only one genus of South Asian mammal, *Spermophilus* (literally, seed-eater), belongs to this category. Usually, the phagonyms end with the suffixes –*phagus*, -eating; -*thēras*, hunter; and –*vorus*, -eating.

3 Generic and Specific Names Explained

Order Proboscidea Illiger, 1811
Family Elephantidae Gray, 1821

1. ***Elephas maximus* Linnaeus, 1758**
Asian Elephant

Elephas: Gr. *elephas* elephant (Gr. *ele* arch; Gr. *phas* huge; literally Gr. ivory).

maximus: Gr. *maximus* greatest or largest (super. >*magnus* great).

South and Southeast Asia. Besides the nominate subspecies, *E. m. maximus* Sri Lankan Elephant, endemic to Sri Lanka, another subspecies is known from the region.

Subspecies

- *indicus*: L. adj. *indicus* Indian, belonging to India (*-icus* belonging to, pertaining to (L./Gr.)).

  Indian Elephant. Bangladesh, Bhutan, India, and Nepal.

OEN Asiatic Elephant (J. Syn. *asiaticus*), Bengal Elephant (J. Syn. *bengalensis*), Ceylonese Elephant (J. Syn. *ceylanicus*), Deccan Elephant (J. Syn. *dakhunensis*), Eastern Ceylon Elephant (J. Syn. *sinhaleyus*), Marsh Elephant (J. Syn. *vilaliya*), Ceylon Marsh Elephant (J. Syn. *vilaliya*)

VN Hind. *Hathi* (male), *Hathini* (female); Gond. *Yenu*; Tel. *Yenugu*; Kan. *Ani*; Mal. *Ani*; Tam. *Ani*, *Yanei*; Sinh. *Aliya* (female), *Atha* (male), *Athini* (female), *Hora-aliya* (rogue); Tam. (SL) *Yanei*

Order Sirenia Illiger, 1811
Family Dugongidae Gray, 1821

2. ***Dugong dugon* (Müller, 1776)**
Dugong

Dugong: Mod. L. *dugong* from the Malayan name *duyong* for the species.

dugon: possibly also from the same origin as the generic name.

Warm coastal waters from the western Pacific Ocean to the eastern coast of Africa.

OEN 'Sea Pig'

VN Sinh. *Mudu Ura*, *Talla maga*; Tam. (SL) *Caddadt-pandri*

Order Scandentia Wagner, 1855
Family Tupaiidae Gray, 1825

3. *Anathana ellioti* (Waterhouse, 1850)
Madras Treeshrew

Anathana: Mod. L. *anathana* >Tamil *mungil anathaan* (for Bamboo Squirrel).

ellioti: Eponym after Sir Walter Elliot (b. 1803–d. 1887), Indian civil servant and amateur naturalist, (*-i*, commemorating (dedication, eponym) (L.)).

Endemic to India.

OEN Elliot's Treeshrew, Pale Treeshrew (J. Syn. *pallida*), Wrougthon's Treeshrew (J. Syn. *wroughtoni*)
VN Tam. *Mungil anathaan*

4. *Tupaia belangeri* (Wagner, 1841)
Northern Treeshrew

Tupaia: Mod. L. *tupaia* >Malay *tupai* (for various small squirrel-like creatures).

belangeri: Eponym after Charles Belanger (b. 1805–d. 1881), a French traveler, (*-i*, commemorating (dedication, eponym) (L.)).

South and Southeast Asia; Bangladesh, Bhutan, India (North East), and Nepal.

OEN Assam Tree Shrew (J. Syn. *assamensis*), Sikkim Tree Shrew
VN Lepch. *Kalli-tang Zhing*

5. *Tupaia nicobarica* (Zelebor, 1869)
Nicobar Treeshrew

Tupaia: Mod. L. *tupaia* >Malay *tupai* (for various small squirrel-like creatures).

nicobarica: L. adj. *nicobarica* Nicobarese or belonging to Nicobar Islands, India (*-ica* belonging to, pertaining to (L./Gr.)).

Endemic to Nicobar Islands, India. Besides the nominate subspecies, *T. n. nicobarica* Great Nicobar Island Treeshrew, another subspecies, is known from the region.

Subspecies

- *surda* – L. *surdus* silent, still.

  Little Nicobar Island Tree Shrew. Endemic to Little Nicobar Island.

Order Primates Linnaeus, 1758
Family Lorisidae Gray, 1821

6. *Loris tardigradus* (Linnaeus, 1758)
Red Slender Loris

Loris: Mod. L. *loris* from French *loris* for the species (from Dutch *loeris* meaning simpleton).

tardigradus: L. *tardigradus* slow-walker (L. adj. *tardus* slow or limping, L. *gradus* step).

Endemic to Sri Lanka. Besides the nominate subspecies, *L. t. tardigradus* Dry Zone Slender Loris, another subspecies, is known.

Subspecies

- *nycticeboides*: L. adj. *Nycticeboides* nycticebus-like (from Mod. L. *nycticebus* night-monkey; L. genitive *nuktos* night >Mod. L. *nox* or Gr. *nux*; Gr. *kēbos* long-tailed monkey, here simply 'a monkey'), (*-oides* (>Gr. *eidos*) apparent shape).

  Horton Plains Slender Loris. Endemic to Sri Lanka.

OEN Ceylon Sloth (Subsp. *tardigradus, nycticeboides*), Ceylon Loris (Subsp. *tardigradus, nycticeboides*), Ceylon Mountain Slender Loris (Subsp. *nycticeboides*)

VN Sinh. *Unhapuluva*, *Rath Unahapuluwa* (Subsp. *tardigradus*, *nycticeboides*); Tam. (SL) *Thevangu* (Subsp. *tardigradus*, *nycticeboides*)

7. *Loris lydekkerianus* Cabrera, 1908
Grey Slender Loris

Loris: Mod. L. *loris* from French *loris* for the species (from Dutch *loeris* meaning simpleton).

lydekkerianus: eponym after Richard Lydekker (b. 1849–d. 1915), English zoologist and paleontologist (*-anus*, belonging to, pertaining to (L.)).

Endemic to India and Sri Lanka. Besides the nominate subspecies, *L. l. lydekkerianus* Mysore Grey Slender Loris, three more subspecies are known.

Subspecies

- *malabaricus*: L. adj. *malabaricus* meaning from Malabar – a region in southern India lying between the Western Ghats and the Arabian Sea, corresponding to the Malabar District of the Madras Presidency in British India (presently includes areas under five northern districts of Kerala), (*-icus*, belonging to, pertaining to (location, toponym) (L.)).

  Malabar Grey Slender Loris. Endemic to India.

- *grandis*: L. *grandis* large.
  Highland Slender Loris. Endemic to Sri Lanka.
- *nordicus*: L. adj. *nordicus* northern (Fr. *nordique*, *nord* north), (*-icus* belonging to, pertaining to (L./Gr.)).
  Northern Ceylonese Slender Loris. Endemic to Sri Lanka.

OEN Lydekker's Slender Sloris (Subsp. *lydekkerianus*), Gray Slender Loris (Subsp. *lydekkerianus*), Ceylon Sloth (Subsp. *grandis*, *nordicus*), Ceylon Loris (Subsp. *grandis*, *nordicus*)

VN Tam. *Tevangar* (Subsp. *lydekkerianus*, *malabaricus*), *Kattu-papa*, *Kattu-pullaye* (Subsp. *lydekkerianus*, *malabaricus*); Kan. *Kada-papa*, *Adavi-papa* (Subsp. *lydekkerianus*, *malabaricus*); Kurg. *Hunnimunna*, *Singalika*, *Kard-munishya* (Subsp. *malabaricus*); Mar. *Wanur-manushya* (Subsp. *lydekkerianus*); Tel. *Arawi-papa*, *Dewantsi pilli* (Subsp. *lydekkerianus*); Dekh. *Sharminda* (Subsp. *lydekkerianus*); Hind. *Sharminda* (Subsp. *lydekkerianus*); Sinh. *Unhapuluva*, *Alu Unahapuluwa* (Subsp. *lydekkerianus*); Sinh. *Kalu Unhapuluva* (Subsp. *grandis*, *nordicus*); Tam. (SL) *Thevangu*, *Kadu-papa* (Subsp. *grandis*)

8. *Nycticebus bengalensis* (Lacépède, 1800)
Bengal Slow Loris

Nycticebus: Mod. L. *nycticebus* night-monkey (L. genitive *nuktos* night >Mod. L. *nox* or Gr. *nux*; Gr. *kēbos* long-tailed monkey, here simply 'a monkey').
bengalensis: L. adj. *bengalensis* meaning from Bengal, India (*-ensis*, geographic, occurrence in (location, toponym) (L.)).
Northeast India and Southeast Asia.

OEN Slow-paced Lemur

VN Beng. *Lajja banar*, *Lajjawoti banar*; Hind. *Sharmindah billi*, *Sharmindi billi*

Family Cercopithecidae Gray, 1821
Subfamily Cercopithecinae Gray, 1821

9. *Macaca silenus* (Linnaeus, 1758)
Lion-tailed Macaque

Macaca: Mod. L. *macaca* macaque (>Fr. *macaque* or Portugese *macaco* meaning 'a monkey').
silenus: Gr. Myth. Silenus, companion of the roman god *Bacchus*, one among the *Sileni* (plural), for gods of the woods.
Endemic to India; Karnataka, Kerala, and Tamil Nadu (in the Western Ghats).

OEN Lion Monkey, Liontail Macaque, Wanderoo

VN Mal. *Nella-manthi, Chingala*; Kan. *Singalika*; Kurg. *Karingode*; Tam. *Kurankarangu, Karupu Korangu*; Tel. *Kondamachu*; Beng. *Nil-bandar*; Hind. *Siah-bandar, Shia-bandar*

10. *Macaca sinica* (Linnaeus, 1771)
Toque Macaque

Macaca: Mod. L. *macaca* macaque (>Fr. *macaque* or Portugese *macaco* meaning 'a monkey').
sinica: L. adj. *sinica* meaning Chinese or from China (a misnomer, as the species is endemic to Sri Lanka), (*-ica* belonging to, pertaining to (L./Gr.)).

Endemic to Sri Lanka. Beside the nominate subspecies, *M. s. sinica* Dry Zone Toque Macaque known from dry zones of Central, Eastern, North Central, North Western, Northern, Sabaragamuwa, Southern, and Uva Provinces, two more subspecies are known from the region.

Subspecies

- *aurifrons*: Mod. L. *aurifrons* gold-fronted (L. *aurum*, *auri* gold; L. *frons* forehead, front).
  Wet Zone Toque Macaque. Wet zones of Central, Sabaragamuwa, Southern, Uva, and Western Provinces.
- *opisthomelas*: Mod. L. *opisthomelas* black-backed (Gr. *opisthe* behind, rear; Gr. *melas* black).
  Highland Toque Macaque. Central Province.

OEN Red Monkey (Subsp. *sinica*, *aurifrons*, *opisthomelas*), Mountain Toque Monkey (Subsp. *opisthomelas*)

VN Sinh. *Rilawa* (Subsp. *sinica, aurifrons, opisthomelas*); Tam. (SL) *Kurangu* (Subsp. *sinica, aurifrons, opisthomelas*), *Siru Kurangu* (Subsp. *sinica*), *Sirra Kurangu* (Subsp. *sinica*)

11. *Macaca mulatta* (Zimmermann, 1780)
Rhesus Macaque

Macaca: Mod. L. *macaca* macaque (>Fr. *macaque* or Portugese *macaco* meaning 'a monkey').
mulatta: L. *mulus* a mule, from where *'mulatto'* is derived, which pertains to the offspring of a black person and a white person.

South Asia, Southeast Asia, and southern China.

OEN Rhesus Monkey, Indian Rhesus Monkey, Red Rhesus Monkey (J. Syn. *fulvus*), McMahon's Rhesus Macaque (J. Syn. *mcmahoni*), Nepal Rhesus Monkey (J. Syn. *nipalensis*), Oniops (J. Syn. *oniops*), Bengal Monkey (J. Syn. *rhesus*), Tibetan Rhesus Monkey (J. Syn. *vestita*), Kashmir Rhesus Monkey (J. Syn. *villosus*)

VN Beng. *Morkot*; Bhut. *Piyu*; Hind. *Bandar*; Lepch. *Marcut-banur, Banur, Suhu*; Kash. *Wandar, Puriz, Punj, Ponj*

12. *Macaca radiata* (E. Geoffroy, 1812)
Bonnet Macaque

Macaca: Mod. L. *macaca* macaque (>Fr. *macaque* or Portugese *macaco* meaning 'a monkey').
radiata: L. *radiatus* furnished with rays, a disc (*-ata*, provided with (L.)).

Endemic to India; beside the nominate subspecies, *M. r. radiata* Dark-bellied Bonnet Macaque, found in most of the peninsular India south of the river Godavari, another subspecies is known from the region.

Subspecies

- *diluta*: L. *dilutus* weak, diluted (L. *diluere* to dilute).
  Pale-bellied Bonnet Macaque. Southern parts of Tamil Nadu, Kerala, and Puducherry.

OEN Bonnet Monkey, Madras Monkey, Agasthyamalai Bonnet Macaque (Subsp. *diluta*), Southern Bonnet Macaque (Subsp. *diluta*), Light-bellied Bonnet Macaque (Subsp. *diluta*), Northern Bonnet Macaque (Subsp. *radiata*)

VN Kurg. *Much*; Kan. *Munga, Mang, Kodaga, Koti, Kapi, Maungya, Kemp Munga*; Mal. *Koranga, Vella munthi*; Mar. *Makadu, Makad, Lal-manga, Wanur, Kerda*; Tam. *Korangu, Vella-manthi*; Tel. *Koti*; Toda *Kodan*; Hind. *Bandar, Bandra*; Dekh. *Bandar, Bandra*; Hind. *Bandar*

13. *Macaca fascicularis* (Raffles, 1821)
Crab-eating Macaque

Macaca: Mod. L. *macaca* macaque (>Fr. *macaque* or Portugese *macaco* meaning 'a monkey').
fascicularis: Mod. L. *fascicularis* small-banded (L. *fasciculus* a band from L. *fascia*; Mod. L. *-aris*, a suffix meaning pertaining to or in allusion of).

Bangladesh, Nicobar Islands, and Southeast Asia. Two subspecies are known from the region.

Subspecies

- *aureus*: L. *aureus* golden (*aurum* gold; *-eus* having the quality of).
  Burmese Long-tailed Macaque. Bangladesh (marginally).
- *umbrosus*: L. adj. *umbrosus* (>L. *umbra* shade, dark) shady, of shade, of twilight.
  Nicobar Long-tailed Macaque. Nicobar Islands.

OEN Long-tailed Macaque, Golden Macaque (Subsp. *aureus*), Nicobar Crab-eating Macaque (Subsp. *umbrosus*)

VN Nic. *Makphoum*

14. *Macaca arctoides* (I. Geoffroy, 1831)
Stump-tailed Macaque

Macaca: Mod. L. *macaca* macaque (>Fr. *macaque* or Portugese *macaco* meaning 'a monkey').
arctoides: Mod. L. *arctoides* bear-like (Gr. *arktos* a bear), (*-oides* (>Gr. *eidos*) apparent shape).

Northeast India and Southeast Asia.

OEN Bear Macaque, Brown Stump-tailed Macaque (S. Syn. *speciosa*)
VN Naga *Chantee*

15. *Macaca assamensis* (McClelland, 1839)
Assam Macaque

Macaca: Mod. L. *macaca* macaque (>Fr. *macaque* or Portugese *macaco* meaning 'monkey').
assamensis: L. adj. *assamensis* meaning from Assam, India, (*-ensis*, geographic, occurrence in (location, toponym) (L.)).

Himalayas, southern China, and Southeast Asia. Besides the nominate subspecies, *M. a. assamenis* Eastern Assam Macaque from the Himalayas of northeast India, another subspecies is known.

Subspecies

- *pelops*: L. *pelops* (Gr. *pélops*) dark face, eye.
  Western Assam Macaque. Himalayas east of Uttarakhand, Nepal, and to Sundarbans in Bangladesh.

OEN Assamese Macaque, Eastern Assamese Macaque (Subsp. *assamensis*), Himalayan Macaque (Subsp. *pelops*), Western Assamese Macaque (Subsp. *assamensis*), Problematic Assamese Macaque (J. Syn. *problematicus*)
VN Lepch. *Sahu*; Bhut. *Pio*; Pahar. *Bandar*; *Panah Bandar* (near Darjeeling)

16. *Macaca leonina* (Blyth, 1863)
Northern Pig-tailed Macaque

Macaca: Mod. L. *macaca* macaque (>Fr. *macaque* or Portugese *macaco* meaning 'a monkey').
leonina: L. *leoninus* belonging to lion (>Gr. *léònina* tawny-coloured, *léòn* lion).

Northeast India and Southeast Asia.

OEN Northern Pigtail Macaque, Northern Pigtail Monkey, Burmese Pig-tailed Macaque (J. Syn. *andamanensis*), Blyth's Pig-tailed Macaque (J. Syn. *blythii*)
VN Naga *Kangh*

17. *Macaca munzala* Sinha, Datta, Madusudhan and Mishra, 2005
Arunachal Macaque

Macaca: Mod. L. *macaca* macaque (>Fr. *macaque* or Portugese *macaco* meaning 'a monkey').

munzala: Mod. L. *munzala* from *mun zala* ('monkey of the deep forest') in the dialect spoken by Dirang Monpa.

Endemic to South Asia; Himalayas in India (Arunachal Pradesh) and Bhutan.

VN Mon. *Mun zala*

18. *Macaca leucogenys* Li, Zhao and Fan, 2015
White-cheeked Macaque

Macaca: Mod. L. *macaca* macaque (>Fr. *macaque* or Portugese *macaco* meaning 'a monkey').

leucogenys: Mod. L. *leucogenys* white-cheeked (Gr. *leukogenus* white-cheek, from Gr. *leukos* white; L. *genys* (>Gr. *genus*) cheek or jaw).

Himalayas in India (Arunachal Pradesh).

Subfamily Colobinae Jerdon, 1867

19. *Semnopithecus vetulus* (Erxleben, 1777)
Purple-faced Langur

Semnopithecus: Mod. L. *semnopithecus* 'sacred-ape' (Gr. *semnos* solemn, sacred, august; Gr. *pithecus* ape).

vetulus: L. *vetulus* old, a little old man.

Endemic to Sri Lanka. Beside the nominate subspecies, *S. v. vetulus* Southern Lowland Wet Zone Purple-faced Langur known from wet zones of Sabaragamuwa, Southern and Western Provinces, three more subspecies are known from the region.

Subspecies

- *nestor*: Gr. Myth. *Nestor*, grey-haired wise old king of Pylos at the siege of Troy – a character in Homer's *Iliad*.

  Southern Lowland Wet Zone Purple-faced Langur. Central, North Western, Sabaragamuwa and Western Provinces.
- *monticola*: L. *monticola* mountain-dweller, mountaineer (*mons*, *montis* mountain; L. *-cola* inhabitant >L. *colere* to dwell).

  Montane Purple-faced Langur. Central, Sabaragamuwa, and Uva Provinces.
- *philbricki*: Eponym after A. N. Philbrick (dates not known), a manager of the Tea Estate at Mousakande, Sri Lanka where W.W.A. Phillips worked, (*-i*, commemorating (dedication, eponym) (L.)).

  Dry Zone Purple-faced Langur. Central, Eastern, North Central, Northern and Uva Provinces.

OEN Southern Purple-faced Leaf-monkey (Subsp. *vetulus*), Black Wanderoo (Subsp. *vetulus*), Western Purple-faced Leaf-monkey (Subsp. *nestor*), Wanderoo (Subsp. *nestor*), Highland Purple-faced Leaf-monkey (Subsp. *monticola*), Bear Monkey (Subsp. *monticola*), Highland Wanderoo (Subsp. *monticola*), Northern Purple-faced Leaf-monkey (Subsp. *philbricki*), Black Wanderoo (Subsp. *philbricki*)

VN Sinh. *Kalu Vandhura* (Subsp. *vetulus, philbricki*), *Vandhura* (Subsp. *nestor*), Maha Vandhura (Subsp. *monticola*); Tam. (SL) *Mundi*

20. *Semnopithecus entellus* (Dufresne, 1797)
Northern Plains Gray Langur

Semnopithecus: Mod. L. *semnopithecus* 'sacred-ape' (Gr. *semnos* solemn, sacred, august; Gr. *pithecus* ape).
entellus: Mod. L. *entellus* from Gr. *éntéllo* to command, species is held in veneration in India.

Endemic to South Asia; India and Bangladesh (introduced).

OEN Bengal Hanuman Langur, Bengal Langur, Indian Langur, Indian Entellus

VN Hind. *Hanuman, Langur*; Kan. *Musya*; Guj. *Vandra*; Mar. *Makur, Wanur*

21. *Semnopithecus johnii* (Fischer, 1829)
Nilgiri Langur

Semnopithecus: Mod. L. *semnopithecus* 'sacred-ape' (Gr. *semnos* solemn, sacred, august; Gr. *pithecus* ape).
johnii: Eponym after Rev. dr. Christoph Samuel John (b. 1747–d. 1813), a medical missionary at Tranquebar (now Tharangambadi) in Tamil Nadu, India, (*-ii*, commemorating (dedication, eponym) (L.)).

Endemic to India; Western Ghats of Karnataka, Kerala and Tamil Nadu.

OEN Nilgiri Black Langur, Nilgiri Leaf Monkey, Black Leaf Monkey, Hooded Leaf Monkey, Indian Hooded Leaf Monkey, John's Langur, Malabar Langur, John's Monkey

VN Toda *Turuni, Kodan, Pershk*; Baduga *Korangu*; Kurumba *Korangu*; Mal. *Karing Korangu*; Kurg. *Kari-Mushya*; Tam. *Mandi*

22. *Semnopithecus schistaceus* Hodgson, 1840
Central Himalayan Langur

Semnopithecus: Mod. L. *semnopithecus* 'sacred-ape' (Gr. *semnos* solemn, sacred, august; Gr. *pithecus* ape).

schistaceus: Mod. L. *schistaceus* slate-grey (>Mod. L. *schistus* slate), (-*aceus*, pertaining to, having the nature of (L.)).

Himalayas, in Bhutan, China, India, Nepal, Pakistan, and possibly in Afghanistan.

OEN Nepal Gray Langur, Black-faced Langur, Pale-armed Himalayan Langur, Himalayan Langur

VN Bhut. *Kubup*, *Propyaka*, *Piopyaka*; Mallaha *Langur*; Pahar. *Derdca*; Lepch. *Kamba Suhu*, *Sahu Kaboo*; Kuma. *Gooni*

23. *Semnopithecus hypoleucos* Blyth, 1841
Dark-legged Malabar Langur

Semnopithecus: Mod. L. *semnopithecus* 'sacred-ape' (Gr. *semnos* solemn, sacred, august; Gr. *pithecus* ape).

hypoleucos: Mod. L. *hypoleucos* somewhat white, from Gr. *hupoleukos* whitish (>Gr. *hupo* beneath; Gr. *leukos* white).

Endemic to India; Goa, Karnataka and Kerala.

OEN Black-footed Langur, Malabar Sacred Langur

VN Kan. *Musya*; Mal. *Karing Korangu*; Tam. *Mandi*

24. *Semnopithecus dussumieri* I. Geoffroy, 1843
Southern Plains Gray Langur

Semnopithecus: Mod. L. *semnopithecus* 'sacred-ape' (Gr. *semnos* solemn, sacred, august; Gr. *pithecus* ape).

dussumieri: Eponym after Jean-Jacques Dussumier (b. 1792–d. 1883), a French collector, traveler, and trader, (-*i*, commemorating (dedication, eponym) (L.)).

Endemic to India.

OEN Malabar Langur, Dark-armed Malabar Langur, Dussumier's Malabar Langur, Dussumier's Sacred Langur, Satpura Langur (J. Syn. *achates*), Western Hanuman Langur (J. Syn. *achates*), Jog Langur (J. Syn. *iulus*), Nagarhole Langur (J. Syn. *elissa*), Shernelly Langur (J. Syn. *priamellus*)

VN Mar. *Makur*, *Wanur*

25. *Semnopithecus priam* Blyth, 1844
Coromandel Gray Langur

Semnopithecus: Mod. L. *semnopithecus* 'sacred-ape' (Gr. *semnos* solemn, sacred, august; Gr. *pithecus* ape).

priam: Gr. Myth. *Priam*, King of Troy – a major character in Homer's *Iliad*.

Endemic to South Asia. Beside the nominate subspecies, *S. p. priam* Coromandel Gray Langur, endemic to south India, another subspecies is known from the region.

Subspecies

- *thersites:* Gr. Myth. *Thersites*, a soldier in Greek army in Trojan war – a character in Homer's *Iliad*.

  Tufted Gray Langur. Endemic to Sri Lanka.

OEN Gray Langur (Subsp. *thersites*), Grey Wanderoo (Subsp. *thersites*), Pale-footed Langur (S. Syn. *pallipes*)

VN Tel. *Gandangi*; Sinh. *Vandhura, Konde Vandhura, Kala Vandhura, Maha Vandhura, Elli Vandhura, Eli-wadura*; Tam. (SL) *Mundi, Periya Mundi, Mundi-kurangu*

26. *Semnopithecus ajax* (Pocock, 1928)
Himalayan Gray Langur

Semnopithecus: Mod. L. *semnopithecus* 'sacred-ape' (Gr. *semnos* solemn, sacred, august; Gr. *pithecus* ape).
ajax: Gr. Myth. *Ajax*, a hero of the Trojan War – a major character in Homer's *Iliad*.

Endemic to India; Himachal Pradesh.

OEN Kashmir Gray Langur, Dark-armed Himalayan Langur, Himalayan Langur, Chamba Sacred Langur, Dark-eyed Himalayan Langur, Himalayan Grey Langur, Western Himalayan Langur

27. *Semnopithecus hector* (Pocock, 1928)
Lesser Hill Langur

Semnopithecus: Mod. L. *semnopithecus* 'sacred-ape' (Gr. *semnos* solemn, sacred, august; Gr. *pithecus* ape).
hector: Gr. Myth. *Hector*, prince of Troy – a major character in Homer's *Iliad*.

Endemic to South Asia; in the Himalayas of Bhutan, India, and Nepal.

OEN Lesser Langur, Terai Gray Langur, Gray Langur, Hanuman Langur, Terai Sacred Langur

28. *Trachypithecus pileatus* (Blyth, 1843)
Capped Langur

Trachypithecus: 'Shaggy-ape' (Gr. *trachus* rough or shaggy, in allusion to longer hair; Gr. *pithecus* ape).
pileatus: L. *pileatus* -capped (L. *pileus* felt-cap), (*-atus*, provided with (L.)).

South Asia and Southeast Asia. Beside the nominate subspecies, *T. p. pileatus* Blond-bellied Capped Langur, occurring in most of northeast India, three more subspecies are known from the region.

Subspecies

- *brahma*: Mod. L. *brahma* for Brahma, a supreme Hindu deity.

  Buff-bellied Capped Langur. Endemic to India, Arunachal Pradesh.

- *durga*: Mod. L. *durga* for Durga, a major Hindu goddess.
  Orange-bellied Capped Langur. Bangladesh and northeast India.
- *tenebricus*: L. *tenebricus* dark, gloomy.
  Tenebrous Capped Langur. Endemic to South Asia, Bhutan and northeast India.

OEN Bonneted Langur (nom. form), Assam Golden Headed Langur (nom. form), Capped Leaf Monkey (nom. form), Capped Monkey (nom. form), Yellow-bellied Capped Langur (Subsp. *pileatus*), Blond-bellied Langur (Subsp. *pileatus*), White-bellied Capped Langur (Subsp. *brahma*), Buff-bellied Capped Leaf Monkey (Subsp. *brahma*), Buff-bellied Langur (Subsp. *brahma*), Orange-bellied Capped Leaf Monkey (Subsp. *durga*), Orange-bellied Langur (Subsp. *durga*), Dusky Capped Langur (Subsp. *tenebricus*), Tenebrous Capped Leaf Monkey (Subsp. *tenebricus*), Silver-backed Capped Langur (J. Syn. *argentatus*), Brown Capped Langur (J. Syn. *saturatior*)

29. *Trachypithecus phayrei* (Blyth, 1847)
Phayre's Leaf Monkey

Trachypithecus: 'Shaggy-ape' (Gr. *trachus* rough or shaggy, in allusion to longer hair; Gr. *pithecus* ape).
phayrei: Eponym after Arthur Purves Phayre (b. 1812–d. 1885), a British Indian army officer and first commissioner of British Burma and naturalist, (-*i*, commemorating (dedication, eponym) (L.)).

Only the nominate subspecies *T. p. phayrei* occurs in northeast India and Bangladesh.

OEN Phayre's Langur (nom. form), Phayre's Leaf-Monkey (nom. form)

30. *Trachypithecus geei* (Khajuria, 1956)
Gee's Golden Langur

Trachypithecus: 'Shaggy-ape' (Gr. *trachus* rough or shaggy, in allusion to longer hair; Gr. *pithecus* ape).
geei: Eponym after Edward Prichard Gee (b. *ca.* 1910–d. 1966), a British tea planter, who first observed this species, (-*i*, commemorating (dedication, eponym) (L.)).

Endemic to South Asia; Himalayas of Assam in India and adjoining Bhutan.

OEN Golden Leaf Monkey (nom. form), Southern Gee's Golden Langur (nom. form), Southern Golden Leaf Monkey (nom. form), Northern Gee's Golden Langur (J. Syn. *bhutanensis*), Northern Golden Leaf Monkey (J. Syn. *bhutanensis*)

Family Hylobatidae Gray, 1871

31. *Hoolock hoolock* (Harlan, 1834)
Hoolock Gibbon

Hoolock: Mod. L. >Burmese *Huluk*, the native name for gibbon.
hoolock: Mod. L. >Burmese *Huluk*, the native name for gibbon.

South and Southeast Asia. Beside the nominate subspecies, *H. h. hoolock* Hoolock Gibbon, occurring in northeast India, another subspecies is known from the region.

Subspecies

- *mishmiensis*: L. adj. *mishmiensis* from Mishmi Hills, Arunachal Pradesh, India, (*-ensis*, geographic, occurrence in (location, toponym) (L.)).

  Mishmi Hills Hoolock Gibbon. Endemic to India, Arunachal Pradesh.

OEN Hoolock (nom. form), White-browed Gibbon (nom. form), Western Hoolock (nom. form), Western Hoolock Gibbon (nom. form), Brown Gibbon (J. Syn. *fuscus*), White Whiskered Gibbon (J. Syn. *scyritus*)

VN Assam. *Holook*, *Huluk*; Khas. *Huluk*; Shyl. *Holook*; Dima. *Holook*; Hind. *Uluk*

32. *Hoolock leuconedys* Groves, 1967
Eastern Hoolock Gibbon

Hoolock: Mod. L. >Burmese *Huluk*, the native name for gibbon.
leuconedys: Mod. L. *leuconedys* pale-ventered (Gr. *leukos* white; Gr. *nedys* stomach).

South and Southeast Asia; northeast India.

Order Rodentia
Suborder Sciuromorpha Brandt, 1855
Family Sciuridae Hemprich, 1820
Subfamily Ratufinae Moore, 1959

33. *Ratufa macroura* (Pennant, 1769)
Grizzled Giant Squirrel

Ratufa: Mod. L. *ratufa* >*ratuphar*, the local name of this squirrel in northern Bihar, India.
macroura: Mod. L. *macroura* long-tailed (Gr. *makros* long; Gr. *-ouros* -tailed (Gr. *oúrá* tail).

Endemic to South Asia. Besides the nominate subspecies, *R. m. macroura* Highland Ceylon Giant Squirrel from Sri Lanka, two subspecies are known from the region.

Subspecies

- *melanochra*: Mod. L. *melanochra* >Gr. *melanokhrōs* black-skinned (Gr. *melas* black; Gr. *khrōs* colour).
  Western Ceylon Giant Squirrel. Endemic to Sri Lanka.
- *dandolena*: Mod. L. *dandolena* after Sinhalese name *dandolena* for the species.
  Grizzled Giant Squirrel. Endemic to India (southern Western Ghats) and Sri Lanka.

OEN Sri Lankan Giant Squirrel (Subsp. *macroura*), Highland Grizzled Giant Squirrel (Subsp. *macroura*), Long-tailed Giant Squirrel (Subsp. *macroura*), Highland Rock-Squirrel (Subsp. *macroura*), Grizzled Hill Squirrel (Subsp. *melanochra*), Western Ceylon Giant Squirrel (Subsp. *melanochra*), Black and Yellow Giant Squirrel (Subsp. *melanochra*), Black Rock-Squirrel (Subsp. *melanochra*), Grizzled Indian Squirrel (Subsp. *dandolena*), Common Ceylon Giant Squirrel (Subsp. *dandolena*), Rock Squirrel (Subsp. *dandolena*), White-legged Squirrel (J. Syn. *albipes*)

VN Sinh. *Kallo Dandolena* (Subsp. *macroura*, *melanochra*), *Dandolena* (Subsp. *dandolena*); Tam. (SL) *Mali-Anil*

34. *Ratufa indica* (Erxleben, 1777)
Indian Giant Squirrel

Ratufa: Mod. L. *ratufa* >*ratuphar*, the local name of this squirrel in northern Bihar, India.

indica: L. adj. *indica* belonging to India, (*-ica* belonging to, pertaining to (L./Gr.)).

Endemic to India. Besides the nominate subspecies, *R. i. indica* Indian Giant Squirrel from Gujarat, Karnataka, and Maharashtra, two subspecies are known from the region.

Subspecies

- *maxima*: Gr. *maximus* greatest (super. >*magnus* great).
  Malabar Giant Squirrel. Endemic to the Western Ghats of Kerala and Tamil Nadu.
- *centralis*: L. *centralis* central, in the middle (L. *centrum* middle).
  Central Indian Giant Squirrel. Endemic to forests of Andhra Pradesh, Chhattisgarh, Goa, Gujarat, Jharkhand, Karnataka, Kerala, Madhya Pradesh, Maharashtra, Orissa and Tamil Nadu.

OEN Red Squirrel (Subsp. *maxima*); Central Indian Red Squirrel (Subsp. *centralis*); Bombay Red Squirrel (J. Syn. *bombayus*); Elphinstone's Red Squirrel (J. Syn. *elphinstonei*)

VN Hind. *Jangli Gilheri*, *Karrat*; Bihar. *Ratuphar*; Beng. *Kat berral*, *Rasu*, *Ratuphar*; Kol *Kondeng*; Gond. *Per-warsti*; Tel. *Bet-Udata*; Mar. *Shekra*; Kan. *Kes-annalu*

35. *Ratufa bicolor* (Sparrman, 1778)
Black Giant Squirrel

Ratufa: Mod. L. *ratufa* >*ratuphar*, the local name of this squirrel in northern Bihar, India.
bicolor: L. adj. *bicolor* having two colours (L. *bi* two; L. *color* colour).

Subspecies

- *gigantea*: L. *giganteus* gigantic (L. *gigas* giant).
McClelland's Giant Squirrel. Himalayas in Bangladesh, Bhutan, India (northeast) and Nepal.

OEN Black Hill Squirrel
VN Bhut. *Shingsham*; Lepch. *Le-hyuk*

Subfamily Sciurinae Fischer de Waldheim, 1817

36. *Belomys pearsonii* (Gray, 1842)
Hairy-footed Flying Squirrel

Belomys: Mod. L. *belomys* beautiful mouse (L. *bellus* beautiful; Gr. *mȳs* mouse).
pearsonii: Eponym after Dr. John Thomas Pearson (dates not known), fellow medical student of J.E. Gray, subsequently collected in India, (-*ii*, commemorating (dedication, eponym) (L.)).
China, Southeast Asia, and South Asia. Himalayas in Bhutan, India (northeast) and Nepal.

OEN Tufted-eared Flying Squirrel (J. Syn. *trichotis*), Upper Assam Flying Squirrel (J. Syn. *villosus*)

37. *Biswamoyopterus biswasi* Saha, 1981
Namdapha Flying Squirrel

Biswamoyopterus: Mod. L. *biswamoyopterus* 'winged-Biswamoy', the generic name is an eponym after Dr. Biswamoy Biswas (b. 1923–d. 1994), renowned ornithologist and former Director, Zoological Survey of India in conjunction of L. -*pterus* winged (Gr. *ptero* wing).
biswasi: Eponym after S. Biswas (dates not known), a scientist at Zoological Survey of India, (-*i*, commemorating (dedication, eponym) (L.)).
Endemic to India; Namdapha Tiger Reserve, Arunachal Pradesh, India.

38. *Eoglaucomys fimbriatus* (Gray, 1837)
Small Kashmir Flying Squirrel

Eoglaucomys: Mod. L. *eoglaucomys* Eastern blue-green mouse (L. *eos* East or pertaining to East; Gr. *glauco* blue green colour; Gr. *mȳs* mouse) *fimbriatus*: Med. L. *fimbriatus* bordered (L. *fimbriae* border or fringe; -*atus*, provided with (L.)).

Endemic to South Asia. Besides the nominate subspecies, *E. f. fimbriatus* Small Kashmir Flying Squirrel, from Himalayas in India (Jammu and Kashmir, Himachal Pradesh, and Uttarakhand) and Pakistan (Khyber Pakhtunkhwa), another subspecies is known from the region.

Subspecies

- *baberi*: Eponym after Zahir-ud-din Muhammad Babur (also spelt as Baber) (b. 1483–d. 1530), a conqueror from central Asia who laid the foundation to Mughal Dynasty, a naturalist and a keen chronicler, his work '*Baburnama*' includes accounts of numerous animals, one such being this species, with which it was confused (-*i*, commemorating (dedication, eponym) (L.)).

  Small Afghan Flying Sqiurrel. Afghanistan (Badakhshan and Nuristan), India (Jammu nd Kashmir) and Pakistan (Punjab).

OEN Kashmir Pygmy Flying Squirrel (Subsp. *fimbriatus*), Kashmir Flying Squirrel (Subsp. *fimbriatus*), Grey Flying Squirrel (Subsp. *fimbriatus*), Afghan Pygmy Flying Squirrel (Subsp. *baberi*), Afghan Flying Squirrel (Subsp. *baberi*)

39. *Eupetaurus cinereus* Thomas, 1888
Woolly Flying Squirrel

Eupetaurus: Mod. L. *eupetaurus* true flying squirrel (L. *eu* (>Gr. *eÿ*) true or well; L. *petaurus* (>Gr. *petauron*) a springboard or a perch).
cinereus: L. *cinereus* ash-coloured (L. *cinis*, *cineris* ashes), (-*eus* having the quality of).

Himalayas in Bhutan (West), India (Jammu and Kashmir and Sikkim) and Pakistan (Khyber Pakhtunkhwa).

40. *Hylopetes alboniger* (Hodgson, 1836)
Parti-coloured Flying Squirrel

Hylopetes: Mod. L. *hylopetes* forest flying squirrel (L. *hylo* (Gr. *ÿlè*) forest, wood; Gr. *petēs* winged, flyer).
alboniger: Mod. L. *alboniger* white and black (L. adj. *albus* white; L. *niger* black).

Eastern Himalayas, in Bhutan, India, and Nepal.

OEN Particolored Pygmy Flying Squirrel (nom. form), Black and White Flying Squirrel (nom. form), Turnbull's Flying Squirrel (J. Syn. *turnbulli*)

VN Bhut. *Piyam-piyu*; Lepch. *Khim*

41. *Petaurista petaurista* (Pallas, 1766)
Red Giant Flying Squirrel

Petaurista: Mod. L. *petaurista* flying squirrel (L. *petaurus* (>Gr. *petauron*) a springboard or a perch; *-ista* L. suffix denoting ability); alternatively from Gr. *petauristís* 'rope-dancer'.

petaurista: Mod. L. *petaurista* flying squirrel (L. *petaurus* (>Gr. *petauron*) a springboard or a perch; *-ista* L. suffix denoting ability); alternatively from Gr. *petauristís* 'rope-dancer'.

Southeast Asia, China, and South Asia. One subspecies is known from the region.

Subspecies

- *albiventer*: Mod. L. *albiventer* bearing white belly (L. *albus* white; L. *venter* belly).

  White-bellied Flying Squirrel. Himalayas in Afghanistan, Bangladesh, Bhutan, India, and Nepal.

OEN Brown Flying Squirrel (nom. form), Birrel's Flying Squirrel (J. Syn. *birrelli*), Western Himalayan Flying Squirrel (J. Syn. *fulvinus*), Himalayan Flying Squirrel (J. Syn. *inornatus*), White-bellied Flying Squirrel (J. Syn. *inornatus*)

VN Kash. *Rusi-gugar*

42. *Petaurista magnificus* (Hodgson, 1836)
Himalayan Giant Flying Squirrel

Petaurista: Mod. L. *petaurista* flying squirrel (L. *petaurus* (>Gr. *petauron*) a springboard or a perch; *-ista* L. suffix denoting ability); alternatively from Gr. *petauristís* 'rope-dancer'.

magnificus: L. adj. *magnificus* magnificent.

South Asia and China (Tibet). Besides the nominate subspecies, *P. m. magnificus* Himalayan Giant Flying Squirrel, from the Himalayas in Bhutan, India (Sikkim and West Bengal) and Nepal, another subspecies is known from the region.

Subspecies

- *hodgsoni*: Eponym after Brian Houghton Hodgson (b. 1801–d. 1894), a British civil servant, naturalist and ethnologist stationed in Nepal, (*-i*, commemorating (dedication, eponym) (L.)).

  Hodgson's Giant Flying Squirrel. Himalayas in India (Arunachal Pradesh, Sikkim, and West Bengal) and Nepal (Central Nepal).

OEN Himalayan Flying Squirrel (nom. form), Red-bellied Flying Squirrel (nom. form)

VN Lepch. *Biyom*

43. *Petaurista philippensis* (Elliot, 1839)
South Indian Giant Flying Squirrel

Petaurista: Mod. L. *petaurista* flying squirrel (L. *petaurus* (>Gr. *petauron*) a springboard or a perch; *-ista* L. suffix denoting ability); alternatively from Gr. *petauristís* 'rope-dancer'.

philippensis: L. adj. *philippensis* meaning from the Philippines, (*-ensis*, geographic, occurrence in (location, toponym) (L.)). Erroneous, type locality later reallocated as near Madras, India (in original description no exact location was given but the paper in which it was dealt was about Mammals of South Mahratta Country).

South Asia, Southeast Asia, and China. The nominate subspecies occurs in peninsular India and Sri Lanka.

OEN Large Brown Flying Squirrel (nom. form), Indian Giant Flying Squirrel (nom. form), Brown Flying Squirrel (nom. form), Large Ceylon Flying Squirrel (J. Syn. *lanka*), Large Grey Flying Squirrel (J. Syn. *lanka*), Ceylon Grey Flying Squirrel (J. Syn. *lanka*)

VN Mar. *Pakya*; Kol *Oral*; Sinh. *Hambawa*; Tam. (SL) *Parravanil*, *Vemba*, *Paravai-anil*

44. *Petaurista elegans* (Müller, 1840)
Spotted Giant Flying Squirrel

Petaurista: Mod. L. *petaurista* flying squirrel (L. *petaurus* (>Gr. *petauron*) a springboard or a perch; *-ista* L. suffix denoting ability); alternatively from Gr. *petauristís* 'rope-dancer'.

elegans: L. adj. *elegans* elegant, handsome or fine.

Southeast Asia, South Asia and China (south). One subspecies is known from the region.

Subspecies

- *caniceps*: Mod. L. *caniceps* grey-headed (L. *canus* grey; L. *-ceps* -headed (L. *caput* head)).

  Grey-headed Flying Squirrel. Himalayas in Bhutan, India (Arunachal Pradesh, Sikkim, and West Bengal) and Nepal.

OEN Lesser Giant Flying Squirrel (Subsp. *caniceps*), Grey Flying Squirrel (J. Syn. *senex*), Gorkha Flying Squirrel (J. Syn. *gorkhali*)

VN Lepch. *Biyom-chimbo*

45. *Petaurista nobilis* (Gray, 1842)
Noble Giant Flying Squirrel

Petaurista: Mod. L. *petaurista* flying squirrel (L. *petaurus* (>Gr. *petauron*) a springboard or a perch; *-ista* L. suffix denoting ability); alternatively from Gr. *petauristís* 'rope-dancer'.

nobilis: L. adj. *nobilis* noble.

Endemic to South Asia. Besides the nominate subspecies, *P. n. nobilis* Noble Giant Flying Squirrel, from the Himalayas in India (Sikkim and West Bengal) and Nepal, another subspecies is known from the region.

Subspecies

- *singhei*: Eponym after Jigme Singye Wangchuck (b. 1955), the king of Bhutan, (*-i*, commemorating (dedication, eponym) (L.))

  Bhutan Giant Flying Squirrel. Himalayas in Bhutan and India (Arunachal Pradesh).

OEN Gray's Giant Flying Squirrel (Subsp. *nobilis*)

46. *Petaurista mechukaensis* Choudhury, 2007
Mechuka Giant Flying Squirrel

Petaurista: Mod. L. *petaurista* flying squirrel (L. *petaurus* (>Gr. *petauron*) a springboard or a perch; *-ista* L. suffix denoting ability); alternatively from Gr. *petauristís* 'rope-dancer'.

mechukaensis: L. adj. *mechukaensis* meaning from the Mechuka Valley, West Siang District, Arunachal Pradesh, India, (*-ensis*, geographic, occurrence in (location, toponym) (L.)).

Endemic to India.

47. *Petaurista mishmiensis* Choudhury, 2009
Mishmi Hills Giant Flying Squirrel

Petaurista: Mod. L. *petaurista* flying squirrel (L. *petaurus* (>Gr. *petauron*) a springboard or a perch, *-ista* L. suffix denoting ability); alternatively from Gr. *petauristís* 'rope-dancer'.

mishmiensis: L. adj. *mishmiensis* from Mishmi Hills, Arunachal Pradesh, India, (*-ensis*, geographic, occurrence in (location, toponym) (L.)).

Endemic to India.

48. *Petaurista siangensis* Choudhury, 2013
Mebo Giant Flying Squirrel

Petaurista: Mod. L. *petaurista* flying squirrel (L. *petaurus* (>Gr. *petauron*) a springboard or a perch; *-ista* L. suffix denoting ability); alternatively from Gr. *petauristís* 'rope-dancer'.

siangensis: L. adj. *siangensis* from Siang basin, Siang District, Arunachal Pradesh, India, (*-ensis*, geographic, occurrence in (location, toponym) (L.)).

Endemic to India.

49. *Petinomys fuscocapillus* (Jerdon, 1847)
Travancore Flying Squirrel

Petinomys: Mod. L. *petinomys* flying squirrel (L. *petino*, >Gr. *petauron* a springboard or a perch; Gr. *mȳs* mouse).

fuscocapillus: Mod. L. *fuscocapillus* dusky-headed (L. adj. *fuscus* dark or black; L. *-capillus* headed (L. *capillus* hair of the head)).

Endemic to South Asia; India (Western Ghats in Karnataka, Kerala and Tamil Nadu) and Sri Lanka.

OEN Small Travancore Flying Squirrel (nom. form *fuscocapillus*), Ceylon Small Flying Squirrel (J. syn. *layardi*), Small Ceylon Flying Squirrel (J. syn. *layardi*), Small Flying Squirrel (J. syn. *layardi*)

VN Sinh. *Podi Hambawa* (J. syn. *layardi*); Tam. (SL) *Parravani*, *Paravai-anil* (J. syn. *layardi*)

Subfamily Callosciurinae Pocock, 1923

50. *Callosciurus erythraeus* (Pallas, 1799)
Pallas's Squirrel

Callosciurus: Mod. L. *callosciurus* beautiful squirrel (L. *callo* (Gr. *kalos*) beautiful; L. *sciurus* squirrel).

erythraeus: L. *erythraeus* red-coloured (Gr. *eruthros* red; *-eus* having the quality of).

Southeast Asia, South Asia, and China. Besides the nominate subspecies, *C. e. erythraeus* Pallas's Squirrel, from the Himalayas in Bangladesh, Bhutan, and India (northeast), two subspecies are known from the region.

Subspecies

- *erythrogaster*: Mod. L. *erythrogaster* red-bellied (Gr. *eruthros* red; Gr. *gastēr* belly).

  Red-bellied Squirrel. Himalayas in India (Assam, Manipur, Mizoram, and Nagaland).
- *intermedia*: L. *intermedius* intermediate (Med. L. *intermediatus* intermediate).

  Intermediate Squirrel. Himalayas in India (Meghalaya and Tripura).

OEN Mountain Red-bellied Squirrel (Subsp. *erythrogaster*), Gray's Squirrel (J. Syn. *punctatissimus*), Bhutan Red-bellied Squirrel (J. Syn. *bhutanensis*), Naga Squirrel (J. Syn. *nagarum*), Crump's Squirrel (J. Syn. *crumpi*), Wells' Squirrel (J. Syn. *wellsi*), Anderson's Chestnut-bellied Squirrel (J. Syn. *aquilo*)

51. *Callosciurus pygerythrus* (I. Geoffroy Saint–Hilaire, 1832)
Irrawady Squirrel

Callosciurus: Mod. L. *callosciurus* beautiful squirrel (L. *callo* (Gr. *kalos*) beautiful; L. *sciurus* squirrel).

pygerythrus: Mod. L. *pygerythrus* red-backed (Mod. L. *pygi* (>Gr. *pugé*) buttock or rump; L. *erythrus* (Gr. *eruthrós*) red).

Southeast Asia (western), South Asia (northeastern) and China (southern). Three subspecies are known from the region.

Subspecies

- *lokroides*: Mod. L. *lokroides* resembling lokriah (*lokriah* >Nep. name of Orange-bellied Himalayan Squirrel; *-oides* (>Gr. *eidos*) apparent shape).
  Hoary-bellied Himalayan Squirrel. Himalayas in Bhutan, India (northeast) and Nepal.
- *blythi*: Eponym after Edward Blyth (b. 1810–d. 1873), an English zoologist and pharmacist, curator of Museum at the Royal Asiatic Society of Bengal, (*i*, commemorating (dedication, eponym) (L.)).
  Blyth's Squirrel. Himalayas in Bangladesh (Chittagong and Sylhet) and India (Assam).
- *stevensi*: Eponym after H. Stevens (dates not known), an English army officer, who collected the type from Beni-Chang, Abor-Miri Hills, northern frontier of Upper Assam, (*-i*, commemorating (dedication, eponym) (L.)).
  Steven's Squirrel. Himalayas in India (Assam).

OEN Albino Hoary-bellied Himalayan Squirrel (Subsp. *lokroides*), Hoary-bellied Gray Squirrel (Subsp. *lokroides*)

52. *Dremomys lokriah* (Hodgson, 1836)
Orange-bellied Himalayan Squirrel

Dremomys: Mod. L. *dremomys* fast running squirrel (L. *dromos* running (from L. *drameïn* to run), Gr. *mŷs* mouse).

lokriah: Mod. L. *lokriah* >Nep. *lokriah*, the local name of this squirrel.

South Asia (northeast), China (Tibet and southwest) and Southeast Asia (west). Besides the nominate subspecies, *D. l. lokriah* Orange-bellied Himalayan Squirrel, from the Himalayas in Bangladesh, Bhutan, India (northeast) and Nepal, two subspecies are known from the region.

Subspecies

- *macmillani*: Eponym after S.A. Macmillan (d. 1915), a British army officer and naturalist, who collected accompanied Col. G.C. Shortridge on his Chindwin trip, (*-i*, commemorating (dedication, eponym) (L.)).
  Macmillan's Squirrel. Himalayas in India (Meghalaya, Nagaland and Tripura).
- *garonum*: L. adj. *garonum* meaning from Garo Hills, Meghalaya, India, (*-unum*, belonging to, pertaining to (location, toponym) (L.)).
  Garo Hills Squirrel. Himalayas in India (Meghalaya).

OEN Orange-bellied Ground Squirrel (Subsp. *lokriah*), Himalayan Ground Squirrel (Subsp. *lokriah*), Orange-bellied Gray Squirrel (Subsp. *lokriah*), Red-thighed Squirrel (Subsp. *lokriah*), Long-snouted Nepal Squirrel (Subsp. *lokriah*), Long-snouted Bhutan Squirrel (J. Syn. *bhotia*), Nepal Squirrel (J. Syn. *subflaviventris*)

VN Nep. *Lokria*; Bhut. *Zhamo*; Lepch. *Killi*, *Killi-tingdon*

53. *Dremomys pernyi* (Milne-Edwards, 1867)
Pernyi's Long-nosed Squirrel

Dremomys: Mod. L. *dremomys* fast running squirrel (L. *dromos* running (from *drameïn* to run), Gr. *mÿs* mouse).

pernyi: Eponym after Paul-Hubert Perny (b. 1818–d. 1907), a French missionary, (-*i*, commemorating (dedication, eponym) (L.)).

South Asia (northeast), China and Southeast Asia. Besides the nominate subspecies, *D. p. pernyi* Pernyi's Long-nosed Squirrel, from the Himalayas in India (northeast), one subspecies is known from the region.

Subspecies

- *howelli*: Eponym after E.B. Howell (dates not known), a British army officer and naturalist posted in Burma, who collected the type specimen from Ma Chang Kai, S.W. of Tengyueh, (-*i*, commemorating (dedication, eponym) (L.)).

  Howell's Long-nosed Squirrel. Himalayas in India (Arunachal Pradesh).

OEN Perny's Ground Squirrel (Subsp. *pernyi*), Howell's Ground Squirrel (Subsp. *howelli*)

54. *Dremomys rufigenis* (Blanford, 1878)
Red-cheeked Squirrel

Dremomys: Mod. L. *dremomys* fast running squirrel (L. *dromos* running (from *drameïn* to run), Gr. *mÿs* mouse).

rufigenis: Mod. L. *rufigenis* red-cheeked (L. *rufi* (from *rufus*) red; L. *genys* cheek or jaw).

Southeast Asia, China (southwestern) and South Asia (northeast). India (Himalayas in Arunachal Pradesh and Nagaland).

OEN Asian Red-cheeked Ground Squirrel (Subsp. *rufigenis*)

55. *Funambulus palmarum* (Linnaeus, 1766)
Three-striped Palm Squirrel

Funambulus: Mod. L. *funambulus* 'rope-walker' (L. *funis* rope or cord; L. v. *ambulo* I walk).

palmarum: L. *palmarum* (>L. *palmarius*) pertaining to palm trees.

Endemic to South Asia; India and Sri Lanka. Besides the nominate subspecies, *F. p. palmarum* Three-striped Palm Squirrel, from India (southern West Bengal to most of peninsular India), four subspecies are known from the region.

Subspecies

- *brodiei*: Eponym after A.O. Brodie (dates not known), a British officer in Ceylon and naturalist, (-*i*, commemorating (dedication, eponym) (L.)).

  Northern Ceylon Palm Squirrel. Endemic to Sri Lanka.

- *robertsoni*: Eponym after Laurence Roberston (dates not known), a British civil servant posted at Junagadh, a naturalist who was Honorary Treasurere of the Bombay Natural History Society when the species was described, (*-i*, commemorating (dedication, eponym) (L.)).
  Robertson's Palm Squirrel. Endemic to India (Chhattisgarh, Gujarat and Madhya Pradesh).
- *bellaricus*: L. adj. *bellaricus* meaning from Bellary, Karnataka, (*-icus*, belonging to, pertaining to (location, toponym) (L.)).
  Bellary Palm Squirrel. Endemic to India (Karnataka and Maharashtra).
- *matugamaensis*: L. adj. meaning from the Matugama, Kalutara, Western Province, Sri Lanka, (*–ensis*, geographic, occurrence in (location, toponym) (L.)).
  Matugama Palm Squirrel. Endemic to Sri Lanka (Western Province).

OEN Common Palm Squirrel (Subsp. *palmarum*), Common Striped Squirrel (Subsp. *palmarum*), Striped Squirrel (Subsp. *brodiei*, J. Syn. *favonicus*, J. Syn. *kelaarti*, J. Syn. *olympius*), 'Tree-Rat' (Subsp. *brodiei*, J. Syn. *favonicus*, J. Syn. *kelaarti*, J. Syn. *olympius*), Khurree (J. Syn. *penicillatus*), Bengal Palm Squirrel (J. Syn. *bengalensis*), South Indian Common Palm Squirrel (J. Syn. *comorinus*), Western Ceylon Palm Squirrel (J. Syn. *favonicus*), Submontane Ceylon Palm Squirrel (J. Syn. *favonicus*), Eastern Ceylon Palm Squirrel (J. Syn. *kelaarti*), Lowland Ceylon Palm Squirrel (J. Syn. *kelaarti*), Highland Ceylon Palm Squirrel (J. Syn. *olympius*)

VN Hind. *Gil'heri*; Beng. *Beral*, *Lakki*; Kan. *Alalu*; Tel. *Vudata*; Wad. *Urta*; Tam. *Anil*; Sinh. *Lena*; Tam. (SL) *Anil*, *Sinna Anil*

56. *Funambulus tristriatus* (Waterhouse, 1837)
Jungle Striped Squirrel

Funambulus: Mod. L. *funambulus* 'rope-walker' (L. *funis* rope or cord; L. v. *ambulo* I walk).

tristriatus: Mod. L. *tristriatus* three-striped (L. *tri* (*tres*) >Sans. *tri* three; L. adj. *striatus* striped (*stria* furrow or channel)).

Endemic to India. Besides the nominate subspecies, *F. t. tristriatus* Jungle Striped Squirrel, from the Western Ghats of Karnataka, Maharashtra, and Tamil Nadu, two subspecies are known from the region.

Subspecies

- *wroughtoni*: Eponym after Robert Charles Wroughton (b. 1849–d. 1921), an English officer of Indian Forest Service in Bombay Presidency, member of the Bombay Natural History Society, described many new taxa resulting from Mammal Survey of British India, (*-i*, commemorating (dedication, eponym) (L.)).
  Wroughton's Striped Squirrel. Western Ghats in Karnataka (south) and Kerala.
- *numarius*: L. *numarius* of or belonging to money.
  Prater's Striped Squirrel. Western Ghats in Maharashtra.

OEN Jungle Palm Squirrel (Subsp. *tristriatus*), Western Ghats Striped Squirrel (Subsp. *tristriatus*), Dussumier's Palm Squirrel (J. Syn. *dussumieri*), Annandale's Jungle Palm Squirrel (J. Syn. *annandalei*), Thomas's Jungle Palm Squirrel (J. Syn. *thomasi*)

57. *Funambulus sublineatus* (Waterhouse, 1838)
Dusky-striped Squirrel

Funambulus: Mod. L. *funambulus* 'rope-walker' (L. *funis* rope or cord; L. v. *ambulo* I walk).

sublineatus: Mod. L. *sublineatus* faint-lined (L. *sub* slight; L. *-lineatus* -lined (*linea*, from *līnum* flax, for line or cord)).

Endemic to South Asia. Besides the nominate subspecies, *F. s. sublineatus* Indian Dusky-striped Squirrel. Endemic to India from the Western Ghats of Karnataka, Kerala and Tamil Nadu, another subspecies is known from the region.

Subspecies

- *obscurus*: L. *obscurus* dark, dusky.
  Sri Lankan Dusky-striped Squirrel. Endemic to Sri Lanka.

OEN Dusky Palm Squirrel (Subsp. *sublineatus*), Neelgherry Striped Squirrel (Subsp. *sublineatus*), Ceylon Dusky-striped Jungle Squirrel (Subsp. *obscurus*), Little Striped Jungle Squirrel (Subsp. *obscurus*) Ceylon Dusky-striped Squirrel (J. Syn. *kathleenae*), Three-lined Dusky-striped Squirrel (J. Syn. *trilineatus*), Delessert's Striped Squirrel (J. Syn. *delesserti*)

VN Sinh. *Podi-Lena*; Tam. (SL) *Sinna Anil*

58. *Funambulus layardi* (Blyth, 1849)
Layard's Striped Squirrel

Funambulus: Mod. L. *funambulus* 'rope-walker' (L. *funis* rope or cord; L. v. *ambulo* I walk).

layardi: Eponym after E.L. Layard (b. 1824–d. 1900), civil servant in Ceylon, (*-i*, commemorating (dedication, eponym) (L.)).

Endemic to Sri Lanka.

OEN Layard's Palm Squirrel, Striped Jungle Squirrel (nom. form *layardi*), Flame-striped Jungle Squirrel (nom. form *layardi*, J. Syn. *signatus*), Western Flame-striped Jungle Squirrel (J. Syn. *signatus*)

VN Sinh. *Mookula Lena*; Tam. (SL) *Karupu Anil*

59. *Funambulus pennantii* Wroughton, 1905
Northern Palm Squirrel

Funambulus: Mod. L. *funambulus* 'rope-walker' (L. *funis* rope or cord; L. v. *ambulo* I walk).

pennantii: Eponym after T. Pennant (b. 1726–d. 1798), British naturalist, (*-ii*, commemorating (dedication, eponym) (L.)).

South Asia and Iran. Besides the nominate subspecies, *F. p. pennantii* Northern Palm Squirrel. Endemic to India from the peninsular region, four subspecies are known from the region.

Subspecies

- *lutescens*: Mod. L. *lutescens* somewhat yellowish (>L. *luteus* saffron-yellow) (*cf.* L. *lutescens* muddy-coloured >*lutum*, *luti* mud).
  Gujarat Palm Squirrel. Endemic to South Asia; India (northwestern) and Pakistan (Punjab).
- *argentescens*: L. *argentescens* silverish (L. *argenteus* of silver, silvery, *argentum* silver).
  Pakistan Palm Squirrel. Endemic to South Asia; India (Gujarat and Rajasthan) and Pakistan.
- *chhattisgarhi*: L. adj. *chhattisgarhi* belonging to Chhattisgarh, (-*i* commemorating (L./Gr.)).
  Chhattisgarh Palm Squirrel. Endemic to South Asia; Bangladesh and India (central and eastern India).
- *gangutrianus*: L. adj. *gangutrianus* belonging to Gangotri, (-*anus* belonging to, pertaining to (L./Gr.)).
  Gangotri Palm Squirrel. Endemic to South Asia; Bhutan, India (northern) and Nepal.

OEN Five-striped Squirrel (Subsp. *pennantii*), Sind Banyan Squirrel (Subsp. *argentescens*)
VN Hind. *Gil'heri*; Beng. *Beral*, *Lakki*; Tel. *Vudata*

60. *Tamiops macclellandi* (Horsfield, 1839)
Himalayan Striped Squirrel

Tamiops: Mod. L. *tamiops* bearing an aspect of *tamias* chipmunk (Gr. *Tamias* chipmunk – a treasurer or the one who stores; Gr. *óps* aspect).

macclellandi: Eponym after Dr. John McClelland (b. 1805–d. 1875), geologist and amateur naturalist, worked in Assam, (-*i*, commemorating (dedication, eponym) (L.)).

Southeast Asia, South Asia (northeast) and China. Himalayas in Bhutan, India, and Nepal.

OEN Small Himalayan Squirrel
VN Lepch. *Kalli Gangdin*

Subfamily Xerinae Osborn, 1910

61. *Spermophilopsis leptodactylus* (Lichtenstein, 1823)
Long-clawed Ground Squirrel

Spermophilopsis: Mod. L. *spermophilopsis* spermophilus-like, ground squirrel–like (L. *spermophilus* ground squirrel; Gr. -*opsis* -appearance).

leptodactylus: Mod. L. *leptodactylus* slender-fingered (Gr. *leptos* fine or slender; L. *dactylus* (Gr. *daktulōs*) finger or toe).

Endemic to Central Asia and marginally in South Asia (northwest). One subspecies occurs in the region.

Subspecies

- *bactrianus*: L. adj. *bactrianus* Bactrian, belonging to Bactria – an ancient country of central Asia, roughly corresponding to modern Turkistan and Balkh, northern Afghanistan), (*-anus*, geographic, belonging to, pertaining to (L.) geographic (location, toponym)).
  Bactrian Long-clawed Ground Squirrel. Afghanistan.

OEN Thin-clawed Ground Squirrel (nom. form)

62. *Marmota himalayana* (Hodgson, 1841)
Himalayan Marmot

Marmota: Mod. L. *marmota* marmot (Ital. *marmotta* >*murmont* of local dialect which is >*muris* (genitive singular of *mӯs* mouse) and *montis* (genitive singular of *mons*)).

himalayana: Mod. L. *himalayana* from Himalayas (Sans. *Himalaya* an abode of snow (*him* snow, *alaya* abode), (*-ana*, belonging to, pertaining to (L.) geographic (location, toponym).

South Asia (northwest) and China. Himalayas in India and Nepal.

OEN White Marmot (nom. form); Tibet Marmot (J. Syn. *tibetanus*); Red Marmot (J. Syn. *hemachalanus*)

VN Bhut. *Chibi*, *Chipi*; Lepch. *Lho*, *Pot sammiong*, *Sammiong*; Kash. *Brin*, *Drun*; Tib. *Kadia-piu*

63. *Marmota caudata* (Geoffroy, 1844)
Long-tailed Marmot

Marmota: Mod. L. *marmota* marmot (Ital. *marmotta* >*murmont* of local dialect which is >*muris* (genitive singular of *mӯs* mouse) and *montis* (genitive singular of *mons*)).

caudata: L. *caudata* tailed, having a (long/short) tail (L. *cauda* tail).

Central Asia, China, and South Asia (northwest). Afghanistan (Hindukush) and India (Himalayas in Jammu and Kashmir).

Subspecies

- *aurea*: L. *aureus* golden (*aurum* gold).
  Golden Marmot. Pakistan (Khyber Pakhtunkhwa).

OEN Red Marmot (nom. form), Bicolored Marmot (J. Syn. *dichrous*), Littledale's Marmot (J. Syn. *littledalei*), Stirling's Marmot (J. Syn. *stirlingi*)

64. *Spermophilus fulvus* (Lichtenstein, 1823)
Yellow Ground Squirrel

Spermophilus: Mod. L. *spermophilus* seed-lover (Gr. *sperma* seed; Gr. *philos* loving).
fulvus: L. *fulvus* tawny, yellow.
Central Asia and South Asia (northwest).

OEN Yellow Suslik (nom. form), Large-toothed Suslik (nom. form)

Family Gliridae Muirhead, 1819
Subfamily Leithiinae Lydekker, 1896

65. *Dryomys nitedula* (Pallas, 1778)
Forest Dormouse

Dryomys: Mod. L. *dryomys* dormouse (Gr. *druos* (genitive singular of *drӱs*) oak or tree; Gr. *mӱs* mouse).
nitedula: L. *nitedula* dormouse.
Europe, Central Asia and South Asia (northwest). One subspecies occurs in the region.

Subspecies

- *pictus*: L. *pictus* painted (*pingere* to paint).
  Painted Forest Dormouse. Afghanistan and Pakistan.

OEN Oak Dormouse (nom. form)

66. *Dryomys niethammeri* Holden, 1996
Niethhammer's Forest Dormouse

Dryomys: Mod. L. *dryomys* dormouse (Gr. *druos* (genitive singular of *drӱs*) oak or tree, *mӱs* mouse).
niethammeri: Eponym Jochen Neithammer (dates not known), mammalogist, an expert on dormice, (*-i*, commemorating (dedication, eponym) (L.)).
Endemic to Pakistan; northeast Balochistan.

Family Dipodidae Fischer, 1817
Subfamily Allactaginae Vinogradov, 1925

67. *Allactaga elater* (Lichtenstein, 1828)
Small Five-toed Jerboa

Allactaga: Mod. L. *allactaga* >Mongol. *alak-daagha* (*alak* variegated, *daagha* colt); in allusion to its variegated coat of the type species of the genus.
elater: Mod. L. *elater* (Gr. *elatēr*) one who elates, jumps.
Middle East and South Asia (northwest). Afghanistan and Pakistan.

OEN Bactrian Small Five-toed Jerboa (J. Syn. *bactriana*), Afghan Small Five-toed Jerboa (J. Syn. *indica*)

68. *Allactaga williamsi* Thomas, 1897
William's Jerboa

Allactaga: Mod. L. *allactaga* >Mongol. *alak-daagha* (*alak* variegated, *daagha* colt); in allusion to its variegated coat of *Dipus jaculus* type species of the genus.
williamsi: Eponym after Major W.H. Williams (dates not known), H.M. Consul at Van (=Artemita), in Kurdistan (= Turkey), who collected the type, (*-i*, commemorating (dedication, eponym) (L.)).

Middle East and South Asia (northwest). Afghanistan (Bamian and Kabul Provinces).

OEN Ellerman's Jerboa (J. Syn. *caprimulga*)

69. *Allactaga hotsoni* Thomas, 1920
Hotson's Five-toed Jerboa

Allactaga: Mod. L. *allactaga* >Mongol. *alak-daagha* (*alak* variegated, *daagha* colt); in allusion to its variegated coat of *Dipus jaculus* type species of the genus.
hotsoni: Eponym after John Ernest Buttery Hotson (b. 1877–d. 1944), an Indian civil servant, subsequently the Governor of the Bombay Presidency, naturalist, collected extensively in Balochistan and Persia (*-i*, commemorating (dedication, eponym) (L.)).

Middle East and South Asia (northwest). Afghanistan and Pakistan (Balochistan).

Subfamily Cardiocraniinae Vinogradov, 1925

70. *Salpingotulus michaelis* (Fitzgibbon, 1966)
Balochistan Pygmy Jerboa

Salpingotulus: Mod. L. *salpingotulus* dim. of *salpingotus* trumpet-eared pygmy jerboa (Gr. *salpingos* trumpet; L. *otus* ear), (*-ulus,* diminutive (comparison), somewhat (adj.) (L.)).
michaelis: Eponym after Michael FitzGibbon (dates not known), an English animal importer, in whose consignment from Balochistan, the types of this species were discovered.

Endemic to Pakistan; Balochistan Province.

OEN Dwarf Three-toed Jerboa (nom. form), Thomas's Pygmy Jerboa (J. Syn. *thomasi*)

Subfamily Dipodinae G. Fischer, 1817

71. *Jaculus blanfordi* (Murray, 1884)
Blanford's Jerboa

Jaculus: L. *jaculus* that which is thrown, a dart; in allusion to animal's dart-like leaps.

blanfordi: Eponym after William Thomas Blanford (b. 1832–d. 1905), an English geologist and naturalist, collected in many parts of India, also instrumental in establishing Indian Museum, (*-i*, commemorating (dedication, eponym) (L.)).

Iran and South Asia (northwest). Afghanistan and Pakistan (Balochistan).

Subfamily Sicistinae Allen, 1901

72. *Sicista concolor* (Büchner, 1892)
Chinese Birch Mouse

Sicista: Mod. L. *sicista* from Tartar name *sikistan* 'gregarious mouse'.

concolor: L. *concolor* uniform, similar in colour, plain.

China (southwest) and South Asia (northwest). Himalayas in India (Jammu and Kashmir) and Pakistan (Khyber Pakhtunkhwa).

OEN Leathem's Birch Mouse (J. Syn. *leathemi*), Yellow Birch Mouse (J. Syn. *flavus*)

Superfamily Muroidea Illiger, 1811
Family Platacanthomyidae Alston, 1876

73. *Platacanthomys lasiurus* Blyth, 1859
Malabar Spiny Dormouse

Platacanthomys: Mod. L. *platacanthomys* flat-spine bearing mouse (Gr. *platōs* broad, flat; Gr. *akantha* spine, thorn; Gr. *mȳs* mouse); in allusion to the flattened spines mingled with fur.

lasiurus: Mod. L. *lasiurus* hairy-tailed (Gr. *lásios* hairy; Gr. *oúrá* tail); in allusion to its hairy tail.

Endemic to India; Western Ghats in Karnataka, Kerala and Tamil Nadu.

OEN Malabar Spiny Tree Mouse (nom. form), Long-tailed Spiny Mouse (nom. form), Pepper Rat (nom. form)

Family Spalacidae Gray, 1821
Subfamily Rhizomyinae Winge, 1887

74. *Cannomys badius* (Hodgson, 1841)
Bay Bamboo Rat

Cannomys: Mod. L. *cannomys* bamboo rat (L. *canna* (Gr. *kannē*) reed, bamboo; Gr. *mȳs* mouse).

badius: L. *badius* chestnut-coloured.

Southeast Asia, China (southwest) and South Asia (northeast). Himalayas in Bangladesh, India (northeast) and Nepal.

OEN Lesser Bamboo Rat (nom. form), Bay Tree Mouse (nom. form)

75. *Rhizomys pruinosus* Blyth, 1851
Hoary Bamboo Rat

Rhizomys: Mod. L. *rhizomys* root rat (L. *rhizo* (>Gr. *rhiza*) root; Gr. *mŷs* mouse); after its preference to roots of bamboo.
pruinosus: L. *pruinosus* frosty, cold (L. *pruina* hoar-frost).
Southeast Asia, China (south and southwest) and South Asia (northeast). Himalayas (northeast) in India (Manipur, Meghalaya and Nagaland).

Family Calomyscidae Vorontsov and Potapova, 1979

76. *Calomyscus baluchi* Thomas, 1920
Baluchi Mouse-like Hamster

Calomyscus: Mod. L. *calomyscus* beautiful mouse (Gr. *kalos* beautiful; Gr. *myscus* mouse).
baluchi: Mod. L. *baluchi* from Balochistan.
Endemic to South Asia; Afghanistan and Pakistan.

OEN Baluchi Brush-tailed Mouse (nom. form), Baluchi Calomyscus (nom. form), Kabul Mouse-like Hamster (J. Syn. *mustersi*), Muster's Mouse-like Hamster (J. Syn. *mustersi*)

77. *Calomyscus hotsoni* Thomas, 1920
Hotson's Mouse-like Hamster

Calomyscus: Mod. L. *calomyscus* beautiful mouse (Gr. *kalos* beautiful; Gr. *myscus* mouse).
hotsoni: Eponym after John Ernest Buttery Hotson (b. 1877–d. 1944), an Indian civil servant, subsequently the Governor of the Bombay Presidency, naturalist, collected extensively in Balochistan and Persia, (-*i*, commemorating (dedication, eponym) (L.)).
Endemic to Pakistan; Balochistan.

OEN Hotson's Brush-tailed Mouse (nom. form), Mekran Mouse-like Hamster (nom. form), Hotson's Calomyscus (nom. form)

78. *Calomyscus elburzensis* Goodwin, 1938
Goodwin's Mouse-like Hamster

Calomyscus: Mod. L. *calomyscus* beautiful mouse (Gr. *kalos* beautiful; Gr. *myscus* mouse).
elburzensis: L. adj. *elburzensis* meaning from the Elburz (=Alborz) mountains, Iran, (-*ensis*, geographic, occurrence in (location, toponym) (L.)).
Middle East (Iran), Central Asia (Turkmenistan) and South Asia (northwest). Afghanistan (Herat Province).

OEN Godwin's Brush-tailed Mouse (nom. form), Elburz Mouse-like Hamster (nom. form),Godwin's Calomyscus (nom. form)

Family Cricetidae Fischer, 1817
Subfamily Cricetinae Fischer, 1817

79. *Cricetulus migratorius* (Pallas, 1773)
Little Grey Hamster

Cricetulus: Mod. L. *cricetulus* dim. of *cricetus* hamster (Ital. *criceto*), (-*ulus*, diminutive (comparison), somewhat (adj.) (L.)).
migratorius: Med. L. *migratorius* migrating (L. *migrator, migratoris* migrant, wanderer (L. *migrare* to migrate)), (-*ius*, having the nature of (L./Gr.)).

Europe, Middle East, Central Asia, China, and South Asia (northwest). Two subspecies occur in the region.

Subspecies

- *cinerascens*: Mod. L. *cinerescens* ashen (>L. *cinis, cineris* ashes, >*cinerescere* to turn to ashes).
  Wagner's Little Grey Hamster. Afghanistan and Pakistan.
- *fulvus*: L. *fulvus* tawny, yellow.
  Sandy Hamster. Afghanistan and India (Jammu and Kashmir).

OEN Grey Dwarf Hamster (nom. form), Sandy Little Grey Hamster (Subsp. *fulvus*)

80. *Cricetulus alticola* Thomas, 1917
Ladakh Hamster

Cricetulus: Mod. L. *cricetulus* dim. of *cricetus* hamster (Ital. *criceto*), (-*ulus*, diminutive (comparison), somewhat (adj.) (L.)).
alticola: Med. L. *alticola* inhabitant of high altitudes (L. *altus* high, *colo* to dwell, to inhabit), as the animal lives in elevated habitats.

China (southwest and Tibet) and South Asia (northwest and north). India (Ladakh in Jammu and Kashmir) and Nepal (Western Nepal).

OEN Tibetan Dwarf Hamster (J. Syn. *tibetanus*)

Subfamily Arvicolinae

81. *Alticola roylei* (Gray, 1842)
Royle's Vole

Alticola: Med. L. *alticola* inhabitant of high altitudes (L. *altus* high, *colo* to dwell, to inhabit), as the animal lives in elevated habitats.
roylei: Eponym after Dr. John Forbes Royle (b. 1799–d. 1858), an Indian-born British botanist, surgeon, and collector, (-*i*, commemorating (dedication, eponym) (L.)).

Endemic to India; Himalayas in Himachal Pradesh and Uttarakhand.

OEN Himalayan Vole (nom. form), Wells' Vole (J. Syn. *cautus*)

82. *Alticola stoliczkanus* (Blanford, 1875)
Stoliczka's Vole

Alticola: Med. L. *alticola* inhabitant of high altitudes (L. *altus* high, *colo* to dwell, to inhabit), as the animal lives in elevated habitats.
stoliczkanus: Eponym after Dr. Ferdinand Stoliczka (b. 1838–d. 1874), a British paleontologist, who participated in Yarkand Missions.

China (Tibet) and South Asia (north). Himalayas in India (Jammu and Kashmir and Sikkim) and Nepal (Eastern Nepal).

OEN Stoliczka's Mountain Vole (nom. form), Abbott's Mountain Vole (J. Syn. *acrophilus*), Ladhak Mountain Vole (J. Syn. *stracheyi*), Strachey's Mountain Vole (J. Syn. *stracheyi*), Short-tailed Mountain Vole (J. syn. *cricetulus*), Bhatnagar's Mountain Vole (J. Syn. *bhatnagari*)

83. *Alticola argentatus* (Severtzov, 1879)
Silver Mountain Vole

Alticola: Med. L. *alticola* inhabitant of high altitudes (L. *altus* high, *colo* to dwell, to inhabit), as the animal lives in elevated habitats.
argentatus: L. *argentatus* ornamented with silver (L. *argentum* silver).

China, Central Asia, Middle East (northeast) and South Asia (northwest). Pamirs in Afghanistan and Himalayas in India (Jammu and Kashmir) and Pakistan (Khyber Pakhtunkhwa).

OEN Severtzov's Mountain Vole (nom. form), Gilgit Vole (J. syn. *blanfordi*), Lahul Vole (J. Syn. *lahulius*), Small-toothed Vole (J. Syn. *parvidens*)

84. *Alticola albicauda* (True, 1894)
White-tailed Mountain Vole

Alticola: Med. L. *alticola* inhabitant of high altitudes (L. *altus* high, *colo* to dwell, to inhabit), as the animal lives in elevated habitats.
albicauda: Mod. L. *albicauda* white-tailed (L. *albus* white; L. *cauda* tail).

Endemic to India; Himalayas in Jammu and Kashmir.

85. *Alticola montosa* (True, 1894)
Kashmir Mountain Vole

Alticola: Med. L. *alticola* inhabitant of high altitudes (L. *altus* high, *colo* to dwell, to inhabit), as the animal lives in elevated habitats.
montosa: L. *montosus* mountainous.

Endemic to South Asia; Himalayas in India (Jammu and Kashmir) and Pakistan (Khyber Pakhtunkhwa).

OEN Central Kashmir Vole (nom. form)

86. *Blanfordimys afghanus* (Thomas, 1912)
Afghan Vole

Blanfordimys: Mod. L. *blanfordimys* Blanford's mouse (Blanford: William Thomas Blanford (1832–1905), an English geologist and naturalist, collected in many parts of India, also instrumental in establishing Indian Museum, Gr. *mȳs* mouse).
afghanus: Mod. L. *afghanus* belonging to Afghan or Afghanistan, (*-anus*, belonging to, pertaining to (L.) geographic (location, toponym)).
Central Asia and South Asia (northwest). Pamirs in Afghanistan.

87. *Blanfordimys bucharensis* (Vinogradov, 1930)
Bucharian Vole

Blanfordimys: Mod. L. *blanfordimys* Blanford's mouse (Blanford – William Thomas Blanford (1832–1905), an English geologist and naturalist, collected in many parts of India, also instrumental in establishing Indian Museum, Gr. *mȳs* mouse).
bucharensis: L. adj. *bucharensis* from Bukhara, Uzbekistan, (*-ensis*, geographic, occurrence in (location, toponym) (L.)).
Central Asia and South Asia (northwest). Pamirs in Afghanistan.

88. *Ellobius talpinus* (Pallas, 1770)
Northern Mole Vole

Ellobius: Mod. L. *ellobius* round-eared (Gr. *éllóvion* earring), from the rudimentary, somewhat circular, external ears.
talpinus: Mod. L. *talpinus* talpa-like (L. *talpa* European mole, *-inus* pertaining to (L.)).
Central Asia, Russia, South Asia (northwest). Afghanistan and Pakistan (Balochistan).

89. *Ellobius fuscocapillus* (Blyth, 1842)
Afghan Mole Vole

Ellobius: Mod. L. *ellobius* round-eared (Gr. *éllóvion* earring), from the rudimentary, somewhat circular, external ears.
fuscocapillus: Mod. L. *fuscocapillus* dusky-headed (L. adj. *fusco* dark or black; L. *-capillus* headed (L. *capillus* hair of the head).
Central Asia, Middle East (Iran) and South Asia (northwest). Pamirs in Afghanistan.

OEN Southern Mole Vole (nom. form), Quetta Mole Vole (nom. form)

90. *Eothenomys melanogaster* (Milne-Edwards, 1871)
Pere David's Vole

Eothenomys: Mod. L. *eothenomys* early mouse (Gr. *íós* dawn; Gr. *then* from; Gr. *mȳs* mouse).
melanogaster: Gr. *melanogastēr* black-belly (Gr. *melanos* black; Gr. *gastēr* belly).

China (south and southwest), Southeast Asia and South Asia (northeast). Himalayas in Arunachal Pradesh.

OEN Formosan Black-bellied Vole (nom. form)

91. *Hyperacrius wynnei* (Blanford, 1881)
Murree Vole

Hyperacrius: Mod. L. *hyperacrius* high-altitude vole (Gr. *oíýperákrios* inhabitants of the heights); due to its high altitude range.
wynnei: Eponym after Arthur Beavor Wynne (b. 1835–d. 1906), an officer with the Indian Geological Survey, worked in Bombay and Punjab presidencies, (-*i*, commemorating (dedication, eponym) (L.)).

Endemic to South Asia; Himalayas in India (Jammu and Kashmir) and Pakistan (Khyber Pakhtunkhwa and Punjab).

OEN Swat Vole (J. Syn. *traubi*)

92. *Hyperacrius fertilis* (True, 1894)
Subalpine Kashmir Vole

Hyperacrius: Mod. L. *hyperacrius* high-altitude vole (Gr. *oíýperákrios* inhabitants of the heights); due to its high altitude range.
fertilis: L. *fertilis* productive, fertile (L. *ferō* carry, bear).

Endemic to South Asia; Himalayas in India (Jammu and Kashmir) and Pakistan (Khyber Pakhtunkhwa).

OEN True's Vole (nom. form), Aitchison's Vole (J. Syn. *aitchisoni*), Short-eared Vole (J. Syn. *brachelix*), Phillips' Vole (J. Syn. *zygomaticus*)

93. *Microtus ilaeus* Thomas, 1912
Kazkhstan Vole

Microtus: Mod. L. *microtus* small-eared (Gr. *mikros* small; Gr. *otus* ear).
ilaeus: Specific epithet meaning unclear; presumed L. origin *ileum* groin, flank, lower part of the body.

China (northwest), Central Asia and South Asia (northwest). Afghanistan.

94. *Neodon sikimensis* (Hodgson, 1849)
Sikkim Vole

Neodon: Mod. L. *neodon* new-tooth (Gr. *neos* new; Gr. *ódoús*, *ódôn* tooth).
sikimensis: L. adj. *sikimensis* meaning from Sikkim, India (-*ensis*, geographic, occurrence in (location, toponym) (L.)).

China (Tibet) and South Asia (Himalayas). Himalayas in Bhutan, India (Sikkim and West Bengal) and Nepal.

OEN Sikkim Mountain Vole (nom. form), Sikkim Pine Vole (nom. form)
VN Nep. *Phalchua*; Tib. *Sing Phuchi*; Kiran. *Cheek yu*

95. ***Neodon juldaschi*** **(Severtzov, 1879)**
Juniper Vole

Neodon: Mod. L. *neodon* new-tooth (Gr. *neos* new; Gr. *ódoús*, *ódôn* tooth).
juldaschi: Origin not clear, could be an eponym, Juldasch is an Uzbek name (-*i*, commemorating (dedication, eponym) (L.)).

China (Tibet and northwest), Central Asia and South Asia (northwest). Pamirs in Afghanistan and Himalayas in India (Jammu and Kashmir) and Pakistan (Khyber Pakhtunkhwa).

OEN Juniper Mountain Vole (nom. form), Carruther's Vole (J. Syn. *carruthersi*), Carruther's Juniper Vole (J. Syn. *carruthersi*)

96. ***Phaiomys leucurus*** **Blyth, 1863**
Blyth's Vole

Phaiomys: Mod. L. *Phaiomys* grey-mouse (Gr. *phaiós* grey; Gr. *mȳs* mouse).
leucurus: L. *leucurus* (Gr. *leukouros*) white-tailed.

China (western and Tibet) and South Asia (north). Himalayas in India (Jammu and Kashmir and Himachal Pradesh) and Nepal (Eastern and Western Nepal).

OEN Blyth's Mountain Vole (nom. form), Blanford's Vole (J. Syn. *blythi*), Whitehead's Vole (J. Syn. *petulans*), Everest Vole (J. Syn. *everesti*)

Family Muridae Illiger, 1811
Subfamily Deomyinae Thomas, 1888

97. ***Acomys dimidiatus*** **(Cretzschmar, 1826)**
Arabian Spiny Mouse

Acomys: Mod. L. *acomys* spiny mouse (Gr. *ákê* a sharp point, (>*acantha* thorn or prick); Gr. *mȳs* mouse); owing to its spiny fur.
dimidiatus: L. *dimidiatus* divided (L. *dimidius* halved; L. *di*- apart; L. *medius* middle).

Egypt, Arabian Peninsula, Middle East, and South Asia (west). Pakistan (Balochistan and Sindh).

OEN Sinai Spiny Mouse (nom. form), Yellowish Spiny Mouse (J. Syn. *flavidus*)

Subfamily Gerbillinae Gray, 1825

98. ***Gerbillus nanus*** **Blanford, 1875**
Balochistan Gerbil

Gerbillus: L. *gerbillus* gerbil (dim. of *gerbua* or *jerboa*, >Arab. *yarbū* the flesh of the back and loins, an oblique descending muscle); in allusion to its strong muscles or its hind legs.

nanus: L. *nanus* dwarf (Gr. *nanos* dwarf).

Africa (north), Arabian Peninsula, Middle East, and South Asia (west). Afghanistan, India (western) and Pakistan.

OEN Baluchistan Dwarf Gerbil (nom. form), Dwarf Jerboa-Rat (nom. form), Baluchistan Dipodil (nom. form), Indian Dwarf Gerbil (J. Syn. *indus*)

99. *Gerbillus gleadowi* Murray, 1886
Little Hairy-footed Gerbil

Gerbillus: L. *gerbillus* gerbil (dim. of *gerbua* or *jerboa*, >Arab. *yarbū* the flesh of the back and loins, an oblique descending muscle); in allusion to its strong muscles or its hind legs.

gleadowi: Eponym after F. Gleadow (dates not known), a naturalist and painter, worked in north India, (*-i*, commemorating (dedication, eponym) (L.)).

Endemic to South Asia; India (Gujarat and Rajasthan) and Pakistan (Punjab and Sindh).

100. *Gerbillus aquilus* Schlitter and Stezer, 1972
Swarthy Gerbil

Gerbillus: L. *gerbillus* gerbil (dim. of *gerbua* or *jerboa*, >Arab. *yarbū* the flesh of the back and loins, an oblique descending muscle); in allusion to its strong muscles or its hind legs.

aquilus: L. adj. *aquilus* dark-coloured.

Middle East (Iran) and South Asia (west). Afghanistan and Pakistan (Balochistan).

OEN Eastern Iran Gerbil (nom. form), Eastern Swarthy Gerbil (J. Syn. *subsolanus*)

101. *Meriones meridianus* (Pallas, 1773)
Mid-day Jird

Meriones: Mod. L. *meriones* jird (Gr. *mèrós* thigh); in allusion to the development of its hind legs.

meridianus: L. *meridianus* of noon, southern (L. *meridies* south); in allusion to its activity during the middle of the day.

Europe (south), Central Asia, Russia, China, Middle East (Iran) and South Asia (northwest). Afghanistan (Konduz Province).

OEN Mid-day Gerbil (nom. form), Chinese Jird (nom. form)

102. *Meriones libycus* Lichtenstein, 1823
Libyan Jird

Meriones: Mod. L. *meriones* jird (Gr. *mèrós* thigh); in allusion to the development of its hind legs.

libyicus: L. adj. *libycus* Libyan, belonging to Libya, (*-icus* belonging to, pertaining to (L./Gr.)).

Africa (north), Arabian Peninsula, Middle East, Central Asia, China (Tibet and west) and South Asia (northwest). Afghanistan and Pakistan (Balochistan).

OEN Red-tailed Jerboa-Rat (J. Syn. *erythrourus*), Afghan Jird (J. Syn. *afghanus*)

103. *Meriones crassus* Sundevall, 1842
Sundevall's Jird

Meriones: Mod. L. *meriones* jird (Gr. *mèrós* thigh); in allusion to the development of its hind legs.
crassus: L. *crassus* thick, heavy, dense.

Africa (north), Arabian Peninsula, Middle East, Central Asia and South Asia (northwest). Afghanistan and Pakistan (Balochistan and Khyber Pakhtunkhwa).

OEN Swinhoe's Jird (J. Syn. *swinhoei*)

104. *Meriones hurrianae* (Jerdon, 1867)
Indian Desert Gerbil

Meriones: Mod. L. *meriones* jird (Gr. *mèrós* thigh); in allusion to the development of its hind legs.
hurrianae: Mod. L. *hurrianae* from Hurriana district (presently Haryana), India, (*-ae*, geographic (location, toponym) (L.)).

Middle East (Iran) and South Asia (west). India (Gujarat, Haryana and Rajasthan) and Pakistan.

OEN Western Desert Gerbil (J. Syn. *collinus*)

105. *Meriones persicus* (Blanford, 1875)
Persian Jird

Meriones: Mod. L. *meriones* jird (Gr. *mèrós* thigh); in allusion to the development of its hind legs.
persicus: L. adj. *persicus* Persian, belonging to Persia (*-icus* belonging to, pertaining to (L./Gr.)).

Middle East, Central Asia and South Asia (northwest). Besides the nominate subspecies, *M. p. persicus* Persian Jird, from Afghanistan and Pakistan (Balochistan), another subspecies is known from the region.

Subspecies

- *baptistae*: Eponym after N.A. Baptista (dates not known), taxidermist and collected in India for the Mammal Survey of British India, (*-ae*, commemorating (dedication, eponym) (L.)).

  Baptista's Persian Jird. Pakistan (Balochistan).

106. *Meriones zarudnyi* Heptner, 1937
Zarudny's Jird

Meriones: Mod. L. *meriones* jird (Gr. *mèrós* thigh); in allusion to the development of its hind legs.
zarudnyi: Eponym after Dr. N.A. Zarudny (b. 1859–d. 1919), curator of Tashkent Museum, Uzbekistan, and collected in Persia (*-i*, commemorating (dedication, eponym) (L.)).

Middle East (Iran), Central Asia (Turkmenistan) and South Asia (northwest). Afghanistan (Faryab Province).

107. *Rhombomys opimus* (Lichtenstein, 1823)
Great Gerbil

Rhombomys: Mod. L. *rhombomys* (Gr. *rómbos* rhombus; Gr. *mӱs* mouse); in allusion to the shape of the upper molars.
opimus: L. adj. *opimus* fat.

Mongolia, China (west and southwest), Central Asia, Middle East, and South Asia (northwest). Afghanistan and Pakistan (Balochistan).

108. *Tatera indica* (Hardwicke, 1807)
Indian Gerbil

Tatera: Etymology unknown.
indica: L. adj. *indica* belonging to India, (*-ica* belonging to, pertaining to (L./Gr.)).

Middle East and South Asia. Besides the nominate subspecies, *T. i. indica* Indian Gerbil, from Afghanistan, Bangladesh, India, Nepal, and Pakistan, another subspecies is known from the region.

Subspecies

- *cuvieri*: Eponym after M.F. Cuvier (b. 1773–d. 1838), a French zoologist and younger brother of Georges Cuvier, author of a monumental work on *Gerbillus*, (*-i*, commemorating (dedication, eponym) (L.)).
  South Indian Gerbil. India (south) and Sri Lanka.

OEN Antelope Rat (nom. form), Indian Jerboa-Rat (nom. form), Ceylon Gerbil (J. Syn. *ceylonica*), Ceylon Antelope-Rat (J. Syn. *ceylonica*), Sand-Rat (J. Syn. *ceylonica*), Kangaroo-Rat (J. Syn. *ceylonica*), Sherrin's Gerbil (J. Syn. *sherrini*), Dunn's Gerbil (J. Syn. *dunni*), Hardwicke's Gerbil (J. Syn. *hardwickei*)

VN Beng. *Jhenku indur*; Hind. *Harna mus*; Kan. *Billa ilei*; Tel. *Yeri-yelka*; Sinh. *Wer-miya*; Tam. (SL) *Vel-elli*

Subfamily Murinae Illiger, 1811

109. *Apodemus draco* (Barrett-Hamilton, 1900)
South China Field Mouse

Apodemus: Mod. L. *apodemus* (Gr. *ápódèmos*, from *apodēmein* to go abroad, be abroad) abroad; meaning inhabiting fields.

draco: L. *draco* dragon; no explanation provided, probably due to its origin in south China.

China and South Asia (northeast India).

OEN South China Wood Mouse

110. *Apodemus pallipes* Barrett-Hamilton, 1900
Himalayan Field Mouse

Apodemus: Mod. L. *apodemus* (Gr. *ápódèmos*, from *apodēmein* to go abroad, be abroad) abroad; meaning inhabiting fields.
pallipes: Mod. L. *pallipes* pale-footed (L. *pallidus* pale; L. *pes* foot, >Gr. *pous* foot).

South Asia (Himalayas and Mountains of Afghanistan) and Central Asia (marginally).

OEN Hazara Field Mouse (J. Syn. *pentax*)

111. *Apodemus latronum* Thomas, 1911
Sichuan Field Mouse

Apodemus: Mod. L. *apodemus* (Gr. *ápódèmos*, from *apodēmein* to go abroad, be abroad) abroad; meaning inhabiting fields.
latronum: Mod. L. *latronum* barking (L. *latro* bark (>Gr. *lātris*)).

China (parts of southern and central) and India (northeast).

OEN Big-eared Field Mouse (nom. form)

112. *Apodemus rusiges* Miller, 1913
Kashmir Field Mouse

Apodemus: Mod. L. *apodemus* (Gr. *ápódèmos*, from *apodēmein* to go abroad, be abroad) abroad; meaning inhabiting fields.
rusiges: No explanation provided; possibly after L. *russeus* reddish (L. *russus* red), owing to the coloration of the dorsal fur being somewhat reddish.

Endemic to South Asia (Himalayas in India and Pakistan, and also from southern mountain ranges in Pakistan).

OEN Gray Field Mouse (J. Syn. *griseus*)

113. *Apodemus gurkha* Thomas, 1924
Himalayan Wood Mouse

Apodemus: Mod. L. *apodemus* (Gr. *ápódèmos*, from *apodēmein* to go abroad, be abroad) abroad; meaning inhabiting fields.
gurkha: Mod. L. *gurkha* from Gorkha, Nepal, where its type locality is located.

Endemic to Nepal.

OEN Himalayn Field Mouse (nom. form)

114. ***Bandicota indica*** **(Bechstein, 1800)**
Greater Bandicoot Rat

Bandicota: L. *bandicota* >*bandicoot*, from Tel. *pandi-kokku*, meaning pig rat.

indica: L. adj. *indica* belonging to India, (*-ica* belonging to, pertaining to (L./Gr.)).

South Asia and Southeast Asia. Besides the nominate subspecies, *B. i. indica* Greater Bandicoot Rat, from Bangladesh, India and Sri Lanka, two more subspecies are known from the region.

Subspecies

- *nemorivaga*: L. *nemorivaga* forest roaming, wood roaming (L. *nemus*, *nemoris* wood; L. *vagari* to wander).
  Forest Greater Bandicoot Rat. Himalayas in Bhutan, India (northeast) and Nepal.
- *malabarica*: L. adj. *malabarica* meaning from Malabar – a region in southern India lying between the Western Ghats and the Arabian Sea corresponding to the Malabar District of the Madras Presidency in British India (presently includes areas under five northern districts of Kerala), (*-ica*, belonging to, pertaining to (location, toponym) (L.)).
  Malabar Greater Bandicoot Rat. Endemic to the Western Ghats in India.

OEN Large Bandicoot Rat (nom. form), Bandicoot (nom. form), Indian Bandicoot (Subsp. *indica*), Elliot's Field-Rat (J. Syn. *elliotanus*), Bengal Bandicoot (J. Syn. *elliotanus*), Small Nepal Water Rat (J. Syn. *hydrophillus*), Gigantic Bandicoot Rat (J. Syn. *gigantea*), Large-footed Bandicoot Rat (J. Syn. *macropus*), Malabar Bandicoot (Subsp. *malabrica*), Greater Bandicoot Rat (J. Syn. *maxima*)

VN Hind. *Ghus*, *Ghous*; Beng. *Ikria*, *Ikara*; Sans. *Indur*; Kan. *Heggin*; Tel. *Pandi-kokku*; Sinh. *Uru-miya*; Tam. (SL) *Peritchelli*

115. ***Bandicota bengalensis*** **(Gray and Hardwicke, 1833)**
Lesser Bandicoot Rat

Bandicota: L. *bandicota* >*bandicoot*, from Tel. *pandi-kokku*, meaning pig rat.

bengalensis: L. adj. *bengalensis* meaning from Bengal, India, (*-ensis*, geographic, occurrence in (location, toponym) (L.)).

South Asia and Southeast Asia (Myanmar, and partially in Thailand and China). Besides the nominate subspecies, *B. b. bengalensis* Lesser Bandicoot Rat from Bangladesh, Bhutan, India, Nepal and Sri Lanka, another subspecies is known from the region.

Subspecies

- *wardi*: Eponym after Colonel A.E. Ward (dates not known), an amateur zoologist, sportsman and collector, active in the Kashmir Himalayas, (*-i*, commemorating (dedication, eponym) (L.)). The types were collected by C.M. Crump and presented by Col. A.E. Ward.

Wroughton's Lesser Bandicoot Rat. Himalayas in India (northwest) and Pakistan.

OEN Indian Mole-rat (nom. form), Barclay's Field-rat (J. Syn. *barclayanus*), Bengal Field-Rat (J. Syn. *blythianus*), Deccan Field-rat (J. Syn. *deccanensis*), Ceylon Mole Rat (J. Syn. *gracilis*), Lesser Bandicoot (J. Syn. *gracilis*), Paddy-field rat (J. Syn. *gracilis*, *insularis*), Northern Ceylon Mole Rat (J. Syn. *insularis*), Kok Bandicoot (J. Syn. *kok*), Konkan Mole Rat (J. Syn. *lordi*), Nepal Rat (J. Syn. *plurimammis*), Southern India Field Rat (J. Syn. *providens*), Sindh Mole Rat (J. Syn. *sindicus*), Terai Mole Rat (J. Syn. *tarayensis*)

VN Sinh. *Wel-miya*; Tam. (SL) *Kurumb'elli*, *Viel'elli*, *Nell'elli*, *Agillarne*

116. *Berylmys bowersi* (Anderson, 1879)
Bower's White-toothed Rat

Berylmys: Mod. L. *berylmys* crystalline mouse (L. *beryllus* beryl; Gr. *mȳs* mouse).

bowersi: Eponym after Alexander Bowers (d. 1887), a British naval captain and merchantman stationed at Burma, collected from northern Burma and Yunnan, (*-i*, commemorating (dedication, eponym) (L.)).

Southeast Asia, China (south) and South Asia (northeast India).

OEN Bower's Rat (nom. form), Bower's Berylmys (nom. form), Wells' White-toothed Rat (J. Syn. *wellsi*)

117. *Berylmys mackenziei* (Thomas, 1916)
Mackenzie's White-toothed Rat

Berylmys: Mod. L. *berylmys* crystalline mouse (L. *beryllus* beryl; Gr. *mȳs* mouse).

mackenziei: Eponym after J. M. D. Mackenzie (dates not known), an officer with Burmese Forest Service, collected from Burma for the Mammal Survey of British India, (*-i*, commemorating (dedication, eponym) (L.)).

South Asia (northeast India) and very localized in China, Myanmar and Vietnam.

OEN Mackenzie's Rat (nom. form), Kenneth's White-toothed Rat (inappropriately given name).

118. *Berylmys manipulus* (Thomas, 1916)
Manipur White-toothed Rat

Berylmys: Mod. L. *berylmys* crystalline mouse (L. *beryllus* beryl; Gr. *mȳs* mouse).

manipulus: L. adj. *manipulus* belonging to Manipur, India.
South Asia (northeast India) and Southeast Asia (Myanmar).

OEN Manipur Rat (nom. form), Naga Hills Manipur Rat (J. Syn. *kekrimus*)

119. *Chiropodomys gliroides* (Blyth, 1856)
Pencil-tailed Tree Mouse

Chiropodomys: Mod. L. *chiropodomys* (Gr. *kheir, kheiros* hand; Gr. *podos* foot; Gr. *mŷs* mouse).
gliroides: Mod. L. *gliroides* glires-like (L. *glires* squirrel); *–oides* (>Gr. *eidos*) apparent shape.
Southeast Asia and South Asia (northeast India).

OEN Pencillate-tailed Tree-mouse (nom. form), Cherrapunjee Tree Mouse (nom. form), Pegu Tree Mouse (J. Syn. *peguensis*)

120. *Cremnomys cutchicus* Wroughton, 1912
Cutch Rock-rat

Cremnomys: Mod. L. *cremnomys* rock rat (Gr. *cremnos* a steep bank; Gr. *mŷs* mouse).
cutchicus: L. adj. *cutchicus* meaning from Kutch, Gujarat, India, (*-icus*, belonging to, pertaining to (location, toponym) (L.)).
Endemic to India.

OEN Cutch Rock Rat (nom. form), Indian Desert Rock rat (nom. form), Kathiawar Rock-rat (J. Syn. *medius*), Mt. Abu Rock-rat (J. Syn. *rajput*), Dull-coloured Rock-rat (J. Syn. *coenosus*), Mysore Rock-rat (J. Syn. *australis*), Sivasamudram Rock-rat (J. Syn. *siva*)

121. *Cremnomys elvira* (Ellerman, 1947)
Large Rock-rat

Cremnomys: Mod. L. *cremnomys* rock rat (Gr. *cremnos* a steep bank; Gr. *mŷs* mouse).
elvira – Mod. L. *elvira* a female given name, originally means 'the white'.
Endemic to India.

OEN Large Rock Rat (nom. form), Elvira Rat (nom. form)

122. *Dacnomys millardi* Thomas, 1916
Millard's Rat

Dacnomys: Mod. L. *Dacnomys* biting mouse (Gr. *dakno* to bite; Gr. *mŷs* mouse); owing to its unsually large maxillary toothrow.
millardi: Eponym after Dr. Walter Samuel Millard (b. 1864–d. 1952), a British naturalist, Honorary Secretary of the Bombay Natural History Society, and along with Oldfield Thomas commenced the Mammal Survey of British India, (*-i*, commemorating (dedication, eponym) (L.)).

South Asia (northeast India and Nepal) and China.

OEN Large-toothed Giant Rat (nom. form), Wroughton's Large-toothed Rat (J. Syn. *wroughtoni*)

123. *Diomys crumpi* Thomas, 1917
Crump's Mouse

Diomys: Mod. L. *Diomys* God's mouse (Gr. *dios* god; Gr. *mӯs* mouse); as Oldfield Thomas described this new genus and new species based on an extracted skull, supposedly collected on Mt. Paresnath, Bihar.
crumpi: Eponym after C. A. Crump (dates not known), who collected in Deccan for the Mammal Survey of British India, (-*i*, commemorating (dedication, eponym) (L.)).

Endemic to South Asia (India – Jharkhand and Manipur, and Nepal – Western Nepal).

124. *Golunda ellioti* Gray, 1837
Indian Bush-rat

Golunda: Mod. L. *golunda* from Kan. *gulandi*, the native name for Indian Bush-Rat.
ellioti: Eponym after Sir Walter Elliot (b. 1803–d. 1887), Indian civil servant and amateur naturalist (-*i*, commemorating (dedication, eponym) (L.)).

South Asia and Iran. Besides the nominate subspecies, *G. e. ellioti* Indian Bush-rat from Bangladesh, Bhutan, India, Nepal, Pakistan, and Sri Lanka, another subspecies is known from the region.

Subspecies

- *newara*: Mod. L. *newara* from Nuwara Eliya (historically Newara Eliya), Central Province, Sri Lanka, (a toponym).
  Ceylon Highland Bush-rat. Endemic to Sri Lanka.

OEN Coffee Rat (Subsp. *ellioti*, *newara*), Bombay Bush-rat (J. Syn. *bombax*), Assam Bush-rat (J. Syn. *coenosa*), Coffee Bush-rat (J. Syn. *coffaeus*), Coorg Bush-rat (J. Syn. *coraginis*), Gujarat Bush-rat (J. Syn. *gujerati*), Pakistan Bush-rat (J. Syn. *limitaris*), Himalayan Bush-rat (J. Syn. *myothrix*), Newara Eliya Bush-rat (J. Syn. *newara*), Ambala Bush-rat (J. Syn. *paupera*), Northern Indian Bush-rat (J. Syn. *watsoni*)
VN Kan. *Gulandi*; Yan. *Sora Panji-gadur*; Wad. *Gulat-yelka*; Sinh. *Copie-watte Miya*, *Miya*; Tam. (SL) *Copie-elli*, *Sarak'elli*

125. *Hadromys humei* (Thomas, 1886)
Hume's Rat

Hadromys: Mod. L. *hadromys* stout-mouse (Gr. *hadrōs* thick, stout; Gr. *mӯs* mouse).

humei: Eponym after Allan Octavian Hume (b. 1829–d. 1912), a civil servant, political reformer, ornithologist, and horticulturalist in British India, one of the founders of the Indian National Congress and co-founders of the Bombay Natural History Society, (*-i*, commemorating (dedication, eponym) (L.)).

Endemic to India (Manipur and Assom).

OEN Manipur Bush Rat (nom. form)

126. *Leopoldamys edwardsi* (Thomas, 1882)
Edward's Long-tailed Giant Rat

Leopoldamys: Mod. L. *Leopoldamys* Leopold's mouse (Leopold and Gr. *mŷs* mouse).
edwardsi: Eponym after Alphonse Milne-Edwards (b. 1835–d. 1900), a French mammalologist, ornithologist, and carcinologist, (*-i*, commemorating (dedication, eponym) (L.)).

China, Southeast Asia, and South Asia (northeast India).

OEN Edward's Rat (nom. form), Edward's Giant Rat (nom. form), Lister's Giant Rat (J. Syn. *listeri*), Garo Hills Giant Rat (J. syn. *garonum*)

127. *Leopoldamys sabanus* (Thomas, 1887)
Long-tailed Giant Rat

Leopoldamys: Mod. L. *Leopoldamys* Leopold's mouse (Leopold and Gr. *mŷs* mouse).
sabanus: Mod. L. *sabanus* from Sabah peninsula, Malaysia, (*-anus*, geographic, belonging to, pertaining to (L.) geographic (location, toponym)).

Southeast Asia and South Asia (Bangladesh).

OEN Noisy Rat (nom. form)

128. *Madromys blanfordi* (Thomas, 1881)
White-tailed Wood Rat

Madromys: Mod. L. *Madromys* Madras mouse (Madras and Gr. *mŷs* mouse); owing to the fact that the type was collected from Cuddapah (=Kadapa), Madras Presidency.
blanfordi: Eponym after William Thomas Blanford (b. 1832–d. 1905), an English geologist and naturalist, collected in many parts of India, also instrumental in establishing Indian Museum, (*-ii*, commemorating (dedication, eponym) (L.)).

South Asia (Bangladesh, India and Sri Lanka).

OEN Blanford's Rat (nom. form), White-tailed Rat (nom. form), Blanford's Indian Rock Rat (nom. form), Blanford's Madromys (nom. form)

VN Sinh. *Miya*; Tam. (SL) *Vell'elli*

129. ***Micromys minutus*** **(Pallas, 1771)**
Eurasian Harvest Mouse

Micromys: Mod. L. *micromys* small mouse (Gr. *mikros* small; Gr. *mȳs* mouse).

minutus: L. *minutus* small, little (*minuere* to make smaller).

Europe and Asia (temperate regions), China, Japan and South Asia (northeast India).

Subspecies

- *erythrotis*: Mod L. *erythrotis* red-eared (Gr. *eruthros* red; Gr. *-ōtis* -eared).

  Cherrapunji Red-eared Mouse. Northeast India (Meghalaya).

130. ***Millardia meltada*** **(Gray, 1837)**
Soft-furred Metad

Millardia: Generic *nomen* is an eponym after Dr. Walter Samuel Millard (b. 1864–d. 1952), a British naturalist, Honorary Secretary of the Bombay Natural History Society, and along with Oldfield Thomas commenced the Mammal Survey of British India, (*-ia*, commemorating (dedication, eponym) (L./Gr.)).

meltada: Mod. L. *meltada* metad from the Wadarri *mettade*, a name by which this species in known.

Endemic to South Asia (Bangladesh, India, Nepal, Pakistan and Sri Lanka).

OEN Soft-furred Field Rat (nom. form), Indian Soft-furred Field Rat (nom. form), Indian Soft-furred Metad (nom. form), Comber's Field Rat (J. Syn. *comberi*), Dunn's Field Rat (J. Syn. *dunni*), Liston's Field Rat (J. Syn. *listoni*), Pallid Field Rat (J. Syn. *pallidior*), Singur Field Rat (J. Syn. *singuri*)

VN Yan. *Metta-yelka*; Wad. *Mettade*; Kan. *Kera ilei*

131. ***Millardia gleadowi*** **(Murray, 1885)**
Sand-coloured Metad

Millardia: Generic *nomen* is an eponym after Dr. Walter Samuel Millard (b. 1864–d. 1952), a British naturalist, Honorary Secretary of the Bombay Natural History Society, and along with Oldfield Thomas commenced the Mammal Survey of British India, (*-ia*, commemorating (dedication, eponym) (L./Gr.)).

gleadowi: Eponym after F. Gleadow (dates not known), a naturalist and painter, worked in north India, (*-i*, commemorating (dedication, eponym) (L.)).
Endemic to South Asia (northwest India and Pakistan).

132. *Millardia kondana* Mishra and Dhanda, 1975
Kondana Rat

Millardia: Generic *nomen* is an eponym after Dr. Walter Samuel Millard (b. 1864–d. 1952), a British naturalist, Honorary Secretary of the Bombay Natural History Society, and along with Oldfield Thomas commenced the Mammal Survey of British India, (*-ia*, commemorating (dedication, eponym) (L./Gr.)).
kondana: Mod. L. *kondana* from the Kondana fort (also known as Sinhgad fort), Pune, Maharashtra, India, (*-ana*, belonging to, pertaining to (L.) geographic (location, toponym)).
Endemic to India (Maharashtra).

OEN Large Metad (nom. form)

133. *Mus musculus* Linnaeus, 1758
House Mouse

Mus: Gr. *mŷs* mouse.
musculus: Mod. L. *musculus* small mouse (Gr. *mŷs* mouse, *-ulus,* diminutive (comparison); somewhat (adj.) (L.)).
Cosmopolitan, introduced to many countries.

Subspecies

- *castaneus*: L. *castaneus* chestnut-brown (L. *castanea* chestnut).
  Southwestern Asian House Mouse. Bangladesh, Bhutan, India, Nepal and Sri Lanka.
- *domesticus*: L. *domesticus* belonging to the house, domestic (L. *domus* house).
  Western House Mouse. Afghanistan, India (northwest), Nepal and Pakistan.
- *bactrianus*: L. adj. *bactrianus* Bactrian, belonging to Bactria – an ancient country of central Asia, roughly corresponding to modern Turkistan and Balkh, northern Afghanistan), (*-anus*, geographic, belonging to, pertaining to (L.) geographic (location, toponym)).
  Asiatic House Mouse. Afghanistan, India, and Pakistan.

OEN East European House Mouse (Subsp. *musculus*), Persian House Mouse (Subsp. *bactrianus*), Sandy Mouse (Subsp. *bactrianus*), House-Mouse (Subsp. *castaneus*), Nepal House Mouse (J. Syn. *dubius*), Blyth's House Mouse (J. Syn. *gerbillinus*), Hill Mouse (J. Syn. *homorous*), Himalayan House Mouse (J. Syn. *homorous*), Ceylon House Mouse

(J. Syn. *manei*), Blanford's House Mouse (J. Syn. *sublimis*), Theobald's House Mouse (J. Syn. *theobaldi*), Long-haired Mouse (J. Syn. *tytleri*), Common Indian Mouse (J. Syn. *urbanus*)

VN Sinh. *Kossattu'miya*; Tam. (SL) *Sund'elli*, *Sit'elli*

134. *Mus platythrix* Bennett, 1832
Brown Spiny Mouse

Mus: Gr. *mŷs* mouse.
platythrix: Mod. L. *platythrix* broad-spined (Gr. *platōs* broad; Gr. *thríx* hair); in allusion to broad spine-like hairs on its body.
Endemic to India.

OEN Flat-haired Mouse (nom. form), Karwar Spiny Mouse (J. Syn. *bahadur*), Coorg Hill Spiny Mouse (J. Syn. *grahami*), Coorg Lowland Spiny Mouse (J. Syn. *hannyngtoni*)

VN Kan. *Kal ilei*; Wad. *Legyade*, *Kal-yelka*; Tel. *Gijeli-gadu*

135. *Mus booduga* (Gray, 1837)
Common Indian Field Mouse

Mus: Gr. *mŷs* mouse.
booduga: Mod. L. *booduga* from *buduga* native name of the species in Madras Presidency.
South Asia and Southeast Asia (Myanmar).

OEN Little Indian Field Mouse (nom. form), Indian Field Mouse (nom. form), Small Spiny Mouse (J. Syn. *lepidus*), White-bellied Field Mouse (J. Syn. *albidiventris*), Ceylon Field Mouse (J. Syn. *fulvidiventris*)

VN Tel. *Chitta-burkani*, *Chita-yeluka*, *Chita-ganda*; Sinh. *Podi Wel-miya*; Tam. (SL) *Sund'elli*, *Sit'elli*

136. *Mus saxicola* Elliot, 1839
Brown Spiny Mouse

Mus: Gr. *mŷs* mouse.
saxicola: Mod. L. *saxicola* stone-dweller (>L. *saxum* stone; L. *-cola* dweller (>L. *colere* to dwell)).
Endemic to South Asia. India, Nepal, and Pakistan. Besides the nominate subspecies *M. s. saxicola* Elliot's Spiny Mouse that is endemic to India, two more subspecies are found in the region.

Subspecies

- *sadhu*: L. adj. *sadhu* (from Sans. *sādhu* an ascetic, holy man), perhaps owing to its drab grey pelage in allusion to the ash-smeared ascetics.
  Wroughton's Spiny Mouse. Endemic to South Asia; India (northwest) and Pakistan.

- *gurkha*: Mod. L. *gurkha*, no explanation as to why this nomen has been given; perhaps owing to its collector C.A. Crump, who collected extensively in Nepal and Himalayas in the Kumaon region of India.

  Kumaon Spiny Mouse. Endemic to South Asia, India (northwest Himalayas and also in Bihar, Jharkhand, and Rajasthan), and Nepal.

OEN Dusky Spiny Mouse (J. Syn. *spinulosus*), Ramnad Spiny Mouse (J. Syn. *ramnadensis*), Cutch Spiny Mouse (J. Syn. *cinderella*), Small Spiny Mouse (J. Syn. *pygmaeus*), Khumbu Spiny Mouse (J. Syn. *khumbuensis*)

137. *Mus cervicolor* Hodgson, 1845
Fawn-coloured Mouse

Mus: Gr. *mÿs* mouse.
cervicolor: Mod. L. *cervicolor* fawn-coloured (L. *cervus* deer; L. *color* colour).

South Asia and Southeast Asia (Myanmar). Bhutan, India, Nepal, and Pakistan.

OEN Fawn Field Mouse (nom. form), Nepal Field Mouse (J. Syn. *strophiatus*), Little Rabbit-mouse (J. Syn. *cunicularis*), Imphal Field Mouse (J. Syn. *imphalensis*)

138. *Mus terricolor* Blyth, 1851
Earth-coloured Mouse

Mus: Gr. *mÿs* mouse.
terricolor: L. *terricolor* earth-coloured (L. *terra* earth; L. *color* colour).

Endemic to South Asia. Bangladesh, India, Nepal, and Pakistan.

OEN Earthy Field Mouse (nom. form), Beavan's Field Mouse (J. Syn. *beavanii*), Northern Field Mouse (J. Syn. *dunni*)

139. *Mus famulus* Bonhote, 1898
Bonhote's Mouse

Mus: Gr. *mÿs* mouse.
famulus: L. *famulus* servile.

Endemic to India (Western Ghats of Kerala and Tamil Nadu).

OEN Servant Mouse (nom. form)

140. *Mus phillipsi* Wroughton, 1912
Wroughton's Small Spiny Mouse

Mus: Gr. *mÿs* mouse.
phillipsi: Eponym after R.M. Phillips (dates not known), District Superintendent of Police, Dharwar, India, assisted C. A. Crump in

collecting for the Mammal Survey of British India, (*-i*, commemorating (dedication, eponym) (L.)).

Endemic to India. Nepal record needs verification.

OEN Grey Small Spiny Mouse (J. Syn. *surkha*), Sivasamudram Small Spiny Mouse (J. Syn. *siva*)

141. *Mus cookii* Ryley, 1914
Ryley's Spiny Mouse

Mus: Gr. *mȳs* mouse.

cookii: Eponym after J. Pemberton Cook (b. 1865–d. 1924), worked for the Burma Teak Company and collected from Moulmein for the Mammal Survey of British India, (*-ii*, commemorating (dedication, eponym) (L.)).

Southeast Asia and South Asia (northeast India).

OEN Cook's Mouse (nom. form), Naga Spiny Mouse (J. Syn. *nagarum*), Palni Hills Spiny Mouse (J. Syn. *palnica*), Darjeeling Spiny Mouse (S. Syn. *darjilingensis*)

142. *Mus mayori* (Thomas, 1915)
Mayor's Mouse

Mus: Gr. *mȳs* mouse.

mayori: Eponym after Major E.W. Mayor (dates not known), collected in Sri Lanka for the Mammal Survey of British India, (*-i*, commemorating (dedication, eponym) (L.)).

Endemic to Sri Lanka.

OEN Mayor's Rat (nom. form), Mayor's Bicoloured Rat (nom. form), Highland Spiny Rat (nom. form), Highland Coelomys (nom. form), Spiny Rat (nom. form), Bicoloured Spiny Rat (J. Syn. *bicolor*), Bi-coloured Coelomys (J. Syn. *bicolor*), Pocock's Bicoloured Spiny Rat (J. Syn. *pococki*), Pocock's Bicoloured Coelomys (J. Syn. *pococki*)

VN Sinh. *Miya*, *Kelle-miya*; Tam. (SL) *Yelli*, *Kart'elli*

143. *Mus pahari* Thomas, 1916
Sikkim Mouse

Mus: Gr. *mȳs* mouse.

pahari: Mod. L. *pahari* from Hind. *pahari* someone from the hills; owing to the fact that the types were collected in the Himalayas.

Southeast Asia and South Asia (Bhutan and India – northeast). Only the nominate subspecies, *M. p. pahari* Sikkim Mouse, occurs in the region.

OEN Sikkim Hill Mouse (nom. form), Mrs. Jackson's Hill Mouse (J. Syn. *jacksoniae*)

Remarks The accepted common name, as per some authorities, Gairdner's Shrewmouse is applicable only to an extralimital subspecies, *Mus pahari gairdneri* (Kloss, 1920). This name is after Kenneth G. Gairdner (b. 1900–d. 1950), a collector attached to Raffles Museum, Singapore. As the nominate subspecies was described from Sikkim in India, the name Sikkim Mouse is appropriate.

144. *Mus fernandoni* (Phillips, 1932)
Ceylon Spiny Mouse

Mus: Gr. *mӱs* mouse.
fernandoni: Eponym after E.C. Fernando (b. 1900–d. 1966), a collector and taxidermist at the Colombo Museum, Sri Lanka, (*-i*, commemorating (dedication, eponym) (L.)).
Endemic to Sri Lanka.

OEN Spiny Mouse (nom. form), Sri Lanka Spiny Mouse (nom. form)
VN Sinh. *Miya*, *Podi-Miya*; Tam. (SL) *Sund'elli*

145. *Nesokia indica* (Gray and Hardwicke, 1832)
Short-tailed Bandicoot-rat

Nesokia: Mod. L. *nesokia*, >*nesoki*, a local name for the species.
indica: L. adj. *indica* belonging to India, (*-ica* belonging to, pertaining to (L./Gr.)).
Middle East, South Asia, Arabian Peninsula, China, and Egypt. Afghanistan, Bangladesh, India, Nepal, and Pakistan.

OEN Indian Mole Rat (nom. form); Hardwicke's Field Rat (J. Syn. *hardwickei*), Short-tailed Mole Rat (J. Syn. *hardwickei*), Hazara Nesokia (J. Syn. *griffithi*), Eastern Indian Mole Rat (J. Syn. *indicus*), Hutton's Mole Rat (J. Syn. *huttoni*), Sindh Desert Mole Rat (J. Syn. *beaba*), Chitral Mole Rat (J. Syn. *chitralensis*)
VN Kan. *Kok*; Tel. *Galatta Koku*

146. *Niviventer niviventer* (Hodgson, 1836)
Himalayan Niviventer

Niviventer: Mod. L. *niviventer* white-bellied (L. *nivis* snow; L. *ventris* belly); in allusion to the whitish belly.
niviventer: Mod. L. *niviventer* white-bellied (L. *nivis* snow; L. *ventris* belly); in allusion to the whitish belly.
Endemic to South Asia. Bhutan, India, and Nepal.

OEN Himalayan White-bellied Rat, White-bellied House-rat, Lepcha Rat (J. Syn. *lepcha*), Ghose's white-bellied Rat (J. Syn. *monticola*)

147. ***Niviventer fulvescens*** **(Gray, 1847)**
Indomalayan Niviventer

Niviventer: Mod. L. *niviventer* white-bellied (L. *nivis* snow; L. *ventris* belly); in allusion to the whitish belly.
fulvescens: L. *fulvescens* somewhat tawny (L. *fulvus* tawny; *-escens*, somewhat).
Southeast Asia, China, and South Asia (Himalayas in India, Pakistan, and Nepal).

OEN Indomalayan White-bellied Rat, Chestnut White-bellied Rat (J. Syn. *caudiator*), Chestnut Rat (J. Syn. *caudiator*), Himalayan Spiny Field-Mouse (J. Syn. *jerdoni*)

148. ***Niviventer brahma*** **(Thomas, 1914)**
Brahma Niviventer

Niviventer: Mod. L. *niviventer* white-bellied (L. *nivis* snow; L. *ventris* belly); in allusion to the whitish belly.
brahma: Mod. L. *brahma* for Brahma, a supreme Hindu deity.
South Asia (India – Arunachal Pradesh) and Southeast Asia (Myanmar).

OEN Brahma White-bellied Rat (nom. form), Thomas's Chestnut Rat (nom. form)

149. ***Niviventer eha*** **(Wroughton, 1916)**
Smoke-bellied Niviventer

Niviventer: Mod. L. *niviventer* white-bellied (L. *nivis* snow; L. *ventris* belly); in allusion to the whitish belly.
eha: Eponym after E.H. Aitken (b. 1851–d. 1909), Indian civil servant, co-founder of the Bombay Natural History Society, and writer – wrote under the pseudonym Eha.
South Asia (Himalayas in India – Arunachal Pradesh, and Nepal) and Southeast Asia (Myanmar and southwest China).

OEN Smoke-bellied Rat, Little Himalayan Rat, Spectacled Rat

150. ***Niviventer langbianis*** **(Robinson and Kloss, 1922)**
Lang Bian Niviventer

Niviventer: Mod. L. *niviventer* white-bellied (L. *nivis* snow; L. *ventris* belly); in allusion to the whitish belly.
langbianis: Mod. L. *langbianis* meaning from Lang Bian Peak, on Lâm Viên Plateau, Vietnam, (*-is*, geographic, occurrence in (location, toponym) (L.)).

Southeast Asia, China (southwest) and South Asia (marginally in Arunachal Pradesh).

OEN Lang Bian White-bellied Rat, Dark-tailed Rat, Indo-chinese Arboreal White-bellied Rat (J. Syn. *indosinicus*), Indo-Chinese Arboreal Niviventer (J. Syn. *indosinicus*)

151. *Rattus rattus* (Linnaeus, 1758)
House Rat

Rattus: L. *rattus* rat.
rattus: L. *rattus* rat.

Africa, Arabian Peninsula, Middle East, South Asia, Europe, Southeast Asia, China and possibly also in other parts as an introduced taxon.

OEN House Rat (nom. form), Black Rat (nom. form), Roof Rat (nom. form), Ship Rat (nom. form), Black House Rat (nom. form), Hodgson's Rat (J. Syn. *aequicaudalis*), Alexandria Black Rat (J. Syn. *alexandrinus*), Egyptian House Rat (J. Syn. *alexandrinus*), Egyptian Ship Rat (J. Syn. *alexandrinus*), Arboreal Black Rat (J. Syn. *arboreus*), Asiatic Black Rat (J. Syn. *asiaticus*), Alexandria Black Rat (J. Syn. *alexandrinus*), Barren Island Black Rat (J. Syn. *atridorsum*), Brown Tree Rat (J. Syn. *brunneus*), Ceylon Mouse (J. Syn. *ceylonus*), Large-footed Mouse (J. Syn. *crassipes*), Gir House Rat (J. Syn. *girensis*), Striped-bellied Rat (J. Syn. *infralineatus*), Common Ceylon House Rat (J. Syn. *kandianus*), White-bellied Tree Rat (J. Syn. *kandianus*), Bunglow Rat (J. Syn. *kandianus*), Ceylon Highland Rat (J. Syn. *kelaarti*), Common Highland Rat (J. Syn. *kelaarti*), Kelaart's Rat (J. Syn. *kelaarti*), Narmada Rat (J. Syn. *narbadae*), Ceylon Large Tree Rat (J. Syn. *nemoralis*), Indian House Rat (J. Syn. *rufescens*), Common Indian Rat (J. Syn. *rufescens*), Rufescent Tree Rat (J. Syn. *rufescens*), Rufous Rat (J. Syn. *rufescens*), Wroughton's Black Rat (J. Syn. *wroughtoni*)

VN Hind. *Ghar-ka-chua*; Beng. *Demsa Indur*; Kan. *Manei ilei*; Tel. Yeluka; Tam. *Veedtu'elli*; Sinh. *Miya*, *Gas-Miya, Gay-miya*; Tam. (SL) *Elli, Yelli, Kart'yelli, Karpu'yelli, Lite'elli, Veedtu'elli*

152. *Rattus norvegicus* (Berkenhout, 1769)
Brown Rat

Rattus: L. *rattus* rat.
norvegicus: L. adj. *norvegicus* meaning from Norway, (*-icus*, belonging to, pertaining to (location, toponym) (L.)). This appears to be a simple case of latinization of vernacular name of the rat which was common in Great Britain and Ireland.

Europe, Asia (temperate region), Southeast Asia, Japan, Egypt and South Asia (large port cities in India, Pakistan and Sri Lanka).

OEN Norway Rat
VN Sinh. *Loco-miya*; Tam. (SL) *Yelli*

153. *Rattus tanezumi* (Temminck, 1844)
Oriental House Rat

Rattus: L. *rattus* rat.
tanezumi: Mod. L. *tanezumi* rat (from Japanese *ta-nezumi* the rat).
Southeast Asia, China, and South Asia (in the Himalayas in Afghanistan, Bangladesh, Bhutan, India, Nepal, and Pakistan).

OEN Tanezumi Rat, Asian House Rat, Lesser Brown Rat (J. Syn. *brunneusculus*), Common Nepal House Rat (J. Syn. *brunneus*)
VN Beng. *Gachua Indur*

154. *Rattus nitidus* (Hodgson, 1845)
Himalayan Rat

Rattus: L. *rattus* rat.
nitidus: L. *nitidus* shining, glittering (L. *nitere* to shine).
China, Southeast Asia, and South Asia (in the Himalayas in Bhutan and northeast India). Only the nominate subspecies, *R. n. nitidus* Himalayan Field Rat, occurs in the region.

OEN Himalayan Field Rat, Shining Brown Rat, Hodgson's grey-bellied Rat (J. Syn. *horeites*), Guha's Rat (J. Syn. *guhai*)

155. *Rattus pyctoris* (Hodgson, 1845)
Turkestan Rat

Rattus: L. rattus rat.
pyctoris: Mod. L. *pyctoris* flat-nose (Gr. *puktēs* boxer; Gr. *rhis* nose).
South Asia (Himalayas in Afghanistan, Bangladesh, Bhutan, India, Nepal, and Pakistan), China, Central Asia, and Middle East.

OEN Himalayan Rat, Black Nepal Rat (J. Syn. *rattoides*), Turkestan Rat (J. Syn. *turkestanicus*), North Asian Rat (J. Syn. *vicerex*), Kashgar Rat (J. Syn. *shigaricus*), Khumbu Rat (J. Syn. *khumbuensis*), Gilgit Rat (J. Syn. *gilgitianus*)

156. *Rattus exulans* (Peale, 1848)
Polynesian Rat

Rattus: L. *rattus* rat.
exulans: L. *exulans* wandering (L. *exsulare* to be exiled).
Southeast Asia and South Asia (Bangladesh – Chittagong and Slyhet).

OEN Pacific Rat, Little Burmese Rat (J. Syn. *concolor*)

157. *Rattus andamanensis* (Blyth, 1860)
Indochinese Forest Rat

Rattus: L. *rattus* rat.
andamanensis: L. adj. *andamanensis* meaning from the Andaman Islands, India, (-*ensis*, geographic, occurrence in (location, toponym) (L.)).

South Asia (Andaman and Nicobar Islands, and in the Himalayas in Bhutan, India, and Nepal), China and Southeast Asia.

OEN Sikkim Rat (J. Syn. *sikkimensis*), Andaman Brown Rat (J. Syn. *burrulus*), Nicobar Brown Rat (J. Syn. *holchu*)

158. *Rattus palmarum* (Zelebor, 1869)
Zelebor's Nicobar Rat

Rattus: L. *rattus* rat.
palmarum: L. *palmarum* (from *palmarius*), pertaining to palm trees.

Endemic to India (Car Nicobar and Great Nicobar Islands, Andaman and Nicobar Islands).

OEN Car Nicobar Rat, Palm Rat

159. *Rattus burrus* (Miller, 1902)
Miller's Nicobar Rat

Rattus: L. *rattus* rat.
burrus: L. *burrus* old form of the personal name *Pyrrhus* (>Gr. *purrhos* red).

Endemic to India (Great Nicobar, Little Nicobar and Trinkut Islands, Andaman and Nicobar Islands).

OEN Nonsense Rat, Nicobar Archipelago Rat

160. *Rattus stoicus* (Miller, 1902)
Andaman Rat

Rattus: L. *rattus* rat.
stoicus: Mod. L. *stoicus* (Gr. *stōikos*) stoic, enduring, uncomplaining; in allusion to its silent disposition when trapped.

Endemic to India (Henry Lawrence Island, Middle Andaman Island and South Andaman Island, Andaman and Nicobar Islands).

OEN Miller's Long-footed Rat (J. Syn. *taciturnus*), Roger's Andaman Rat (J. Syn. *rogersi*)

161. *Rattus satarae* Hinton, 1918
Sahyadris Forest Rat

Rattus: L. *rattus* rat.
satarae: Mod. L. *satarae* from Satara, Maharashtra, India, (-*ae*, geographic (location, toponym) (L.)).

Endemic to India (patchily in the Western Ghats in Karnataka, Maharashtra, and Tamil Nadu).

162. *Rattus montanus* Phillips, 1932
Sri Lankan Mountain Rat

Rattus: L. *rattus* rat.
montanus: L. *montanus* of the mountains, mountain- (L. *mons*, *montis* mountain).
Endemic to Sri Lanka (Central and Uva Provinces).

OEN Nillu Rat (nom. form), Nellu Rat (nom. form)
VN Sinh. *Miya*, *Kelle'miya*; Tam (SL) *Yelli*, *Kart'elli*

163. *Rattus ranjiniae* Agrawal and Ghosh, 1969
Ranjini's Field Rat

Rattus: L. *rattus* rat.
ranjiniae: Eponym after Ms. P.V. Ranjini (dates not known), collector of the species, (*-ae*, commemorating (dedication), feminine (L.)).
Endemic to India (Kerala).

OEN Kerala Rat, Ranjini's Rat

164. *Srilankamys ohiensis* Phillips, 1929
Ohiya Rat

Srilankamys: Mod. L. *Srilankamys* Sri Lanka mouse (Sri Lanka and Gr. *mŷs* mouse).
ohiensis: L. adj. *ohiensis* from Ohiya, Sri Lanka, (*-ensis*, geographic, occurrence in (location, toponym) (L.)).
Endemic to Sri Lanka (Central, Sabargamuwa and Uva Provinces).

OEN Jungle Rat (nom. form), Ceylon bicolored Rat (nom. form)
VN Sinh. *Kelle Miya*; Tam. (SL) *Kart-elli*

165. *Vandeleuria oleracea* (Bennett, 1832)
Asiatic Long-tailed Climbing Mouse

Vandeleuria: Etymology not known, No explanation was given by J.E. Gray (1842).
oleracea: L. *oleraceus* herbaceous, in allusion to its arboreal habits.
South Asia and Southeast Asia. Only the nominate subspecies, *V. o. oleracea* Indian Long-tailed Tree Mouse, from Bangladesh, Bhutan, India, Nepal and Sri Lanka occur in the region.

OEN Palm Mouse (Subsp. *oleracea*), Hodgson's Tree Mouse (Subsp. *dumeticola*), Bengal Tree Mouse (J. Syn. *marica*), Kumaon Tree Mouse (J. Syn. *modesta*), Powah Mouse (J. Syn. *povensis*), Rufous Tree Mouse

(J. Syn. *rubida*), Gujarat Tree Mouse (J. Syn. *spadicea*), Wroughton's Tree Mouse (J. Syn. *wroughtoni*)

VN Kan. *Marad ilei*; Tel. *Meina-yelka*; Sinh. *Podi-gas-miya, Kossatta-miya*; Tam. (SL) *Sund'elli*

166. *Vandeleuria nilagirica* (Jerdon, 1867)
Nilgiri Long-tailed Tree Mouse

Vandeleuria: Etymology not known, No explanation was given by J.E. Gray (1842).

nilagirica: L. adj. *nilagirica* belonging to Nilgiri (Sans. *nil*, *neel* blue; Sans. *giri* mountain) Hills, Tamil Nadu, India (*-ica* belonging to, pertaining to (L./Gr.)).

Endemic to India (Western Ghats of Karnataka and Tamil Nadu).

OEN Nilgiri Tree Mouse (nom. form)

167. *Vandeleuria nolthenii* Phillips, 1929
Sri Lankan Highland Tree Mouse

Vandeleuria: Etymology not known, No explanation was given by J.E. Gray (1842).

nolthenii: Eponym after A.C. Tutein-Nolthenius (b. 1898–d. 1953), estate owner and collector in Sri Lanka (*-i*, commemorating (dedication, eponym) (L.)).

Endemic to Sri Lanka (Central Province).

OEN Ceylon Highland Tree Mouse, Nolthenius's Long-tailed Climbing Mouse, Ceylon Highland Long-tailed Tree Mouse

VN Sinh. *Podi-gas-miya, Kossatta-miya*; Tam. (SL) *Sund'elli*

Infraorder Hystricognathi Brandt, 1855
Family Hystricidae G. Fischer, 1817

168. *Atherurus macrourus* (Linnaeus, 1758)
Asiatic Brush-tailed Porcupine

Atherurus: Mod. L. *atherurus* brush-tailed (Gr. *ather* the beard of the ear of the corn; Gr. *oúrá* tail); in allusion to the bunch of flattened scaly bristles at the tip of the tail.

macrourus: Mod. L. *macrourus* long-tailed (Gr. *makros* long; Gr. *-ouros* -tailed (Gr. *oúrá* tail)).

Southeast Asia, China (southern and southwest) and South Asia (India). Only the nominate subspecies, *A. m. macrourus* Asiatic Brush-tailed Procupine, in northeast India occurs in the region.

OEN Asiatic Brush-tailed Porcupine (nom. form), Bengal Porcupine (J. Syn. *assamensis*)

169. *Hystrix brachyura* Linnaeus, 1758
Malayan Porcupine

Hystrix: L. *hystrix* (Gr. *ýstriz*) porcupine (apparently from Gr. *ýs* hog and Gr. *trikh* hair).

brachyura: Mod. L. *brachyura* short-tailed (Gr. *brakhus* short; Gr. *-ouros* -tailed (Gr. *oúrá* tail)).

Southeast Asia, China, and South Asia (Bangladesh and India). Three subspecies are known from the region.

Subspecies

- *bengalensis*: L. adj. *bengalensis* meaning from Bengal, India, (*-ensis*, geographic, occurrence in (location, toponym) (L.)).
  Bengal Porcupine. Himalayas in Bangladesh and India (northeast), and Myanmar (west).
- *hodgsoni*: Eponym after Brian Houghton Hodgson (b. 1801–d. 1894), a British civil servant, naturalist, and ethnologist stationed in Nepal (*-i*, commemorating (dedication, eponym) (L.)).
  Nepal Porcupine. Himalayas in China (Tibet), India (northeast) and Nepal.
- *subcristata*: L. adj. *subcristata* slightly crested (L. *sub* slightly, somewhat; L. *cristatus* crested (L. *crista* crest)).
  Chinese Porcupine. China, India (Himalayas in the northeast), and Southeast Asia.

OEN Himalayan Crestless Porcupine (nom. form), Himalayan South-east Asian Porcupine (nom. form), Nepal Crestless Porcupine (J. Syn. *alophus*), Long-tailed Crestless Porcupine (J. Syn. *longicauda*), Mills' Crestless Porcupine (J. Syn. *millsi*)

VN Lepch. *Sa-thung*; Limbú *O'–e*; Nep. *Anchotia dumsi*

170. *Hystrix indica* Kerr, 1792
Indian Crested Porcupine

Hystrix: L. *hystrix* (Gr. *ýstriz*) porcupine (apparently from Gr. *ýs* hog and Gr. *trikh* hair).

indica: L. adj. *indica* belonging to India, (*-ica* belonging to, pertaining to (L./Gr.)).

South Asia, Central Asia (parts), Middle East, Turkey, Arabian Peninsula. Afghanistan, India, Pakistan, Nepal, and Sri Lanka.

OEN Porcupine (nom. form), Rajputana Porcupine (J. Syn. *cuneiceps*), Shaggy Crested Porcupine (Subsp. *hirsutirostris*), Western Asiatic Crested Porcupine (J. Syn. *leucurus*), Malabar Crested Porcupine (J. Syn. *malabarica*), Ceylon Crested Porcupine (J. Syn. *zeylonensis*)

VN Hind. *Sahi*, *Sayal*, *Sial*, *Sarsel*; Beng. *Sajru*; Kan. *Yed*; Mar. *Salendra*; Tel. *Yedu Pandi*; Gond. *Ho-igu*; Guj. *Saori*; Nep. *Dumsi*; Sinh. *Ittawa*, *Pandura-ittawa*; Tam. (SL) *Mullam-pandi*, *Mullup-pandie*

Order Lagomorpha Brandt, 1855
Family Ochotonidae Thomas, 1897

171. *Ochotona roylei* (Ogilby, 1839)
Royle's Pika

Ochotona: Mod. L. *ochotona* >Mongol. *ochodona*, a native name for pika. *roylei*: Eponym after Dr. John Forbes Royle (b. 1799–d. 1858), an Indian-born British botanist, surgeon, and collector (*-i*, commemorating (dedication, eponym) (L.)).

Himalayas of Bhutan, China, India, Nepal, and Pakistan. Beside the nominate subspecies, *O. r. roylei* Royle's Pika, another subspecies is known from the region.

Subspecies

- *nepalensis*: L. adj. *nepalensis* from Nepal (*-ensis*, geographic, occurrence in (location, toponym) (L.)).

  Nepalese Pika. Nepal.

OEN Himalayan Mouse Hare (nom. form), Red-shouldered Pika (Subsp. *nepalensis*), Hodgson's Pika (J. Syn. *hodgsoni*), Ang Dawa Pika (J. Syn. *angdawai*), Mitchell's Pika (J. Syn. *mitchelli*)

VN Kin. *Rang-runt, Rang-duni*

172. *Ochotona rufescens* (Gray, 1842)
Afghan Pika

Ochotona: Mod. L. *ochotona* >Mongol. *ochodona*, a native name for pika. *rufescens*: L. *rufescens* reddish (L. *rufescere* to become reddish >*rufus* red; *-escens*, somewhat).

Afghanistan, Iran, Pakistan and Turkmenistan.

OEN Rufescent Pika (nom. form), Reddish Pika (nom. form), Baluchistan Pika (J. Syn. *vulturna*)

173. *Ochotona curzoniae* (Hodgson, 1858)
Black-lipped Pika

Ochotona: Mod. L. *ochotona* >Mongol. *ochodona*, a native name for pika. *curzoniae*: Eponym after 'Mrs. Curzon' (details not known), (*-ae*, commemorating (dedication), feminine (L.)).

China (Tibetan Plateau), India, Nepal, and Pakistan.

OEN Plateau Pika (nom. form), Curzon's Pika (nom. form)

174. *Ochotona thibetana* (Milne-Edwards, 1871)
Moupin Pika

Ochotona: Mod. L. *ochotona* >Mongol. *ochodona*, a native name for pika.

thibetana: L. adj. *thibetana* meaning from Tibet (*-ana*, belonging to, pertaining to (L.) geographic (location, toponym).

Bhutan, China, India, Nepal, and Pakistan. Only the undermentioned subspecies occurs in South Asia.

Subspecies

- *sikimaria*: L. adj. *sikimaria* belonging to Sikkim, India, (*-aria* pertaining to (L./Gr.)).

  Sikkim Pika. Himalayas in Bhutan, India, and Nepal.

OEN Forest Pika (nom. form)

175. *Ochotona ladacensis* (Günther, 1875)
Ladakh Pika

Ochotona: Mod. L. *ochotona* >Mongol. *ochodona*, a native name for pika.
ladacensis: L. adj. *ladacensis* from Ladakh, India (*-ensis*, geographic, occurrence in (location, toponym) (L.)).

China (Tibetan Plateau), India, and Pakistan.

OEN Stoliczka's Mouse Hare (nom. form)

176. *Ochotona macrotis* (Günther, 1875)
Large-eared Pika

Ochotona: Mod. L. *ochotona* >Mongol. *ochodona*, a native name for pika.
macrotis: Mod. L. *macrotis* big-eared (>Gr. *makrōtēs* long-eared (Gr. *makros* long; Gr. *ous*, *ōtos* ear)).

Central Asia and South Asia. Beside the nominate subspecies, *O. m. macrotis* Large-eared Pika, occurring in the Himalayas of the Afghanistan, India, Nepal, and Pakistan, two more subspecies are known from the region.

Subspecies

- *wollastoni*: Eponym after Alexander Frederick Richmond 'Sandy' Wollaston (b. 1875–d. 1930), a British medical doctor, ornithologist, and botanist (*-i*, commemorating (dedication, eponym) (L.)).

  Wollaston's Large-eared Pika. Endemic to Nepal.
- *auritus*: L. *auritus* long-eared.

  Ladakh Long-eared Pika. Endemic to India, Ladakh, Jammu and Kashmir.

OEN Indian Pika (nom. form), Everest Pika (Subsp. *wollastoni*), Grey Pika (J. Syn. *griseus*)

177. *Ochotona nubrica* Thomas, 1922
Nubra Pika

Ochotona: Mod. L. *ochotona* >Mongol. *ochodona*, a native name for pika.

nubrica: L. adj. *nubrica* from Nubra Valley, Jammu and Kashmir, India, (*-ica* belonging to, pertaining to (L./Gr.)).

Himalayas of Bhutan, China, India, Nepal, and Pakistan. Beside the nominate subspecies, *O. n. nubrica* Nubra Pika, another subspecies is known from the region.

Subspecies

- *lhasaensis*: L. adj. *lhasaensis* from Lhasa, Tibet, (*-ensis*, geographic, occurrence in (location, toponym) (L.)).

  Lhassa Pika. Nepal.

178. *Ochotona forresti* Thomas, 1923
Forrest's Pika

Ochotona: Mod. L. *ochotona* >Mongol. *ochodona*, a native name for pika.
forresti: Eponym after George Forrest (b. 1873–d. 1932), a Scottish botanist, who collected extensively in Yunnan beginning in 1904 (*-i*, commemorating (dedication, eponym) (L.)).

Bhutan, China, India, and Myanmar.

179. *Ochotona himalayana* Feng, 1973
Himalayan Pika

Ochotona: Mod. L. *ochotona* >Mongol. *ochodona*, a native name for pika.
himalayana: Mod. L. *himalayana* from Himalayas (Sans. *Himalaya* an abode of snow (*him* snow, *alaya* abode)), (*-ana*, belonging to, pertaining to (L.) geographic (location, toponym)).

China (Tibet) and Nepal.

Family Leporidae Fischer, 1817

180. *Caprolagus hispidus* (Pearson, 1839)
Hispid Hare

Caprolagus: Mod. L. *caprolagus* (Gr. *kapros* wild boar; L. *lagos* a hare).
hispidus: L. *hispidus* rough, hairy.

Endemic to India and Nepal.

OEN Assam Rabbit (nom. form)

181. *Lepus tolai* Pallas, 1778
Tolai Hare

Lepus: L. *lepus* rabbit, hare (genitive *Leporis* a hare).
tolai: Mod. L. *tolai* >Transbaikalia or Tungus *tolai*, *tulai*, or *tolui* names by which this species is known.

Central Asia and Afghanistan in South Asia.

OEN Habibi's Hare (J. Syn. *habibi*)

182. *Lepus nigricollis* Cuvier, 1823
Black-naped Hare

Lepus: L. *lepus* rabbit, hare (genitive *Leporis* a hare).
nigricollis: L. *nigricollis* black-collared (L. *nigri* (from *niger*) black; Mod. L. *-collis* -throated (>L. *collum* neck).

South Asia, introduced to many islands of Indian ocean region. Beside the nominate subspecies, *L. n. nigricollis* Black-naped Hare, six more subspecies are known from the region.

Subspecies

- *ruficaudatus*: Mod. L. *ruficaudatus* (L. *rufus* rufous; L. *-caudatus* -tailed (L. *cauda* tail)).
  Rufous-tailed Hare. Endemic to South Asia, Bangladesh, Bhutan, India, Nepal, and Pakistan.
- *aryabertensis*: L. adj. *aryabertensis* belonging to Aryavart (Sans. *Āryāvarta* abode of the Aryans, the tract between the Himalaya and the Vindhya ranges, from the Eastern (Bay of Bengal) to the Western Sea (Arabian Sea)), (*-ensis*, geographic, occurrence in (location, toponym) (L.)).
  Nepal Hare. Endemic to Nepal.
- *dayanus*: Eponym after Dr. Day (dates not known), the collector of the types; could be Dr. Francis Day.
  Sindh Hare. Endemic to South Asia; Afghanistan, India, and Pakistan.
- *simcoxi*: Eponym after A.H.A. Simcox, Indian Civil Services (dates not known), Collector of Khandesh, where the type locality is located.
  Khandesh Hare. Endemic to India; Madhya Pradesh and Maharashtra.
- *singhala*: Mod. L. *singhala* meaning from *Sinhala* (= Ceylon = Sri Lanka).
  Ceylon Hare. Endemic to Sri Lanka.
- *sadiya*: Mod. L. *sadiya* from Sadiya, Assam.
  Assam Hare. Endemic to India; Assom.

OEN Indian Hare (Subsp. *nigricollis*), Northern Indian Hare (Subsp. *nigricollis*), Northern Indian Hare (Subsp. *ruficaudatus*), Common Indian Red-tailed Hare (Subsp. *ruficaudatus*), Bengal Hare (Subsp. *ruficaudatus*), Desert Hare (Subsp. *dayanus*), Tytler's Hare (J. Syn. *tytleri*), Joongshai Hare (J. Syn. *joongshaiensis*), Pachmarhi Hare (J. Syn. *mahadeva*), Rajput Hare (J. Syn. *rajput*), Kutch Hare (J. Syn. *cutchensis*)

VN Hind. *Khar-gosh*, *Kharra*, *Lamma*; Beng. *Sasru*; Pers. *Khar-gosh*; Gond. *Molol*; Mar. *Sassa*; Kan. *Malla*; Tel. *Kundelu*, *Chevula Pilli*; Tam. *Musal*; Sinh. *Hawa*; Tam. (SL) *Mussal*, *Muyal*

183. *Lepus oiostolus* Hodgson, 1840
Woolly Hare

Lepus: L. *lepus* rabbit, hare (genitive *Leporis* a hare).
oiostolus: Mod. L. *oiostolus* woolly (Gr. *óiis* sheep fleece; L. *stola* (Gr. *stolē*) robe, vestment).

China and South Asia; India and Nepal. The nominate subspecies does not occur in the region; two subspecies are known from the region.

Subspecies

- *pallipes*: Mod. L. *pallipes* pale-footed (L. *pallidus* pale; L. *pes* foot, >Gr. *pous* foot).
  White-footed Woolly Hare. Himalayas of India and Nepal.
- *hypsibius*: Mod. L. *hypsibius* (Gr. *ÿpsi* on high; Gr. *bios* living); in allusion to the high elevation range of the animal.
  Ladakh Woolly Hare. China (Tibetan Plateau) and Ladakh, India.

OEN Tibetan Hare (nom. form), Pale-footed Hare (Subsp. *pallipes*), Mountain Hare (Subsp. *hypsibius*)
VN Lep. *Rigong*; Bhot. *Rigong*

184. *Lepus tibetanus* Waterhouse, 1841
Desert Hare

Lepus: L. *lepus* rabbit, hare (genitive *Leporis* a hare).
tibetanus: L. adj. *tibetanus* meaning from or belonging to Tibet, (*-anus*, belonging to, pertaining to (L.) geographic (location, toponym)).

China and marginally in South Asia. Beside the nominate subspecies, *L. t. tibetanus* Desert Hare, found in high-altitude areas of Afghanistan, India, and Pakistan, another subspecies is known from the region.

Subspecies

- *craspedotis*: Mod. L. *craspedotis* large-eared (Gr. *craspedote* veil (*cràspedo* hem, fringe); Gr. *otis*, *ōtos* ear); in allusion to its large ears.
  Balochistan Desert Hare. Endemic to Pakistan.

OEN Tibet Hare (nom. form), Biddulph's Hare (J. Syn. *biddulphi*), Large-eared Hare (Subsp. *craspedotis*), Afghan Hare (Subsp. *craspedotis*), Kopet-Dagh Desert Hare (Subsp. *craspedotis*)
VN Bhut. *Ri-bong*; Tib. *Ri-bong*

Order Erinaceomorpha Gregory, 1910
Family Erinaceidae Fischer, 1817

185. *Hemiechinus auritus* (Gmelin, 1770)
Long-eared Hedgehog

Hemiechinus: Mod. L. *hemiechinus* semi-hedgehog (Gr. *hēmi-* half-; *ékhïnos* hedgehog).
auritus: L. *auritus* long-eared.

Central Asia, Middle East, and South Asia. One subspecies is known from the region.

Subspecies

- *megalotis*: Mod. L. *megalotis* great-eared (Gr. *megalos* great; *-ōtis* -eared (*ous*, *ōtos* ear)).

Great-eared Afghan Hedgehog. Afghanistan and Pakistan.

OEN Large-eared Afghan Hedgehog (Subsp. *megalotis*)

186. *Hemiechinus collaris* (Gray, 1830)
Collared Hedgehog

Hemiechinus: Mod. L. *hemiechinus* semi-hedgehog (Gr. *hēmi-* half-; *ékhïnos* hedgehog).
collaris: L. *collaris* of the neck, collar (L. *collum* neck).
Endemic to South Asia; India and Pakistan.

OEN Hardwicke's Hedgehog (nom. form), Indian long-eared Hedgehog (J. Syn. *indicus*), North-Indian Hedgehog (J. Syn. *indicus*), Gray's Hedgehog (J. Syn. *grayii*), Himalayan Hedgehog (J. Syn. *spatangus*)

187. *Paraechinus hypomelas* (Brandt, 1836)
Brandt's Hedgehog

Paraechinus: Mod. L. *paraechinus* similar to the hedgehog (Gr. *para* near to; Gr. *ékhïnos* hedgehog).
hypomelas: Mod. L. *hypomelas* black-ventered (Gr. *hupo* beneath; Gr. *melas* black).
Middle East and South Asia. Along with the nominate subspecies, *P. h. hypomelas* Brandt's Hedgehog, occurring in Afghanistan and Pakistan, another subspecies is known from the region.

Subspecies

- *blanfordi*: Eponym after William Thomas Blanford (b. 1832–d. 1905), an English geologist and naturalist, collected in many parts of India, also instrumental in establishing Indian Museum, (*-i*, commemorating (dedication, eponym) (L.)).
Blanford's Hedgehog. Endemic to Pakistan.

OEN Anderson's Hedhehog (Subsp. *blanfordi*), Afghan Hedgehog (J. Syn. *amir*), Jerdon's Hedgehog (J. Syn. *jerdoni*)

188. *Paraechinus micropus* (Blyth, 1846)
Indian Hedgehog

Paraechinus: Mod. L. *paraechinus* similar to the hedgehog (Gr. *para* near to; Gr. *ékhïnos* hedgehog).
micropus: Mod. L. *micropus* small-footed (Gr. *mikros* small; Gr. *pous* foot).
Endemic to South Asia; Pakistan and India.

OEN South-Indian Hedgehog (nom. form), Northern Pale Hedgehog (nom. form), Black-chinned Hedgehog (J. Syn. *mentalis*), Painted Hedgehog (J. Syn. *pictus*), Intermediate Hedgehog (J. Syn. *intermedius*), Kutch Hedgehog (J. Syn. *kutchicus*)

189. ***Paraechinus nudiventris*** **(Horsfield, 1851)**
Madras Hedgehog

Paraechinus: Mod. L. *paraechinus* similar to hedgehog (Gr. *para* near to; Gr. *ékhïnos* hedgehog).
nudiventris: Med. L. *nudiventris* bare-bellied (L. *nudus* bare; L. *venter, ventris* belly).

Endemic to India; Andhra Pradesh, Kerala and Tamil Nadu.

OEN Southern Indian Hedgehog (nom. form)

Order Soricomorpha Gregory, 1910
Family Soricidae Fischer, 1817

190. ***Crocidura leucodon*** **(Hermann, 1780)**
Bicoloured White-toothed Shrew

Crocidura: Gr. *crocis* or *crocos* the flock or nap on the woolen cloth, a piece of wool; Gr. *oúrá* tail; in allusion to the tail which is covered with short hairs interspersed by long hairs.
leucodon: Gr. *leukodon* white-toothed (*leukos* white; Gr. *ódoús*, *ódôn* tooth).

Europe (eastern, central and southern), Middle East Asia. India (Jammu and Kashmir).

191. ***Crocidura gmelini*** **(Pallas, 1811)**
Gmelin's White-toothed Shrew

Crocidura: Gr. *crocis* or *crocos* the flock or nap on the woolen cloth, a piece of wool; Gr. *oúrá* tail; in allusion to the tail which is covered with short hairs interspersed by long hairs.
gmelini: Eponym after Johann Friedrich Gmelin (b. 1748–d. 1804), a German naturalist, botanist, entomologist, herpetologist, and malacologist, (-*i*, commemorating (dedication, eponym) (L.)).

Middle Asia, China (western), marginally in South Asia. Afghanistan and Pakistan.

OEN Steppic Pygmy Shrew (nom. form)

192. ***Crocidura fuliginosa*** **(Blyth, 1855)**
Southeast Asian Shrew

Crocidura: Gr. *crocis* or *crocos* the flock or nap on the woolen cloth, a piece of wool; Gr. *oúrá* tail; in allusion to the tail which is covered with short hairs interspersed by long hairs.
fuliginosa: Mod. L. *fuliginosus* sooty (>L. *fuligo, fuliginis* soot).

Southeast Asia; northeast India, no exact location known.

OEN Southeast Asian White-toothed Shrew (nom. form), Coal Shrew (nom. form)

193. ***Crocidura horsfieldii*** **(Tomes, 1856)**
Horsfield's Shrew

Crocidura: Gr. *crocis* or *crocos* the flock or nap on the woolen cloth, a piece of wool; Gr. *oúrá* tail; in allusion to the tail which is covered with short hairs interspersed by long hairs.
horsfieldii: Eponym after Dr. Thomas Horsfield (b. 1773–d. 1859), an American physician and naturalist, later curator of the East India Company Museum in London, (*-ii*, commemorating (dedication, eponym) (L.)).

Endemic to South Asia; India and Sri Lanka, possibly also in Nepal.

OEN Shrew-mouse (J. syn. *retusa*), Mouse-like Shrew (J. Syn. *myoides*)
VN Sinh. *Hik-miya, Podi kunu-miya, Kunuhik-miya*; Tam. (SL) *Mungi'elli, Mungi'elli-kutti*

194. ***Crocidura attenuata*** **Milne-Edwards, 1872**
Grey Shrew

Crocidura: Gr. *crocis* or *crocos* the flock or nap on the woolen cloth, a piece of wool; Gr. *oúrá* tail; in allusion to the tail which is covered with short hairs interspersed by long hairs.
attenuata: L. *attenuatus* shortened, abbreviated, thin (*attenuare* to attenuate).

South Asia and Southeast Asia. Himalayas in Bhutan, India, and Nepal.

OEN Asian Grey Shrew (nom. form), Indochinese Shrew (nom. form), King's Shrew (J. syn. *kingiana*), Anderson's Assam Shrew (J. Syn. *rubricosa*)

195. ***Crocidura andamanensis*** **Miller, 1902**
Andaman White-toothed Shrew

Crocidura: Gr. *crocis* or *crocos* the flock or nap on the woolen cloth, a piece of wool; Gr. *oúrá* tail; in allusion to the tail which is covered with short hairs interspersed by long hairs.
andamanensis: L. adj. *andamanensis* meaning from Andaman Islands, India, (*-ensis*, geographic, occurrence in (location, toponym) (L.)).

Endemic to India; South Andaman Island.

OEN Andaman Shrew (nom. form)

196. ***Crocidura nicobarica*** **Miller, 1902**
Nicobar White-toothed Shrew

Crocidura: Gr. *crocis* or *crocos* the flock or nap on the woolen cloth, a piece of wool; Gr. *oúrá* tail; in allusion to the tail which is covered with short hairs interspersed by long hairs.

nicobarica: L. adj. *nicobarica* belonging to Nicobar Islands, India, (-*ica* belonging to, pertaining to (L./Gr.)).
Endemic to India; Nicobar Islands.

OEN Nicobar Shrew (nom. form)

197. *Crocidura pullata* Miller, 1911
Kashmir White-toothed Shrew

Crocidura: Gr. *crocis* or *crocos* the flock or nap on the woolen cloth, a piece of wool; Gr. *oúrá* tail; in allusion to the tail which is covered with short hairs interspersed by long hairs.
pullata: L. *pullatus* clad in black garments.
Endemic to South Asia; India and Pakistan (in the Himalayas).

OEN Dsuky Shrew (nom. form)

198. *Crocidura hispida* Thomas, 1913
Andaman Shrew

Crocidura: Gr. *crocis* or *crocos* the flock or nap on the woolen cloth, a piece of wool; Gr. *oúrá* tail; in allusion to the tail which is covered with short hairs interspersed by long hairs.
hispida: L. *hispida* (>L. *hispidus*) rough, hairy.
Endemic to India; Middle Andaman Island.

OEN Andaman Spiny Shrew (nom. form)

199. *Crocidura pergrisea* Miller, 1913
Pale Grey Shrew

Crocidura: Gr. *crocis* or *crocos* the flock or nap on the woolen cloth, a piece of wool; Gr. *oúrá* tail; in allusion to the tail which is covered with short hairs interspersed by long hairs.
pergrisea: Mod. L. *pergrisea* pale grey (L. *per*- very- (in comp.) (*per* through); Med. L. *griseus* grey (>Old French *gris* grey)).
Endemic to India; known only from type locality – Skoro Loomba, Shigar, Baltistan, Jammu and Kashmir.

OEN Kashmir Rock Shrew (nom. form)

200. *Crocidura rapax* G. Allen, 1923
Chinese White-toothed Shrew

Crocidura: Gr. *crocis* or *crocos* the flock or nap on the woolen cloth, a piece of wool; Gr. *oúrá* tail; in allusion to the tail which is covered with short hairs interspersed by long hairs.

rapax: L. *rapax* rapacious (L. *rapere* to seize).
Southeast Asia (China and Taiwan) and India (Meghalaya).

OEN Great White-toothed Shrew (nom. form)

201. ***Crocidura vorax* G. Allen, 1923**
Voracious Shrew

Crocidura: Gr. *crocis* or *crocos* the flock or nap on the woolen cloth, a piece of wool; Gr. *oúrá* tail; in allusion to the tail which is covered with short hairs interspersed by long hairs.
vorax: L. *vorax* voracious.
Southeast Asia, China (south) and India (northeast).

202. ***Crocidura zarudnyi* Ognev, 1928**
Zarudny's Rock Shrew

Crocidura: Gr. *crocis* or *crocos* the flock or nap on the woolen cloth, a piece of wool; Gr. *oúrá* tail; in allusion to the tail which is covered with short hairs interspersed by long hairs.
zarudnyi: Eponym after Dr. N.A. Zarudny (b. 1859–d. 1919), curator of Tashkent Museum, Uzbekistan, and collected in Persia (-*i*, commemorating (dedication, eponym) (L.)).
Afghanistan, Iran, and Pakistan.

OEN Pale Gray Shrew (nom. form)

203. ***Crocidura miya* Phillips, 1929**
Sri Lankan Long-tailed Shrew

Crocidura: Gr. *crocis* or *crocos* the flock or nap on the woolen cloth, a piece of wool; Gr. *oúrá* tail; in allusion to the tail which is covered with short hairs interspersed by long hairs.
miya: Mod. L. *miya* shrew from Sinhala *miya* musk-rat.
Endemic to Sri Lanka.

OEN Ceylon Long-tailed Shrew (nom. form)
VN Sinh. *Hik-miya, Kunu-miya, Sri Lanka Kunuhik-miya*; Tam. (SL) *Mungi'elli*

204. ***Crocidura jenkinsi* Chakraborty, 1978**
Jenkin's Andaman Spiny Shrew

Crocidura: Gr. *crocis* or *crocos* the flock or nap on the woolen cloth, a piece of wool; Gr. *oúrá* tail; in allusion to the tail which is covered with short hairs interspersed by long hairs.

jenkinsi: Eponym after Dr. Paulina D. Jenkins (b. 1966), a British zoologist, systematist and senior curator of Mammals at the Natural History Museum, London, (*-i*, commemorating (dedication, eponym) (L.)).

Endemic to India; South Andaman Island.

OEN Jenkin's Shrew (nom. form)

205. *Crocidura hikmiya* Meegaskumbura et al., 2007
Sinharaja Shrew

Crocidura: Gr. *crocis* or *crocos* the flock or nap on the woolen cloth, a piece of wool; Gr. *oúrá* tail; in allusion to the tail which is covered with short hairs interspersed by long hairs.

hikmiya: Mod. L. hik*miya* shrew from Sinhala *hik-miya* musk-rat.

Endemic to Sri Lanka.

206. *Crocidura gathornei* Jenkins, 2013
Gathorne's Rock Shrew

Crocidura: Gr. *crocis* or *crocos* the flock or nap on the woolen cloth, a piece of wool; Gr. *oúrá* tail; in allusion to the tail which is covered with short hairs interspersed by long hairs.

gathornei: Eponym after Gathorne Gathorne-Hardy (b. 1933), V Earl of Cranbrook, a British biologist and author of many books on Southeast Asian mammals (*-i*, commemorating (dedication, eponym) (L.)).

Endemic to India; Himalayas in Himachal Pradesh and Uttarakhand.

207. *Feroculus feroculus* (Kelaart, 1850)
Kelaart's Long-clawed Shrew

Feroculus: L. *feroculus* (dim. of *ferox* fierce), somewhat fierce or spirited (*-ulus*, diminutive (comparison), somewhat (adj.) (L.)).

feroculus: L. *feroculus* (dim. of *ferox* fierce), somewhat fierce or spirited (*-ulus*, diminutive (comparison), somewhat (adj.) (L.)).

Endemic to South Asia; India (Kerala and Tamil Nadu (in the Western Ghats)) and Sri Lanka.

OEN Kelaart's Shrew (nom. form), Kelaart's Mole-Shrew (nom. form), Large-footed Shrew (nom. form), Newara Elyia Long-clawed Shrew (nom. form)

VN Sinh. *Hik-miya*, *Kunu-miya*, *Pirihik-miya*; Tam. (SL) *Mungi'elli*

208. *Solisorex pearsoni* Thomas, 1924
Pearson's Long-clawed Shrew

Solisorex: Mod. L. *solisorex* sun-shrew (L. *solus* sun; *sorex* (>Gr. *ÿrax*) shrew).

pearsoni: Eponym after Dr. Joseph Pearson (b. 1881–d. 1971), a marine biologist and Director of the Colombo Museum, Sri Lanka, (*-i*, commemorating (dedication, eponym) (L.)).

Endemic to Sri Lanka.

OEN Mole-shrew (nom. form)

VN Sinh. *Hik-miya*, *Podi kunu-miya*, *Sri Lanka Mahik-miya*; Tam. (SL) *Mungi'elli*

209. *Suncus murinus* (Linnaeus, 1766)
House Shrew

Suncus: Mod. L. *suncus* shrew from Arabic name '*far sunki*'.

murinus: L. *murinus* of mice, mouse- (>L. *mus*, *muris* mouse); in allusion to its mouse-like appearance.

South Asia, Southeast Asia and introduced to parts of Africa, Arabian Peninsula, and other Oceanic Islands.

OEN Asian House Shrew (nom. form), Blue Musk-rat (nom. form), Indian Musk Shrew (nom. form), Mouse-coloured Shrew (nom. form), Shrew-mouse (nom. form), Dark Brown Shrew (J. Syn. *serpentarius*), Dark-brown Shrew (J. Syn. *saturatior*), Dehra Shrew (J. Syn. *tytleri*), Griffith's Shrew (J. Syn. *grifithii*), Hairy-footed Shrew (J. Syn. *soccatus*), Indian Grey Musk-rat (J. Syn. *caereluscens*), Kandyan Shrew (J. Syn. *kandianus*), Large Black Shrew (J. Syn. *grifithii*), Musk Shrew (J. Syn. *caereluscens*), Musk-rat (J. Syn. *caereluscens*), Musk-shrew (J. Syn. *caereluscens*), Nepal Wood Shrew (J. Syn. *nemorivagus*), Rufescent Shrew (J. Syn. *serpentarius*)

VN Hind. *Chachundi*; Kan. *Sondeli*; Sinh. *Hik-miya*, *Podhu Hik-miya*, *Kunu-miya*; Tam. (SL) *Mungi'elli*, *Munjuru*

210. *Suncus etruscus* (Savi, 1822)
Savi's Pygmy Shrew

Suncus: Mod. L. *suncus* shrew from Arabic name '*far sunki*'.

etruscus: Mod. L. *etruscus* Etruscan, from Etruscus, Italy.

Parts of Mediterranean region, Arabian Peninsula, India, and Southeast Asia.

OEN Pygmy White-toothed Shrew (nom. form), Etruscan Shrew (nom. form), White-toothed Pygmy Shrew (nom. form), Small-clawed Pygmy Shrew (J. Syn. *micronyx*), Kumaon Pygmy Shrew (J. Syn. *micronyx*), Nilgiri Pygmy Shrew (J. Syn. *perrotetti*), Nepal Pygmy Shrew (J. Syn. *hodgsoni*), Darjeeling Pygmy Shrew (J. Syn. *hodgsoni*), Himalayan Pygmy Shrew (J. Syn. *hodgsoni*), Black Pygmy Shrew (J. Syn. *atratus*), Naked-footed Pygmy Shrew (J. Syn. *nudipes*), Black-toothed Pygmy Shrew (J. Syn. *melanodon*), Assamese Pygmy Shrew (J. Syn.

assamensis), Nilagiri Pygmy Shrew (J. Syn. *nilagirica*), Bright-yellow Pygmy Shrew (J. Syn. *nitidofulva*), Anderson's Pygmy Shrew (J. Syn. *pygmeoides*), Travancore Pygmy Shrew (J. Syn. *travancorensis*), Ceylon White-toothed Shrew (J. Syn. *kura*)

VN Sinh. *Podi Hik-miya*

211. *Suncus montanus* (Kelaart, 1850)
Sri Lankan Highland Shrew

Suncus: Mod. L. *suncus* shrew from Arabic name '*far sunki*'.
montanus: L. *montanus* of the mountains, mountain- (L. *mons*, *montis* mountain).

Endemic to Sri Lanka.

OEN Ceylon Highland Shrew (nom. form), Hill Shrew (nom. form), Montane Shrew (nom. form), Black Shrew (nom. form), Shrew-mouse (nom. form), Ceylon Rufescent Shrew (J. Syn. *ferruginea*), Ceylon Rufescent Shrew (J. Syn. *ferrugineus*)

VN Sinh. *Hik-miya*, *Kunu-miya*, *Kandu Hik-miya*; Tam. (SL) *Mungi'elli*

212. *Suncus niger* (Horsfield, 1851)
Indian Highland Shrew

Suncus: Mod. L. *suncus* shrew from Arabic name '*far sunki*'.
niger: L. *niger* black, dark coloured, shining black.

Endemic to India; Karnataka, Kerala and Tamil Nadu (in the Western Ghats).

OEN Nilgiri Wood Shrew

213. *Suncus stoliczkanus* (Anderson, 1877)
Anderson's Shrew

Suncus: Mod. L. *suncus* shrew from Arabic name '*far sunki*'.
stoliczkanus: Eponym after Dr. Ferdinand Stoliczka (b. 1838–d. 1874), a British paleontologist, who participated in Yarkand Missions.

Endemic to South Asia; India, Pakistan, and Nepal, and possibly in Bangladesh.

OEN Yellow-throated Shrew (J. Syn. *subfulva*), Bidie's Shrew (J. Syn. *bidiana*), White-cheeked Shrew (J. Syn. *leucogenys*)

214. *Suncus dayi* (Dobson, 1888)
Day's Shrew

Suncus: Mod. L. *suncus* shrew from Arabic name '*far sunki*'.
dayi: Eponym after Dr. Francis Day (b. 1829–d. 1889), a zoologist, known for his work on fishes, Inspector General of Fisheries for India and Burma (-*i*, commemorating (dedication, eponym) (L.)).

Endemic to India; Karnataka, Kerala, and Tamil Nadu (in the Western Ghats).

OEN Grey-toothed Musk Shrew (nom. form)

215. ***Suncus zeylanicus*** **Phillips, 1928**
Sri Lankan Shrew

Suncus: Mod. L. *suncus* shrew from Arabic name '*far sunki*'.
zeylanicus: L. adj. *zeylanicus* belonging to Ceylon (= Sri Lanka) (*-icus* belonging to, pertaining to (L./Gr.)).

Endemic to Sri Lanka.

OEN Ceylon Jungle Shrew (nom. form), Jungle Shrew (nom. form)
VN Sinh. *Kunu-miya, Sri Lanka Kele Hik-miya*; Tam. (SL) *Mungi'elli*

216. ***Suncus fellowesgordoni*** **Phillips, 1932**
Ceylon Pygmy Shrew

Suncus: Mod. L. *suncus* shrew from Arabic name '*far sunki*'.
fellowesgordoni: Eponym after Ms. Marjory Tutein-Nolthenius née Fellowes-Gordon (dates not known), wife of A. C. Tutein-Nolthenius, estate owner and collector in Sri Lanka (*-i*, commemorating (dedication, eponym) (L.)).

Endemic to Sri Lanka.

OEN Sri Lankan Pygmy Shrew (nom. form)
VN Sinh. *Podi hik-miya, Podi kunu-miya, Sri Lanka Podi Hik-miya*; Tam. (SL) *Sinna Mungi'elli, Mungi'elli-kutti, Sundelli*

217. ***Anourosorex squamipes*** **Milne-Edwards, 1872**
Chinese Mole-Shrew

Anourosorex: Mod. L. *anourosorex* shrew without a tail (Gr. *án* without; Gr. *oúrá* tail; L. *sorex* (>Gr. *ÿrax* shrew)); due to its very short tail.
squamipes: Mod. L. *squamipes* scaly-footed (L. *squameus* scaly, from *squama* scale; L. *pes* foot, >Gr. *pous* foot).

China, India, and parts of Southeast Asia.

OEN Chinese Short-tailed Shrew (nom. form)

218. ***Anourosorex assamensis*** **Anderson, 1875**
Assam Mole-Shrew

Anourosorex: Mod. L. *anourosorex* shrew without a tail (Gr. *án* without; Gr. *oúrá* tail; L. *sorex* (>Gr. *ÿrax* shrew)); due to its very short tail.
assamensis: L. adj. *assamensis* meaning from Assam, India (*-ensis*, geographic, occurrence in (location, toponym) (L.)).

Endemic to India; northeast India.

OEN Assam Short-tailed Shrew (nom. form)

219. *Anourosorex schmidi* Petter, 1963
Giant Mole-Shrew

Anourosorex: Mod. L. *anourosorex* shrew without a tail (Gr. *án* without; Gr. *oúrá* tail; L. *sorex* (>Gr. *ÿrax* shrew)); due to its very short tail.
schmidi: Eponym after Dr. Fernand Schmid (b. 1924–d. 1998), an authority on Trichoptera of the world, (-*i*, commemorating (dedication, eponym) (L.)).

Endemic to South Asia; Himalayas in Bhutan and India.

220. *Chimarrogale himalayica* (Gray, 1842)
Himalayan Water Shrew

Chimarrogale: Mod. L. *chimarrogale* montane water-shrew (Gr. *chimorros* mountain-torrent; Gr. *gale* weasel).
himalayica: L. adj. *himalayica* belonging to the Himalaya (*Himalaya* Sans. noun, an abode of snow; *him* snow, *alaya* abode), (-*ica* belonging to, pertaining to (L./Gr.)).

China, Japan, parts of South Asia (Bhutan, India, and Nepal) and Southeast Asia.

VN Bhut. *Choopitsi*; Lepch. *Ooong-lagniyu*

221. *Episoriculus caudatus* (Horsfield, 1851)
Hodgson's Brown-toothed Shrew

Episoriculus: Mod. L. *episoriculus* somewhat like *soriculus* (Gr. *epi* somewhat, *soriculus* Dim. of L. *sorex* (>Gr. *ÿrax*) shrew, -*ulus*, diminutive (comparison); somewhat (adj.) (L.)).
caudatus: L. *caudatus* -tailed (L. *cauda* tail).

China, Southeast Asia (Myanmar) and South Asia (Himalayas in India and Nepal).

OEN Hodgson's Mountain Shrew (nom. form), Sikkim Mountain Shrew (J. Syn. *gracilicauda*)

222. *Episoriculus leucops* (Horsfield, 1855)
Long-tailed Brown-toothed Shrew

Episoriculus: Mod. L. *episoriculus* somewhat like *soriculus* (Gr. *epi* somewhat, *soriculus* Dim. of L. *sorex* (>Gr. *ÿrax*) shrew, -*ulus*, diminutive (comparison); somewhat (adj.) (L.)).
leucops: Mod. L. *lecops* white-faced (Gr. *leukos* white; *ōps* face, eye).

Endemic to Nepal.

OEN Long-tailed Shrew (nom. form)

223. ***Episoriculus macrurus*** **(Blanford, 1888)**
Arboreal Brown-toothed Shrew

Episoriculus: Mod. L. *episoriculus* somewhat like *soriculus* (Gr. *epi* somewhat, *soriculus* Dim. of L. *sorex* (>Gr. *ÿrax*) shrew, *-ulus*, diminutive (comparison); somewhat (adj.)) (L.)).
macrurus: Mod. L. *macrurus* long-tailed (Gr. *makros* long; Gr. *-ouros* -tailed).

China, Southeast Asia (Myanmar and Vietnam) and South Asia (Himalayas in India and Nepal).

224. ***Episoriculus sacratus*** **(Thomas, 1911)**
Sichuan Brown-toothed Shrew

Episoriculus: Mod. L. *episoriculus* somewhat like *soriculus* (Gr. *epi* somewhat, *soriculus* Dim. of L. *sorex* (>Gr. *ÿrax*) shrew, *-ulus*, diminutive (comparison); somewhat (adj.) (L.)).
sacratus: L. *sacratus* sacred (L. *sacer* sacred).

China, and marginally in to South Asia. One subspecies is known from the region.

Subspecies

- *umbrinus*: Mod. L. *umbrinus* dark, shady, umber-colour (>L. *umbra* shade, dark).

  Dark Brown Sichuan Brown-toothed Shrew. China and marginally in India (Assam).

225. ***Episoriculus baileyi*** **Thomas, 1914**
Long-tailed Brown-toothed Shrew

Episoriculus: Mod. L. *episoriculus* somewhat like *soriculus* (Gr. *epi* somewhat, *soriculus* Dim. of L. *sorex* (>Gr. *ÿrax*) shrew, *-ulus*, diminutive (comparison); somewhat (adj.) (L.)).
baileyi: Eponym after Captain Frederick Marshman Bailey (b. 1882–d. 1967), a British army officer, naturalist and a collector (*-i*, commemorating (dedication, eponym) (L.)).

South Asia and Southeast Asia; Himalayas in northeast India and Nepal.

OEN Gruber's Shrew (J. Syn. *gruberi*)

226. ***Episoriculus soluensis*** **Gruber, 1969**
Solu Brown-toothed Shrew

Episoriculus: Mod. L. *episoriculus* somewhat like *soriculus* (Gr. *epi* somewhat, *soriculus* Dim. of L. *sorex* (>Gr. *ÿrax*) shrew, *-ulus*, diminutive (comparison); somewhat (adj.) (L.)).
soluensis: L. adj. *soluensis* meaning from Solu Province (now Solukhumbu district), Nepal, (*-ensis*, geographic, occurrence in (location, toponym) (L.)).

Endemic to Nepal.

227. ***Nectogale elegans*** **Milne-Edwards, 1870**
Web-footed Shrew

Nectogale: Mod. L. *nectogale* swimming-weasel (Gr. *nektōs* swimming; Gr. *gale* weasel); in allusion to its broad-webbed hind foot; it is an excellent swimmer.
elegans: L. adj. *elegans* elegant, handsome or fine.
China, Southeast Asia, and South Asia; Himalayas in Bhutan, India (Sikkim) and Nepal.

OEN Elegant Water Shrew (nom. form), Tibetan Water Shrew (nom. form), Sikkim Water Shrew (J. Syn. *sikhimensis*), Finger-tailed Water Shrew (J. Syn. *sikhimensis*)

228. ***Sorex minutus*** **Linnaeus, 1766**
Eurasian Pygmy Shrew

Sorex: L. *sorex* (>Gr. *ÿrax*) shrew.
minutus: L. *minutus* small, little.
Europe, East and Central Asia, China (parts), and marginally in South Asia; Himalayas in Nepal.

OEN Pygmy Shrew (nom. form), Lesser Shrew (nom. form), Tibetan Shrew (J. Syn. *thibetanus*)

229. ***Sorex bedfordiae*** **Thomas, 1911**
Lesser Striped Shrew

Sorex: L. *sorex* (>Gr. *ÿrax*) shrew.
bedfordiae: Eponym after the Mary, the Duchess of Bedford (b. 1865–d. 1937), an ornithologist, (*-ae*, commemorating (dedication), feminine (L.)).
China, Myanmar, and Nepal.

OEN Bedford's Striped Shrew (nom. form), Lesser Stripe-backed Shrew (nom. form), Nepal Lesser Striped Shrew (J. Syn. *nepalensis*)

230. ***Sorex planiceps*** **Miller, 1911**
Kashmir Pygmy Shrew

Sorex: L. *sorex* (>Gr. *ÿrax*) shrew.
planiceps: Mod. L. *planiceps* flat-headed (L. *planus* plain, level ground; L. *-ceps* –headed (L. *caput* head)).
Endemic to South Asia; Himalayas in India (north) and Pakistan.

OEN Flat Skull Shrew (nom. form)

231. ***Sorex excelsus*** **Allen, 1923**
Chinese Highland Shrew

Sorex: L. *sorex* (>Gr. *ÿrax*) shrew.

excelsus: L. *excelsus* lofty; owing to its high altitudinal habitat.
China and Nepal.

OEN Allen's Shrew (nom. form), Yunnan Shrew (nom. form), Lofty Shrew (nom. form), Highland Shrew (nom. form)

232. *Soriculus nigrescens* (Gray, 1842)
Sikkim Large-clawed Shrew

Soriculus: Dim. of L. *sorex* (>Gr. *ÿrax*) shrew, (*-ulus*, diminutive (comparison); somewhat (adj.) (L.)).
nigrescens: L. *nigrescens* blackish (L. *nigrescere* to become black >L. *niger* black).

China, and marginally in to South Asia. Along with the nominate subspecies *S. n. nigrescens* from the India and Nepal (in the Himalayas), another subspecies is known from the region.

Subspecies

- *minor*: L. *minor* smaller.
  Dobson's Large-clawed Shrew. Endemic to South Asia; Himalayas in Bhutan and India (Assom and Manipur).

OEN Himalayan Shrew (nom. form), Burrowing Forest Shrew (nom. form), Black Indian Shrew (nom. form), Sikkim Brown-toothed Shrew (J. Syn. *pahari*)
VN Bhut. *Ting-Zhing*; Lepch. *Tang-zhing*

Family Talpidae Fischer, 1817

233. *Euroscaptor micrura* (Hodgson, 1841)
Himalayan Mole

Euroscaptor: Mod. L. *euroscaptor* deep-digger (Gr. *eurus* broad or deep; Gr. *scaptor* to dig).
micrura: Mod. L. *micrura* small-tailed (Gr. *mikros* small; Gr. *-ouros* -tailed).
Himalayas in Bhutan, China, India, and Nepal.

OEN Short-tailed Mole (nom. form), Assam Short-tailed Mole (J. Syn. *cryptura*), Himalayan Long-tailed Mole (J. Syn. *macrura*)

234. *Parascaptor leucura* (Blyth, 1850)
Indian Mole

Parascaptor: Mod. L. *parascaptor* shallow-digger (Gr. *para* near to; Gr. *scaptor* to dig).
leucura: Mod. L. *leucura* white-tailed (>Gr. *leukouros* white-tailed).
China, India and Myanmar.

OEN Assam Mole (nom. form), White-tailed Mole (nom. form), Assamese Mole (nom. form)

Order Chiroptera Blumenbach, 1779
Suborder Megachiroptera Dobson, 1875
Family Pteropodidae Gray, 1821

235. *Cynopterus sphinx* (Vahl, 1797)
Greater Short-nosed Fruit Bat

Cynopterus: Mod. L. *cynopterus* 'winged dog' (Gr. *cynos* dog; L. *-pterus* -winged (>Gr. *pteron* wing)); in allusion to its dog-like head.
sphinx: Gr. Myth. *Sphinx*, a monster of varied appearance at Thebes who set riddles to travellers and killed those who could not answer correctly.

South Asia and Southeast Asia. Along with the nominate subspecies *C. s. sphinx* from Bangladesh, India, Nepal, Pakistan and Sri Lanka, two more subspecies are known from the region.

Subspecies

- *scherzeri*: Eponym after Karl Ritter von Scherzer (b. 1821–d. 1903), an Austrian explorer, diplomat, and naturalist, (*-i*, commemorating (dedication, eponym) (L.)).
  Nicobar Short-nosed Fruit Bat. Endemic to Nicobar Islands, India.
- *angulatus*: L. *angulatus* angular, cornered (L. *angulus* angle, corner).
  Miller's Short-nosed Fruit Bat. Southeast Asia, marginally in South Asia, in the Himalayas in Bhutan.

OEN Small Fruit-Bat (Subsp. *sphinx*), Short-nosed Fruit-Bat (Subsp. *sphinx*), Southern Short-nosed Fruit Bat (Subsp. *sphinx*), Gangetic Short-nosed Fruit Bat (J. syn. *gangeticus*), Margined Short-nosed Fruit Bat (J. syn. *marginatus*), Small Fox-bat (J. syn. *marginatus*)

VN Beng. *Cham-gadal*; Hind. *Gadal*; Sinh. *Yak Wawula*, *Thala-vavula*; Tam. (SL) *Mampala Vaval*, *Vava*

236. *Cynopterus brachyotis* (Müller, 1838)
Lesser Short-nosed Fruit Bat

Cynopterus: Mod. L. *cynopterus* 'winged dog' (Gr. *cynos* dog; *-pterus* -winged (>Gr. *pteron* wing)); in allusion to its dog-like head.
brachyotis: Mod. L. *brachyotis* short-eared (Gr. *brakhus* short; Gr. *-otis* -eared (*ōtos* ear)).

South Asia and Southeast Asia. Along with the nominate subspecies, *C. b. brachyotis* Lesser Short-nosed Fruit Bat, from the Himalayas in northeast India, two more subspecies are known from the region.

Subspecies

- *brachysoma*: Mod. L. *brachysoma* short-bodied (Gr. *brakhusoma*, Gr. *brakhus* short; Gr. *sōma* body).
  Andaman Short-nosed Fruit Bat. Endemic to Andaman Islands, India.
- *ceylonensis*: L. adj. *ceylonensis* belonging to Sri Lanka or Sri Lankan, (*-ensis*, geographic, occurrence in (location, toponym) (L.)).
  Sri Lankan Short-nosed Fruit Bat. Endemic to India (south) and Sri Lanka.

OEN Lesser Dog-faced Fruit Bat (Subsp. *brachyotis*), Common Short-nosed Fruit Bat (Subsp. *brachyotis*), Small Fruit-Bat (Subsp. *ceylonensis*), Andaman Short-nosed Fruit Bat (J. Syn. *andamanensis*)

VN Sinh. *Yak Wawula, Heen Thala-vavula, Kotican Wawula*; Tam. (SL) *Vaval*

237. *Eonycteris spelaea* (Dobson, 1871)
Lesser Dawn Bat

Eonycteris: Mod. L. *eonycteris* dawn bat (Gr. *íós* dawn; Gr. *nikterís* bat).

spelaea: Gr. *spēlaion* cave; owing to its habit of dwelling/roosting in caves.

Southeast Asia and South Asia.

OEN Dawn Bat (nom. form), Cave Fruit Bat (nom. form), Cave Dawn Bat (nom. form), Cave Bat (S. syn. *Macroglossus spelaeus*)

238. *Latidens salimalii* Thonglongya, 1972
Salim Ali's Fruit Bat

Latidens: Mod. L. *latidens* concealed teeth (L. *latens* hidden, secret, unknown (>L. *latere* to be concealed); *dens* tooth).

salimalii: Eponym after Dr. Salim Ali (b. 1896–d. 1987), an Indian ornithologist and conservationist, (-*i*, commemorating (dedication, eponym) (L.)).

Endemic to India (in the Western Ghats of Kerala and Tamil Nadu).

239. *Macroglossus sobrinus* Andersen, 1911
Greater Long-nosed Fruit Bat

Macroglossus: Mod. L. *macroglossus* (>Gr. *makroglōssus*) long-tongued (Gr. *makros* long; Gr. *glōssa* tongue).

sobrinus: L. *sobrinus* maternal cousin.

Southeast Asia and South Asia; Himalayas of northeast India.

OEN Hill Long-tongued Fruit Bat (nom. form), Greater Long-nosed Fruit Bat (nom. form), Greater Long-tongued Fruit Bat (nom. form), Greater Nectar Bat (nom. form)

240. *Megaerops niphanae* Yenbutra and Fenten, 1983
Ratanaworabhan's Fruit Bat

Megaerops: Mod. L. *megaerops* like Megaera (Gr. Myth. *Megaera* one of the Three 'Furies'; *óps* aspect).

niphanae: Eponym after Chanthawanich Ratanaworabhan Niphan (dates not known), a Thai Diptera taxonomist, (-*ae*, commemorating (dedication), feminine (L.)).

Southeast Asia and South Asia; India (northeast).

OEN Niphan's Tail-less Fruit Bat (nom. form)

241. *Pteropus giganteus* (Brünnich, 1782)
Indian Flying Fox

Pteropus: Mod. L. *pteropus* wing-footed (Gr. *pterópous* wing-footed), in allusion to the wing membrane that attaches to the back of the second toe

giganteus: L. *giganteus* gigantic (L. *gigas* giant).

South Asia and marginally in Southeast Asia. Along with the nominate subspecies, *P. g. giganteus* Indian Flying Fox, from the south of Himalayas in India, Bangladesh, Nepal, Pakistan and Sri Lanka, two more subspecies are known from the region.

Subspecies

- *ariel*: Mod. L. *ariel* from Med. folklore *Ariel*, a spirit or sylph of the air.
  Maldives Flying Fox. Endemic to Addu Atoll, Maldive Islands.
- *leucocephalus*: Mod. L. *leucocephalus* white-headed (Gr. *leukokephalus* white-headed, from *leukos* white, *kephalus* head).
  Himalayan Flying Fox. Himalayas in Bhutan, India, Nepal, and Pakistan.

OEN Flying-Fox (nom. form), Indian Ternate Bat (J. syn. *medius*), Indian Pteropus (J. syn. *medius*), Wurbagool (J. syn. *medius*), Large Fox-bat (J. Syn. *edwardsi*), Milne-Edward's Fruit Bat (J. Syn. *edwardsi*), Assamese Flying Fox (J. Syn. *assamensis*), Ceylon Flying Fox (J. syn. *kelaarti*)

VN Beng. *Badul*; Dekh. *Gaddal*, *Barbagal*; Kan. *Toggul bawali*; Hind. *Gadal*, *Bar-bagal*; Mar. *War- baggul*; Tel. *Sikat Yelli*, *Siku rayi*; Wad. *Sikatyelle*, Sinh. *Maha-wawula*, *Ma-vavula*, *Locu-wawula*; Tam. (SL) *Periya-vava*, *Vaval*

242. *Pteropus hypomelanus* Temminck, 1853
Variable Flying Fox

Mod. L. *pteropus* wing-footed (Gr. *pterópous* wing-footed), in allusion to the wing membrane that attaches to the back of the second toe

hypomelanus: Mod. L. *hypomelanus* black underneath (Gr. *hupo* beneath; Gr. *melanos* black).

Oceanic islands of India, Indonesia, Malaysia, Maldives, Myanmar, Papua New Guinea, Philippines, Solomon Islands, Thailand and Viet Nam. Three subspecies are known from the region.

Subspecies

- *geminorum*: Mod. L. *geminorum* (L. *geminus* twin, double).
  South Twin Island Flying Fox; Southeast Asia and in the Andaman Islands, India.
- *satyrus*: L. *satyrus* satyr, horned sylvan deity.
  Narcondam Island Flying Fox. Endemic to Narcondam Island, India.
- *maris*: L. *maris* sea.
  Maldive Island Flying Fox. Endemic to Addu Atoll, Maldives.

243. ***Pteropus melanotus*** **Blyth, 1863**
Black-eared Flying Fox

Mod. L. *pteropus* wing-footed (Gr. *pterópous* wing-footed), in allusion to the wing membrane that attaches to the back of the second toe
melanotus: Mod. L. *melanotus* black-eared (Gr. *melas*, *melanos* black; Gr. *otus* ear).

Endemic to Andaman and Nicobar Islands, India. Along with the nominate subspecies, *P. m. melanotos* Blyth's Black-eared Flying Fox from the Nicobar Islands, another subspecies is known from the region.

Subspecies

- *tytleri*: Eponym after Robert Christopher Tytler (b. 1818–d. 1872), a British soldier, naturalist, and photographer, worked as Superintendent of the convict settlement on Port Blair, Andaman, (*-i*, commemorating (dedication, eponym) (L.)).

  Tytler's Black-eared Flying Fox. Endemic to Andaman Islands, India.

OEN Nicobar Black-eared Flying Fox (J. syn. *ncobaricus*)

244. ***Pteropus faunulus*** **Miller, 1902**
Nicobar Flying Fox

Mod. L. *pteropus* wing-footed (Gr. *pterópous* wing-footed), in allusion to the wing membrane that attaches to the back of the second toe
faunulus: From L. *faunus* (Rom. Myth. horned god of the forest, plains, and fields), (*-ulus*, diminutive (comparison); somewhat (adj.) (L.)); in allusion to its habit of living in a dense forest and having pointed ears.

Endemic to India; Nicobar Island, India.

245. ***Rousettus aegyptiacus*** **(E. Geoffroy, 1810)**
Egyptian Rousette

Rousettus: Mod. L. *rousettus* reddish (>Fr. *rousette* from *rousset* reddish; in allusion to the characteristic colour.
aegyptiacus: L. *aegyptiacus* Egyptian, (*-acus*, belonging to, pertaining to (L./Gr.)).

Africa, Arabian Peninsula and marginally in Mediterranean region and South Asia. One subspecies is known from the region.

Subspecies

- *arabicus*: L. adj. *arabicus* Arabian, or from Arabia (*-icus*, belonging to, pertaining to (location, toponym) (L.)).

  Arabian Rousette. Arabian Peninsula and in South Asia (Afghanistan and Pakistan).

246. ***Rousettus leschenaultii*** **(Desmarest, 1820)**
Leschenault's Rousette

Rousettus: Mod. L. *rousettus* reddish (>Fr. *rousette* from *rousset* reddish; in allusion to the characteristic colour.

leschenaultii: Eponym after Jean-Baptiste Louis Claude Théodore Leschenault de La Tour (b. 1773–d. 1826), a French botanist and ornithologist, collected plants and animals in Madras, Bengal, and Ceylon, (*-ii*, commemorating (dedication, eponym) (L.)).

South Asia and Southeast Asia. Besides the nominate subspecies *R. l. leschenaultii* from Bangladesh, Bhutan, India, Nepal, and Pakistan, another subspecies is known from the region.

Subspecies

- *seminudus*: Mod. L. *seminudus* body slightly covered with fur, (L. *semi-* half- (in comp.) (*semis* half); L. *nudus* bare).

  Sri Lankan Rousette. Endemic to Sri Lanka.

OEN Fulvous Fruit Bat (nom. form), Fulvous Rousette (nom. form), Fulvous Fox-bat (nom. form), Nepal Fruit Bat (J. syn. *pyrivorus*), Indian Cynoptere (J. syn. *affinis*), Dusky Fruit Bat (J. Syn. *fusca*), Large Fruit-Bat (Subsp. *seminudus*), Ceylon Fruit Bat (Subsp. *seminudus*), Dog-faced Bat (Subsp. *seminudus*), Obscure Fruit Bat (J. syn. *infuscata*)

VN Sinh. *Wawula, Rath dumburu pala vavula*; Tam. (SL) *Vaval*

247. *Sphaerias blanfordi* (Thomas, 1891)
Blanford's Fruit Bat

Sphaerias: No explanation provided.

blanfordi: Eponym after William Thomas Blanford (b. 1832–d. 1905), an English geologist and naturalist, collected in many parts of India, also instrumental in establishing Indian Museum (*-i*, commemorating (dedication, eponym) (L.)).

Southeast Asia and South Asia.

Suborder Microchiroptera Dobson, 1875
Family Rhinolophidae Bell, 1836

248. *Rhinolophus ferrumequinum* (Schreber, 1774)
Greater Horseshoe Bat

Rhinolophus: Mod. L. *rhinolophus* crest-nosed (>Gr. *rhinolōphos* (Gr. *rhis*, *rhinos* nose; Gr. *lōphos* crest)); in allusion to the complicated nose leaf consisting of three different parts and sella.

ferrumequinum: L. *ferrumequinum* horseshoe (L. *ferrum* iron; *eqinum*, of horses).

Mediterranean region, Middle East, Mid-West, Himalayas of South Asia, China (southeast and eastern) and Japan. Two subspecies are known from the region.

Subspecies

- *tragatus*: Mod. L. *tragatus* in allusion to the well-developed antitragus (a projection of the auricle of the ear extending to the tragus), (L. *tragus* and *-atus*, provided with (L.)).

Nepal Greater Horseshoe Bat. Himalayas in China, India, and Nepal.

- *proximus*: L. *proximus* very near, nearest, most like (super. >L. *propior* nearer).

  Kashmir Horseshoe Bat. Himalayas in Afghanistan, India, and Pakistan.

OEN Larger Horseshoe Bat (nom. form), Dark-brown Leaf-bat (Subsp. *tragatus*), Large Himalayan Horseshoe Bat (J. syn. *regulus*)

249. *Rhinolophus hipposideros* (Bechstein, 1800)
Lesser Horseshoe Bat

Rhinolophus: Mod. L. *rhinolophus* crest-nosed (>Gr. *rhinolōphos* (Gr. *rhis*, *rhinos* nose; Gr. *lōphos* crest)); in allusion to the complicated nose leaf consisting of three different parts and sella.

hipposideros: Gr. *ïppo-sídèros* horseshoe (*ïppos* horse; Gr. *sídèros* iron); in allusion to the anterior part of the complicated noseleaf.

Europe, Mediterranean Region, Middle East, Arabian Peninsula and South Asia (marginally in Western Himalayas). Only one subspecies is known from the region.

Subspecies

- *midas*: Mod. L. *midas* Gr. Myth. Midas, king of Phrygia, favoured by Bacchus with the gift that everything he touched turned to gold.

  Persian Lesser Horseshoe Bat. Middle East and South Asia (Himalayas in Afghanistan, India, and Pakistan).

250. *Rhinolophus affinis* Horsfield, 1823
Intermediate Horseshoe Bat

Rhinolophus: Mod. L. *rhinolophus* crest-nosed (>Gr. *rhinolōphos* (Gr. *rhis*, *rhinos* nose; Gr. *lōphos* crest)); in allusion to the complicated nose leaf consisting of three different parts and sella.

affinis: L. *affinis* related, allied.

South Asia and Southeast Asia. One subspecies is known from the region.

Subspecies

- *himalayanus*: Mod. L. *himalayanus* from Himalayas (Sans. *Himalaya* an abode of snow (*him* snow, *alaya* abode)), (*-anus*, belonging to, pertaining to (L.) geographic (location, toponym)).

  Himalayan Horseshoe Bat. Himalayas of Bangladesh, Bhutan, China, India, and Nepal.

251. *Rhinolophus pusillus* Temminck, 1834
Least Horseshoe Bat

Rhinolophus: Mod. L. *rhinolophus* crest-nosed (>Gr. *rhinolōphos* (Gr. *rhis*, *rhinos* nose; Gr. *lōphos* crest)); in allusion to the complicated nose leaf consisting of three different parts and sella.

pusillus: L. *pusillus* tiny, very small.

South Asia and Southeast Asia. Two subspecies are known from the region.

Subspecies

- *gracilis*: L. *gracilis* slender, elegant, slim; in allusion to its small forearm.

  Andersen's Least Horseshoe Bat. Endemic to India (Andhra Pradesh, Karnataka, Kerala and Tamil Nadu).
- *blythi*: Eponym after Edward Blyth (b. 1810–d. 1873), an English zoologist and pharmacist, curator of Museum at the Royal Asiatic Society of Bengal (*-i*, commemorating (dedication, eponym) (L.)).

  Himalayan Least Horseshoe Bat. Himalayas in China, India, and Nepal.

OEN Blyth's Lesser Horseshoe bat (Subsp. *blythi*)

252. *Rhinolophus trifoliatus* Temminck, 1834
Trefoil Horseshoe Bat

Rhinolophus: Mod. L. *rhinolophus* crest-nosed (>Gr. *rhinolōphos* (Gr. *rhis*, *rhinos* nose; Gr. *lōphos* crest)); in allusion to the complicated nose leaf consisting of three different parts and sella.

trifoliatus: L. *trifoliatus* three-leaved (L. *tri* (*tres*) >Sans. *tri* three; L. adj. *foliatus* leaved from *folium* leaf).

Southeast Asia, and marginally in China and India – Assom and West Bengal.

253. *Rhinolophus luctus* Temminck, 1835
Woolly Horseshoe Bat

Rhinolophus: Mod. L. *rhinolophus* crest-nosed (>Gr. *rhinolōphos* (Gr. *rhis*, *rhinos* nose; Gr. *lōphos* crest)); in allusion to the complicated nose leaf consisting of three different parts and sella.

luctus: L. *luctus* mourning from *lugere* to mourn.

Southeast Asia and South Asia. One subspecies is known from the region.

Subspecies

- *perniger*: L. *perniger* very black.

  Himalayan Woolly Horseshoe Bat. Himalayas of India and Nepal, and also in Chin Hills and Shan in Myanmar.

254. *Rhinolophus rouxii* Temminck, 1835
Rufous Horseshoe Bat

Rhinolophus: Mod. L. *rhinolophus* crest-nosed (>Gr. *rhinolōphos* (Gr. *rhis*, *rhinos* nose; Gr. *lōphos* crest)); in allusion to the complicated nose leaf consisting of three different parts and sella.

rouxii: Eponym after Jean Louis Florent Polydore Roux (b. 1792–d. 1833), curator of the Museum of Marseille, collected in India and Ceylon (*-ii*, commemorating (dedication, eponym) (L.)).

South Asia and possibly also in southern Myanmar and western Yunnan, China. Besides the nominate subspecies *R. r. rouxii* from India and Nepal, another subspecies is known from the region.

Subspecies

- *rubidus*: L. *rubidus* ruddy, red, dark-red.
  Sri Lankan Rufous Horseshoe Bat. Endemic to Sri Lanka.

OEN Temminck's Horseshoe Bat (Subsp. *rouxii*), Rufous Leaf-bat (Subsp. *rouxii*), Kelaart's Horseshoe Bat (J. syn. *rammanika*), Peter's Horseshoe Bat (J. Syn. *petersi*)

VN Sinh. *Kiri Wawula, Kotican Wawula, Borath Ashladan-vavula*; Tam. (SL) *Vava, Vaval*

255. *Rhinolophus lepidus* Blyth, 1844
Blyth's Horseshoe Bat

Rhinolophus: Mod. L. *rhinolophus* crest-nosed (>Gr. *rhinolōphos* (Gr. *rhis, rhinos* nose; Gr. *lōphos* crest)); in allusion to the complicated nose leaf consisting of three different parts and sella.

lepidus: L. *lepidus* charming, elegant.

South Asia and Southeast Asia. Besides the nominate subspecies *R. l. lepidus* from Bangladesh and India, another subspecies is known from the region.

Subspecies

- *monticola*: L. *monticola* mountain-dweller, mountaineer (L. *mons, montis* mountain; L. *-cola* inhabitant >L. *colere* to dwell).

Montane Horseshoe Bat. Himalayas of Afghanistan, India, Nepal, and Pakistan.

256. *Rhinolophus macrotis* Blyth, 1844
Big-eared Horseshoe Bat

Rhinolophus: Mod. L. *rhinolophus* crest-nosed (>Gr. *rhinolōphos* (Gr. *rhis, rhinos* nose; Gr. *lōphos* crest)); in allusion to the complicated nose leaf consisting of three different parts and sella.

macrotis: Mod. L. *macrotis* big-eared (>Gr. *makrōtēs*) long-eared (Gr. *makros* long; Gr. *ous, ōtos* ear)).

South Asia and Southeast Asia. Besides the nominate subspecies, *R. m. macrotis* Big-eared Horseshoe Bat, from Bangladesh, India, and Nepal, another subspecies is known from the region.

Subspecies

- *topali*: Eponym after Dr. György Topal (b. 1931), a well-known bat taxonomist and biologist, former curator of mammals at the Hungarian Museum of Natural History (*-i*, commemorating (dedication, eponym) (L.)).

Topal's Big-eared Horseshoe Bat. Endemic to Pakistan.

257. *Rhinolophus mitratus* Blyth, 1844
Mitred Horseshoe Bat

Rhinolophus: Mod. L. *rhinolophus* crest-nosed (>Gr. *rhinolōphos* (Gr. *rhis*, *rhinos* nose; Gr. *lōphos* crest)); in allusion to the complicated nose leaf consisting of three different parts and sella.

mitratus: Mod. L. *mitratus* mitred, having a hood (Gr. *mítra* hood; *-atus*, provided with (L.)); origin not known, probably due to long hair on head forming a hood.

Endemic to India, known only from the type locality in Chaibasa, Jharkhand.

258. *Rhinolophus subbadius* Blyth, 1844
Little Nepalese Horseshoe Bat

Rhinolophus: Mod. L. *rhinolophus* crest-nosed (>Gr. *rhinolōphos* (Gr. *rhis*, *rhinos* nose; Gr. *lōphos* crest)); in allusion to the complicated nose leaf consisting of three different parts and sella.

subbadius: Med. L. *subbadius* pale chestnut coloured (L. *sub* beneath, somewhat; L. *badius* chestnut-coloured).

South Asia in the Himalayas of Bangladesh, India, and Nepal, and also reported from Yunnan, China and Nam Tamay valley, Myanmar.

OEN Bay Leaf-bat (nom. form), Garo Hills Horseshoe Bat (J. syn. *garoensis*)

259. *Rhinolophus pearsonii* Horsfield, 1851
Pearson's Horseshoe Bat

Rhinolophus: Mod. L. *rhinolophus* crest-nosed (>Gr. *rhinolōphos* (Gr. *rhis*, *rhinos* nose; Gr. *lōphos* crest)); in allusion to the complicated nose leaf consisting of three different parts and sella.

pearsonii: Eponym after Dr. John Thomas Pearson (dates not known), Fellow Medical Student of J.E. Gray, subsequently collected in India (*-ii*, commemorating (dedication, eponym) (L.)).

South Asia (Himalayas in Bhutan, India, and Nepal; probably also in Bangladesh), Southeast Asia, and China (central and southeast).

OEN Pearson's Leaf-bat (nom. form), Pearson's Nose-leaf Bat (nom. form)

260. *Rhinolophus blasii* Peters, 1866
Blasius's Horseshoe Bat

Rhinolophus: Mod. L. *rhinolophus* crest-nosed (>Gr. *rhinolōphos* (Gr. *rhis*, *rhinos* nose; Gr. *lōphos* crest)); in allusion to the complicated nose leaf consisting of three different parts and sella.

blasii: Eponym after Johann Heinrich Blasius (b. 1809–d. 1870), Professor of Natural Sciences and Director of Natural History Museum, Brunswick, Germany (*-ii*, commemorating (dedication, eponym) (L.)).

Palearctic Europe and Africa, Middle East, and marginally in South Asia. One subspecies is known from the region.

Subspecies

- *meyeroehmi*: Eponym after Dr. D. Meyer-Oehme (dates not known), a German mammologist who collected in Middle East (*-i*, commemorating (dedication, eponym) (L.)).

  Meyer-Oehm's Horseshoe Bat. Afghanistan and Pakistan.

OEN Peters's Horseshoe Bat (nom. form), Peak-saddle Horseshoe Bat (nom. form)

261. *Rhinolophus andamanensis* Dobson, 1872
Homfray's Horseshoe Bat

Rhinolophus: Mod. L. *rhinolophus* crest-nosed (>Gr. *rhinolōphos* (Gr. *rhis*, *rhinos* nose; Gr. *lōphos* crest)); in allusion to the complicated nose leaf consisting of three different parts and sella.

andamanensis: L. adj. *andamanensis* meaning from the Andaman Islands, India (*-ensis*, geographic, occurrence in (location, toponym) (L.)).

Endemic to India (Andaman Islands).

262. *Rhinolophus yunanensis* Dobson, 1872
Dobson's Horseshoe Bat

Rhinolophus: Mod. L. *rhinolophus* crest-nosed (>Gr. *rhinolōphos* (Gr. *rhis*, *rhinos* nose; Gr. *lōphos* crest)); in allusion to the complicated nose leaf consisting of three different parts and sella.

yunanensis: L. adj. *yunanensis* meaning from Yunnan, China (*-ensis*, geographic, occurrence in (location, toponym) (L.)).

Southeast Asia and South Asia (marginally in northeast India).

263. *Rhinolophus beddomei* Andersen, 1905
Beddome's Horseshoe Bat

Rhinolophus: Mod. L. *rhinolophus* crest-nosed (>Gr. *rhinolōphos* (Gr. *rhis*, *rhinos* nose; Gr. *lōphos* crest)); in allusion to the complicated nose leaf consisting of three different parts and sella.

beddomei: Eponym after Col. Richard Henry Beddome (b. 1830–d. 1911), British military officer and naturalist, (*-i*, commemorating (dedication, eponym) (L.)).

Endemic to South Asia. Besides the nominate subspecies, *R. b. beddomei* Beddome's Horseshoe Bat from India, another subspecies is known from the region.

Subspecies

- *sobrinus*: L. *sobrinus* maternal cousin.
  Sri Lankan Horseshoe Bat. Endemic to Sri Lanka.

OEN Lesser Woolly Horseshoe Bat (nom. form), Indian Great Horseshoe Bat (Subsp. *beddomei*)

VN Sinh. *Kotican Wawula, Maha Ashladan-vavula*; Tam. (SL) *Vava*

264. *Rhinolophus sinicus* Andersen, 1905
Chinese Horseshoe Bat

Rhinolophus: Mod. L. *rhinolophus* crest-nosed (>Gr. *rhinolōphos* (Gr. *rhis*, *rhinos* nose; Gr. *lōphos* crest)); in allusion to the complicated nose leaf consisting of three different parts and sella.

sinicus: L. adj. *sinicus* Chinese, belonging to China (*-icus* belonging to, pertaining to (L./Gr.)).

South Asia, Southeast Asia, and China; Himalayas in India and Nepal.

OEN Chinese Rufous Horseshoe Bat (nom. form), Little Nepalese Horseshoe Bat (nom. form)

265. *Rhinolophus cognatus* Andersen, 1906
Andaman Horseshoe Bat

Rhinolophus: Mod. L. *rhinolophus* crest-nosed (>Gr. *rhinolōphos* (Gr. *rhis*, *rhinos* nose; Gr. *lōphos* crest)); in allusion to the complicated nose leaf consisting of three different parts and sella.

cognatus: L. adj. *cognatus* related, similar (L. *co* together; L. *gnātus* born).

Endemic to India. Besides the nominate subspecies *R. c. cognatus* from South and Middle Andaman Islands, another subspecies is known from the region.

Subspecies

- *famulus*: L. *famulus* servile.
  Andersen's Horseshoe Bat. Endemic to Middle Andaman and Narcondam Islands.

266. ***Rhinolophus bocharicus*** **Kastchenko and Akimov, 1917**
Central Asian Horseshoe Bat

Rhinolophus: Mod. L. *rhinolophus* crest-nosed (>Gr. *rhinolōphos* (Gr. *rhis*, *rhinos* nose; Gr. *lōphos* crest)); in allusion to the complicated nose leaf consisting of three different parts and sella.

bocharicus: L. adj. *bocharicus* meaning from Bukhara, Uzbekistan (-*icus* belonging to, pertaining to (L./Gr.)).

Central Asia; marginally in South Asia, Afghanistan, and possibly extinct from Pakistan.

OEN Bokhara Horseshoe Bat (nom. form)

267. ***Rhinolophus shortridgei*** **K. Andersen, 1918**
Shortridge's Horseshoe Bat

Rhinolophus: Mod. L. *rhinolophus* crest-nosed (>Gr. *rhinolōphos* (Gr. *rhis*, *rhinos* nose; Gr. *lōphos* crest)); in allusion to the complicated nose leaf consisting of three different parts and sella.

shortridgei: Eponym after Guy Chester Shortridge (b. 1880–d. 1949), collected for the British Museum (Natural History), and later for the Mammal Survey of British India in southern India and Burma (-*i*, commemorating (dedication, eponym) (L.)).

Southeast Asia (west), China (southwest and south) and marginally in South Asia; India (northeast).

268. ***Rhinolophus indorouxii*** **Chattopadhyay, Garg, Kumar, Doss, Ramakrishnan and Kandula, 2012**
Greater Rufous Horseshoe Bat

Rhinolophus: Mod. L. *rhinolophus* crest-nosed (>Gr. *rhinolōphos* (Gr. *rhis*, *rhinos* nose; Gr. *lōphos* crest)); in allusion to the complicated nose leaf consisting of three different parts and sella.

indorouxii: Mod. L. *indorouxii* Indian-Roux's Horseshoe Bat; specific *nomen* is a combination to refer to populations of Rufous Horsehsoe Bat from peninsular India.

Endemic to India.

Family Hipposideridae Lydekker, 1891

269. ***Asellia tridens*** **(E. Geoffroy, 1813)**
Geoffroy's Trident Roundleaf Bat

Asellia: L. adj. *asellia* (used as n.) ass-like (>L. *asellus*, a little ass); in allusion to long, pointed ears.

tridens: L. *tridens* three-pronged, trident (L. *tri*- three-, L. *dens* tooth); in allusion to its trident-shaped posterior noseleaf.

Northern Africa, Middle East (including the Arabian Peninsula) and marginally in South Asia; Afghanistan and Pakistan. One subspecies is known from the region.

Subspecies

- *murraiana*: Eponym after James A. Murray (dates not known), zoologist and curator Karachi Museum, Pakistan (*-ana*, commemorating (dedication, eponym) (L.)).

  Murray's Trident Bat. Afghanistan and Pakistan (in plains and deserts).

OEN Three-toothed Horseshoe Bat (nom. form), Trident Bat (nom. form), Trident Roundleaf Bat (nom. form)

270. *Coelops frithii* Blyth, 1848

East Asian Tail-less Roundleaf Bat

Coelops: Mod. L. *coelops* hollowed (Gr. *coilos* hollow; Gr. *óps* aspect); probably in allusion to large funnel-shaped ears.

frithii: Eponym after R.W.G. Frith (dates not known), Indigo Planter in northeast India, (*-i*, commemorating (dedication, eponym) (L.)).

Southeast Asia, China (southeastern China) and marginally in South Asia; Bangladesh and India (northeast).

OEN Tailless Bat (nom. form), Tail-less Roundleaf Bat (nom. form)

271. *Hipposideros speoris* (Schneider, 1800)

Schneider's Roundleaf Bat

Hipposideros: Gr. *ïppo-sídèros* horseshoe (Gr. *ïppos* horse; Gr. *sídèros* iron); in allusion to the anterior part of the complicated noseleaf.

speoris: Mod. L. *speoris* (Gr. *spēlaion* cave, *rhis*, *rhinos* nose) meaning cave-like nose; in allusion to the deep depression in which nostrils are located.

Endemic to South Asia; India and Sri Lanka.

OEN Schneider's Roundleaf Bat (Subsp. *speoris*), Indian Horseshoe Bat (Subsp. *speoris*), Hairy Roundleaf Bat (J. Syn. *apiculatus*), Pencilled Horseshoe Bat (J. Syn. *apiculatus*), Sykes' Roundleaf Bat (J. Syn. *dukhunensis*)

VN Sinh. *Kiri Wawula, Podi Wawula, Kesketi Pathnehe-vavula*; Tam. (SL) *Sinna Vaval*

272. *Hipposideros diadema* (E. Geoffroy, 1813)

Diadem Roundleaf Bat

Hipposideros: Gr. *ïppo-sídèros* horseshoe (Gr. *ïppos* horse; Gr. *sídèros* iron); in allusion to the anterior part of the complicated noseleaf.

diadema: L. *diadema* diadem (>Gr. *diadēma* diadem, royal head-dress).

Southeast Asia, Australasia and marginally in South Asia (in Nicobar Islands). One subspecies is known from the region.

Subspecies

- *nicobarensis*: L. adj. *nicobarensis* Nicobarese or belonging to Nicobar Islands, India (*-ensis* belonging to, pertaining to (L./Gr.)).
  Nicobar Diadem Roundleaf Bat. Endemic to Nicobar Islands.
- *masoni*: Eponym after Francis Mason (b. 1799–d. 1874), an American missionary and naturalist, who worked in British Burma (=Myanmar), (*-i*, commemorating (dedication, eponym) (L.)).
  Mason's Diadem Roundleaf Bat. Andaman Islands.

OEN Diadem Roundleaf Bat (nom. form), Diadem Horseshoe Bat (nom. form), Crowned Bat (nom. form)

273. *Hipposideros armiger* (Hodgson, 1835)
Great Roundleaf Bat

Hipposideros: Gr. *ïppo-sídèros* horseshoe (Gr. *ïppos* horse; Gr. *sídèros* iron); in allusion to the anterior part of the complicated noseleaf.
armiger: L. *armiger* bearing weapons, shield-bearer.
Southeast Asia, China (southeast and east) and South Asia. India and Nepal (in Himalayas).

OEN Himalayan Roundleaf Bat (nom. form), Large Horseshoe Bat (nom. form), Great Himalayan Roundleaf Bat (nom. form), Great Roundleaf Bat (nom. form)

274. *Hipposideros fulvus* Gray, 1838
Fulvus Roundleaf Bat

Hipposideros: Gr. *ïppo-sídèros* horseshoe (Gr. *ïppos* horse; Gr. *sídèros* iron); in allusion to the anterior part of the complicated noseleaf.
fulvus: L. *fulvus* tawny.
South Asia, and possibly also in Southeast Asia. Besides the nominate subspecies, *H. f. fulvus* from India and Sri Lanka, another subspecies is known from the region.

Subspecies

- *pallidus*: L. *pallidus* pallid, pale.
  Pallid Roundleaf Bat. Afghanistan, India, Nepal, and Pakistan.

OEN Brown Roundleaf Bat (nom. form), Bicoloured Roundleaf Bat (nom. form), Fulvus Roundleaf Bat (nom. form), Mouse-coloured Horseshoe Bat (J. syn. *murinus*), Little Horseshoe Bat (J. syn. *murinus*)
VN Sinh. *Malekaha Pathnehe-vavula*

275. *Hipposideros galeritus* Cantor, 1846
Cantor's Roundleaf Bat

Hipposideros: Gr. *ïppo-sídèros* horseshoe (Gr. *ïppos* horse; Gr. *sídèros* iron); in allusion to the anterior part of the complicated noseleaf.

galeritus: L. *galeritus* hooded (from *galera* helmet, *galerum* bonnet, cap).

Southeast Asia and South Asia. One subspecies is known from the region.

Subspecies

- *brachyotus*: Mod. L. *brachyotus* short-eared (Gr. *brakhus* short; Gr. *-otis* -eared (*ōtos* ear)).

  Short-eared Roundleaf Bat. Endemic to South Asia; Bangladesh, India, and Sri Lanka.

OEN Cantor's Roundleaf Bat (Subsp. *brachyotus*), Dekhan Roundleaf Bat (Subsp. *brachyotus*)

VN Sinh. *Kiri Wawula, Podi Wawula, Kesdiga Pathnehe-vavula*; Tam. (SL) *Vaval*

276. *Hipposideros ater* Templeton, 1848

Dusky Roundleaf Bat

Hipposideros: Gr. *ïppo-sídèros* horseshoe (Gr. *ïppos* horse; Gr. *sídèros* iron); in allusion to the anterior part of the complicated noseleaf.

ater: L. *ater* black, dark, dull black.

Endemic to South Asia. Besides the nominate subspecies *H. a. ater* from India and Sri Lanka, another subspecies is known from the region.

Subspecies

- *nallamalaensis*: L. adj. *nallamalaensis* meaning from Nallamala Hills, Andhra Pradesh (*-ensis*, geographic, occurrence in (location, toponym) (L.)).

  Nallamala Dusky Roundleaf Bat. Endemic to Andhra Pradesh (Nallamala Hills).

OEN Dusky Roundleaf Bat (Subsp. *ater*), Ceylon Little Roundleaf Bat (Subsp. *ater*), Bi-coloured Roundleaf Bat (Subsp. *ater*), Ceylon Bi-coloured Roundleaf Bat (Subsp. *ater*)

VN Sinh. *Kiri Wawula, Depata Pathnehe-vavula*; Tam. (SL) *Sinna Vaval*

277. *Hipposideros lankadiva* Kelaart, 1850

Kelaart's Roundleaf Bat

Hipposideros: Gr. *ïppo-sídèros* horseshoe (Gr. *ïppos* horse; Gr. *sídèros* iron); in allusion to the anterior part of the complicated noseleaf.

lankadiva: Mod. L. *lankadiva*, meaning belonging to Sri Lanka.

Endemic to South Asia. Besides the nominate subspecies *H. l. lankadiva* from Sri Lanka, two subspecies are known from the region.

Subspecies

- *indus*: L. *indus* Indian.
  Indian Roundleaf Bat. Endemic to South Asia; Bangladesh and India.
- *gyi*: Eponym after Khin Maung Gyi (dates not known), retired professor of zoology at University of Mandalay, Myanmar (*-i*, commemorating (dedication, eponym) (L.)).
  Stanley's Roundleaf Bat. India (northeast) and Myanmar.

OEN Sri Lankan Roundleaf Bat (Subsp. *lankadiva*), Large Ceylon Roundleaf Bat (Subsp. *lankadiva*), Great Ceylon Roundleaf Bat (Subsp. *lankadiva*), Indian Roundleaf Bat (Subsp. *indus*), Indian Roundleaf Bat (Subsp. *indus*), Southern India Roundleaf Bat (J. Syn. *schistaceus*), Split Roundleaf Bat (J. Syn. *schistaceus*), Central India Roundleaf Bat (Subsp. *unitus*), Mysore Roundleaf Bat (J. Syn. *mixtus*)

VN Sinh. *Kotican Wawula, Maha Pathnehe-vavula*; Tam. (SL) *Vaval*

278. *Hipposideros cineraceus* Blyth, 1853
Ashy Roundleaf Bat

Hipposideros: Gr. *ïppo-sídèros* horseshoe (Gr. *ïppos* horse; Gr. *sídèros* iron); in allusion to the anterior part of the complicated noseleaf.
cineraceus: L. *cineraceus* ash-grey (L. *cinis*, *cineris* ashes).
Southeast Asia and South Asia; Himalayas in India, Nepal, and Pakistan.

OEN Ashy Horseshoe Bat (nom. form); Ashy Roundleaf Bat (nom. form), Least Roundleaf Bat (J. Syn. *micropus*)

279. *Hipposideros nicobarulae* Miller, 1902
Nicobar Roundleaf Bat

Hipposideros: Gr. *ïppo-sídèros* horseshoe (Gr. *ïppos* horse; Gr. *sídèros* iron); in allusion to the anterior part of the complicated noseleaf.
nicobarulae: L. adj. *nicobarulae* belonging to Nicobar Islands, India (*-ae*, geographic (location, toponym) (L.)).
Endemic to India, Nicobar Islands.

OEN Nicobar Dusky Roundleaf Bat (nom. form), Nicobar Dusky Roundleaf Bat (nom. form)

280. *Hipposideros gentilis* Andersen, 1918
Andersen's Roundleaf Bat

Hipposideros: Gr. *ïppo-sídèros* horseshoe (Gr. *ïppos* horse; Gr. *sídèros* iron); in allusion to the anterior part of the complicated noseleaf.
gentilis: L. *gentilis* of or belonging to the same clan (L. *gens* clan, race).
Southeast Asia and South Asia. Bangladesh, India, and Nepal.

281. *Hipposideros pomona* Andersen, 1918
Pomona Roundleaf Bat

Hipposideros: Gr. *ïppo-sídèros* horseshoe (Gr. *ïppos* horse; Gr. *sídèros* iron); in allusion to the anterior part of the complicated noseleaf.
pomona: Gr. Myth. *Pomona*, goddess of fruit trees, gardens, and orchards.
South Asia. Endemic to India.

282. *Hipposideros grandis* G.M. Allen, 1936
Grand Roundleaf Bat

Hipposideros: Gr. *ïppo-sídèros* horseshoe (Gr. *ïppos* horse; Gr. *sídèros* iron); in allusion to the anterior part of the complicated noseleaf.
grandis: L. *grandis* large; in allusion to its large size.
Southeast Asia and South Asia; Bangladesh and India (northeast and Little Andaman Islands).

Subspecies

- *leptophyllus*: Mod. L. *leptophyllus* slender-leafletted (Gr. *leptos* fine or slender; Gr. *phýllon* leaf); in allusion to its slender supplementary leaflets.
  Dobson's Roundleaf Bat.

283. *Hipposideros durgadasi* Khajuria, 1970
Durga Das's Roundleaf Bat

Hipposideros: Gr. *ïppo-sídèros* horseshoe (Gr. *ïppos* horse; Gr. *sídèros* iron); in allusion to the anterior part of the complicated noseleaf.
durgadasi: Eponym after Durga Das (dates not known), father in-law of H. Khajuria describer of the species (*-i*, commemorating (dedication, eponym) (L.)).
Endemic to India; Karnataka and Madhya Pradesh.

OEN Khajuria's Roundleaf Bat, Khajuria's Roundleaf Bat, Durga Das's Roundleaf Bat

284. *Hipposideros hypophyllus* Kock and Bhat, 1994
Leafletted Roundleaf Bat

Hipposideros: Gr. *ïppo-sídèros* horseshoe (Gr. *ïppos* horse; Gr. *sídèros* iron); in allusion to the anterior part of the complicated noseleaf.
hypophyllus: Mod. L. *hypophyllus* leafletted (Gr. *hupo* beneath; Gr. *phýllon* leaf); in allusion to it bearing a thin supplementary leaflet.
Endemic to India; Karnataka.

OEN Kolar Roundleaf Bat, Kolar Roundleaf Bat, Leafletted Roundleaf Bat

285. *Hipposideros khasiana* Thabah et al., 2006
Khasian Roundleaf Bat

Hipposideros: Gr. *ïppo-sídèros* horseshoe (Gr. *ïppos* horse; Gr. *sídèros* iron); in allusion to the anterior part of the complicated noseleaf.

khasiana: Mod. L. *khasiana* from Khasi Hills, Meghalaya, India, (*-ana*, belonging to, pertaining to (L.) geographic (location, toponym)).
Endemic to India; Meghalaya.

286. *Triaenops persicus* Dobson, 1871
Persian Trident Bat

Triaenops: Mod. L. *triaenops* trident-faced (Gr. *triaena* trident; Gr. *óps* aspect, face); in allusion to the posterior part of the noseleaf which terminates in three pointed projections resembling the three prongs of a trident.
persicus: L. adj. *persicus* Persian, belonging to Persia (*-icus* belonging to, pertaining to (L./Gr.)).
Africa, Arabian Peninsula, India, Iran and Pakistan.

OEN Triple Roundleaf Bat (nom. form), Persian Greater Roundleaf Bat (nom. form), Persian Roundleaf Bat (nom. form)

Family Megadermatidae H. Allen, 1864

287. *Megaderma spasma* (Linnaeus, 1758)
Lesser False Vampire Bat

Megaderma: Mod. L. *megaderma* large-skinned (Gr. *megas* great; Gr. *derma* skin); in allusion to the large wings and interfemoral membrane.
spasma: L. *spasma* sudden burst of energy or activity (>Gr. *spasmós* convulsion).
Southeast Asia and South Asia. Three subspecies are known from the region.

Subspecies

- *horsfieldii*: Eponym after Dr. Thomas Horsfield (b. 1773–d. 1859), an American physician and naturalist, later curator of the East India Company Museum in London, (*-ii* commemorating (dedication, eponym) (L.)).
  Horsfield's False Vampire Bat. Endemic to India.
- *ceylonense*: L. adj. *ceylonense* belonging to Sri Lanka or Sri Lankan, (*-ense*,= geographic, occurrence in (location, toponym) (L.)).
  Sri Lankan False Vampire Bat. Endemic to Sri Lanka.
- *majus*: L. *maius* great.
  Great False Vampire Bat. India (northeast) and Bangladesh.

OEN Ceylon False Vampire Bat (Subsp. *ceylonense*), Long-eared Bat (Subsp. *ceylonense*), Indian Lesser False Vampire (Subsp. *horsfieldii*), Burmese Lesser False Vampire Bat (Subsp. *majus*)

VN Dakh. *Chamgidar*, *Shabparakh*; Kan. *Kankapati*; Sinh. *Tutican Wawula*, *Kandiga Boru Ley-vavula*, *Kotican Wawula*; Tam. (SL) *Vaval*

288. *Lyroderma lyra* (E. Geoffroy, 1810)
Greater False Vampire Bat

Lyroderma: Mod. L. *lyroderma* lyre-skinned (Gr. *lyro* lyre; Gr. *derma* skin); in allusion to the prominent lyre-shaped noseleaf.
lyra: L. *lyra* lyre.

South Asia, Southeast Asia, and China (central and southeast); Afghanistan, Bangladesh, India, Nepal, Pakistan and Sri Lanka.

OEN Lyre-nosed broad-winged bat (nom. form), Indian Greater False Vampire Bat (Subsp. *lyra*), Large-eared Vampire Bat (Subsp. *lyra*), Greater False Vampire (Subsp. *lyra*), Carnatic False Vampire Bat (J. Syn. *carnatica*), Kashmir Vampire Bat (J. Syn. *spectrum*), Grey Greater False Vampire Bat (J. Syn. *schistaceus*)

VN Dakh. *Chamgidar, Shabparakh*; Kan. *Kankapati*; Sinh. *Tutican Wawula, Boru Ley-vavula, Kotican Wawula*; Tam. (SL) *Vaval*

Family Rhinopomatidae Bonaparte, 1838

289. *Rhinopoma hardwickii* Gray, 1831
Lesser Mouse-tailed Bat

Rhinopoma: Mod. L. *rhinopoma* lid-nosed (Gr. *rhis*, *rhinos* nose; Gr. *pōma* lid, cover); in allusion to the valvular nostrils, which open through a narrow transverse slit.
hardwickii: Eponym after Major-General Thomas Hardwicke (b. 1756–d. 1835), an English soldier and naturalist, travelled extensively throughout Indian subcontinent, (*-i*, commemorating (dedication, eponym) (L.)).

Africa, Arabian Peninsula, Iran, South Asia; Afghanistan, Bangladesh, India, and Pakistan.

OEN Indian Rhinopome (nom. form)

290. *Rhinopoma microphyllum* (Brünnich, 1872)
Greater Mouse-tailed Bat

Rhinopoma: Mod. L. *rhinopoma* lid-nosed (Gr. *rhis*, *rhinos* nose; Gr. *pōma* lid, cover); in allusion to the valvular nostrils, which open through a narrow transverse slit.
microphyllum: Mod. L. *microphyllum* small-leafed (Gr. *mikros* small; Gr. *phýllon* leaf).

Africa, Arabian Peninsula, Iran and South Asia. Besides the nominate subspecies *R. m. microphyllum* from Afghanistan and Pakistan, another subspecies is known from the region.

Subspecies

- *kinneari*: Eponym after Sir Norman Boyd Kinnear (b. 1882–d. 1957) British ornithologist at the British Museum (Natural History) and curator of the Bombay Natural History Museum 1907 (*-i*, commemorating (dedication, eponym) (L.)).

Kinnear's Mouse-tailed Bat. Endemic to South Asia; Bangladesh and India.

OEN Egyptian Rhinopome (nom. form)

291. *Rhinopoma muscatellum* Thomas, 1903
Small Mouse-tailed Bat

Rhinopoma: Mod. L. *rhinopoma* lid-nosed (Gr. *rhis*, *rhinos* nose; Gr. *pōma* lid, cover); in allusion to the valvular nostrils, which open through a narrow transverse slit.

muscatellum: Mod. L. *muscatellum,* meaning from Muscat.

Afghanistan, Iran, Iraq, Pakistan, Oman, and United Arab Emirates. One subspecies is known from the region.

Subspecies

- *seianum*: Toponym after Seistan, a region in Balochistan area of Afghanistan, Iran, and Pakistan (*-anum*, geographic (location, toponym); belonging to, pertaining to (L.)).

  Seistan Mouse-tailed Bat. Afghanistan and Pakistan.

Family Emballonuridae Gervais, 1855

292. *Saccolaimus saccolaimus* (Temminck, 1838)
Pouch-bearing Tomb Bat

Saccolaimus: Gr. *saccolaimus* gular sac (Gr. *sákkos* sac; Gr. *laimūs* throat, gullet); in allusion to the well-developed gular sacs.

saccolaimus: Gr. *saccolaimus* gular sac (Gr. *sákkos* sac; Gr. *laimūs* throat, gullet); in allusion to the well-developed gular sacs.

Southeast Asia, Australia (north), and South Asia. One subspecies occurs in the region.

Subspecies

- *crassus*: L. *crassus* thick, heavy, dense.

  Indian Pouch-bearing Tomb Bat. Bangladesh, India, and Sri Lanka.

OEN Pouch-bearing Sheath-tailed Bat (Subsp. *crassus*); Blyth's Tomb Bat (Subsp. *crassus*); White-bellied Bat (Subsp. *crassus*)

VN Sinh. *Podi Wawula, Maha Kepulum-vavula*; Tam. (SL) *Sinna Vaval*

293. *Taphozous perforatus* E. Geoffroy, 1818
Egyptian Tomb Bat

Taphozous: Mod. L. *taphozous* tomb-dwelling (Gr. *táphos* grave, tomb; *zòós* living); in allusion to its preference to tombs to dwell.

perforatus: L. *perforatus* pierced through (L. *per-* through; L. *forare* pierce); in allusion to its tail, that emerges out of the interfemoral membrane on the dorsal side in mid point.

Africa, Arabian Peninsula, and South Asia.

OEN Geoffroy's Tomb Bat (nom. form)

294. *Taphozous longimanus* Hardwicke, 1825
Long-winged Tomb Bat

Taphozous: Mod. L. *taphozous* tomb-dwelling (Gr. *táphos* grave, tomb; *zòós* living); in allusion to its preference to dwell in tombs.
longimanus: Mod. L. *longimanus*, long-handed (L. *longus* long; L. *manus* hand); in allusion to its long wings.
South Asia and Southeast Asia.

OEN Long-winged Bat (nom. form), Long-armed Sheath-tailed Bat (nom. form), Long-armed Bat (nom. form), Fulvous Long-winged Bat (J. syn. *fulvidus*), Short-tailed Long-winged Bat (J. syn. *brevicaudus*), Cantor's Long-winged Bat (J. syn. *cantori*)
VN Sinh. *Podi Wawula, Dikba Kepulum-vavula*; Tam. (SL) *Vaval*

295. *Taphozous nudiventris* Cretzschmar, 1830
Naked-rumped Tomb Bat

Taphozous: Mod. L. *taphozous* tomb-dwelling (Gr. *táphos* grave, tomb; *zòós* living); in allusion to its preference to dwell in tombs.
nudiventris: Med. L. *nudiventris* bare-bellied (L. *nudus* bare; L. *venter, ventris* belly); in allusion to the naked area of the rump.
Africa, Middle East, and South Asia. One subspecies occurs in the region.

Subspecies

- *kachhensis*: L. adj. *kachhensis* meaning from Kutch, Gujarat (*-ensis*, geographic, occurrence in (location, toponym) (L.)).
Dobson's Naked-rumped Tomb Bat. Afghanistan, Bangladesh, India, and Pakistan.

OEN Naked-bellied Tomb Bat (nom. form), Egyptian Sheath-tailed Bat (nom. form), Naked-rumped Bat (nom. form), Kachh Tomb Bat (Subsp. *kachhensis*), Cutch Sheath-tailed Bat (Subsp. *kachhensis*)

296. *Taphozous melanopogon* Temminck, 1841
Black-bearded Tomb Bat

Taphozous: Mod. L. *taphozous* tomb-dwelling (Gr. *táphos* grave, tomb; *zòós* living); in allusion to its preference to dwell in tombs.
melanopogon: Mod. L. *melanopogon* black-bearded (Gr. *melas* black; Gr. *pōgōn* beard); in allusion to the characteristic black beard present in males.
South Asia and Southeast Asia; Bangladesh, India and Sri Lanka.

OEN Black-bearded Sheath-tailed Bat (nom. form), Black-bearded Bat (nom. form), Bicolored Tomb Bat (J. syn. *bicolor*)
VN Sinh. *Podi Wawula, Ravulkalu Kepulum- vavula*; Tam. (SL) *Vaval*

297. *Taphozous theobaldi* Dobson, 1872
Theobald's Tomb Bat

Taphozous: Mod. L. *taphozous* tomb-dwelling (Gr. *táphos* grave, tomb; *zòós* living); in allusion to its preference to dwell in tombs.
theobaldi: Eponym after William Theobald (b. 1829–d. 1908), an English geologist, naturalist, and malacologist who worked in Burma (*-i*, commemorating (dedication, eponym) (L.)).

Southeast Asia and South Asia. One subspecies occurs in the region.

Subspecies

- *secatus*: L. adj. *secatus* meaning to cut (L. *secare* to cut off), in allusion to the fur not extending on to the dorsal side of the wing-membrane and the interfemoral membrane, unlike the nominate subspecies.
  Central Indian Tomb Bat. Endemic to India.

OEN Theobald's Bat (nom. form)

Family Molossidae Gill, 1872

298. *Chaerephon plicatus* (Buchanan, 1800)
Wrinkle-lipped Free-tailed Bat

Chaerephon: Gr. *Chairephōn* a proper name, a loyal friend and follower of Socrates.
plicatus: L. *plicatus* folded, doubled (L. *plicare* to fold).

Southeast Asia, China, and South Asia. Besides the nominate subspecies, *C. p. plicatus* Wrinkle-lipped Free-tailed Bat from Afghanistan and India, another subspecies is known from the region.

Subspecies

- *insularis*: L. *insularis* of an island (L. *insula* island).
  Sri Lankan Wrinkle-lipped Free-tailed Bat. Endemic to Sri Lanka.

OEN Wrinkle-lipped Free-tailed Bat (Subsp. *plicatus*), Wrinkle-lipped Bat (Subsp. *plicatus*), Ceylon Wrinkle-lipped Bat (Subsp. *insularis*), Sri Lankan Wrinkle-lipped Free-tailed Bat (Subsp. *insularis*)
VN Sinh. *Podi Wawula, Podhu Rallithol-vavula*; Tam. (SL) *Vaval*

299. *Otomops wroughtoni* (Thomas, 1913)
Wroughton's Giant Mastiff Bat

Otomops: Mod. L. *otomops* dog-eared (Gr. *otus* ear, L. *mops* (>Ger. *mops* roly-poly, usually a dog)); in allusion to the doglike appearance of bat's face.

wroughtoni: Eponym after Robert Charles Wroughton (b. 1849–d. 1921), an English officer of Indian Forest Service in the Bombay Presidency, member of the Bombay Natural History Society, described many new taxa resulting from Mammal Survey of British India (-*i*, commemorating (dedication, eponym) (L.)).

South Asia and Southeast Asia. India (Karnataka, Goa, and Meghalaya).

OEN Wroughton's Free-tailed Bat (nom. form)

300. *Tadarida teniotis* (Rafinesque, 1814)
European Free-tailed Bat

Tadarida: Etymology not known, probably from *Tadaris* Rafinesque, 1815 a *nomen nudum.*

teniotis – Mod. L. *teniotis* banded-ears (L. *taenia* (Gr. *taina*) headband; Gr. *otis*, *ōtos* ear).

Mediterranean region, Middle East, and South Asia. Afghanistan.

301. *Tadarida aegyptiaca* (E. Geoffroy, 1818)
Egyptian Free-tailed Bat

Tadarida: Etymology not known, probably from *Tadaris* Rafinesque, 1815 a *nomen nudum.*

aegyptiaca: L. *aegyptiaca* Egyptian (-*aca*, belonging to, pertaining to (L./Gr.)).

Africa, Arabian Peninsula, and South Asia. Three subspecies are known from the region.

Subspecies

- *tragatus*: Mod. L. *tragatus* in allusion to the well-developed antitragus (a projection of the auricle of the ear extending to the tragus), (*tragus* and -*atus*, provided with (L.)).

  Dobson's Free-tailed Bat. Endemic to South Asia; Bangladesh and India.
- *thomasi*: Eponym after Oldfield Thomas (b. 1858–d. 1929), an English zoologist who worked in British Museum (Natural History), who worked relentlessly on mammals, and described many new taxa based on specimens collected during the Mammal Survey of British India of the Bombay Natural History Society (-*i*, commemorating (dedication, eponym) (L.)).

  Thomas's Free-tailed Bat. Endemic to South Asia; India, Pakistan and Sri Lanka.
- *sindhica*: L. adj. *sindhica* belonging to Sindh, Pakistan (-*ica* belonging to, pertaining to (L./Gr.)).

  Sindh Free-tailed Bat. Afghanistan and Pakistan.

OEN Geoffroy's Wrinkled-Lipped Bat (nom. form), Dobson's Wrinkle-lipped Bat (Subsp. *tragatus*), Indian Free-tailed Bat (Subsp. *thomasi*), Indian Wrinkle-lipped Bat (Subsp. *thomasi*), Sindh Wrinkle-lipped Bat (Subsp. *sindhica*)

VN Sinh. *Podi Wawula, Mahadive Rallithol-vavula*; Tam. (SL) *Vaval*

Family Vespertilionidae Gray, 1821
Subfamily Vespertilioninae Miller, 1897
Tribe Eptisicini Volleth and Heller, 1994

302. *Arielulus circumdatus* (Temminck, 1840)
Bronze Sprite

Arielulus: Mod. L. *arielulus* somewhat like *ariel* (from Med. folklore *Ariel*, a spirit or sylph of the air), (*-ulus*, diminutive (comparison); somewhat (adj.) (L.)).

circumdatus: L. *circumdatus* surrounded (L. *circumdare* to surround).

Southeast Asia and South Asia. Himalayas in India (northeast) and Nepal.

OEN Large black Pipistrelle (nom. form), Gilded black Pipistrelle (nom. form)

303. *Cassistrellus dimissus* (Thomas, 1916)
Surat Serotine

Cassistrellus: Mod. L. *cassistrellus* helmeted pipistrelle (L. *cassis* wearer of a helmet; L. *pipistrello*, *vispitrello*, dim. of L. *vespertilio*, a bat).

dimissus: L. *dimissus* abandon; originally thought to be an individual of *E. pachyotis*, later found to be larger and distinct basing on a combination of characters.

Southeast Asia and South Asia. Nepal.

OEN Thailand's Serotine; Helmeted Bat

304. *Eptesicus serotinus* (Schreber, 1774)
Serotine

Eptesicus: Mod. L. *eptesicus* 'house-flyer' (Gr. *éptin* from *pétomai* to fly; Gr. *oíkos* house).

serotinus: L. *serotinus* that which comes late (or that which happens in the evening) (L. *sero*, *serus* late).

Europe, Mediterranean region, Middle East, Central Asia, South Asia, China, and Southeast Asia. One subspecies is known from the region.

Subspecies

- *pashtonus*: Toponym after Pashto, Khyber-Pakhtunkhwa, Pakistan (-*anus*, geographic (location, toponym); belonging to, pertaining to (L.)).
  Gaisler's Serotine. Afghanistan and Pakistan.

OEN Silky Bat (nom. form)

305. *Eptesicus pachyomus* Tomes, 1857
Tomes' Serotine

Eptesicus: Mod. L. *eptesicus* 'house-flyer' (Gr. *éptin* from *pétomai* to fly; Gr. *oíkos* house).

pachyomus: L. *pachyomus* (Gr. *pakhus* large, thick; Gr. *mŷs* mouse); No explanation was given, maybe in allusion to its thick and obtuse muzzle or stout mouse-like appearance.

Europe, Mediterranean region, Middle East, Central Asia, South Asia, China, and Southeast Asia. Himalayas in Afghanistan, India, and Pakistan.

OEN Thick-muzzled bat

306. *Eptesicus pachyotis* (Dobson, 1871)
Thick-eared Bat

Eptesicus: Mod. L. *eptesicus* 'house-flyer' (Gr. *éptin* from *pétomai* to fly; Gr. *oíkos* house).

pachyotis: Mod. L. *pachyotis* thick-eared (Gr. *pakhus* thick, dense; Gr. *-ōtis* –eared (*ous*, *ōtos* ear)).

China, Southeast Asia, and South Asia. Himalayas in Bangladesh and India (northeast).

307. *Eptesicus ognevi* Bobrinski, 1918
Ognev's Serotine

Eptesicus: Mod. L. *eptesicus* 'house-flyer' (Gr. *éptin* from *pétomai* to fly; Gr. *oíkos* house).

ognevi: Eponym after Prof. Sergei Ivanovich Ognev (b. 1886–d. 1951), a Russian zoologist who worked on taxonomy and distribution of mammals in Soviet Union and Central Asia (-*i*, commemorating (dedication, eponym) (L.)).

Arabian Peninsula, Middle East, Central Asia, and South Asia. Afghanistan and India.

OEN Bokhara Serotine

308. *Eptesicus gobiensis* Bobrinski, 1926
Gobi Big Brown Bat

Eptesicus: Mod. L. *eptesicus* 'house-flyer' (Gr. *éptin* from *pétomai* to fly; Gr. *oíkos* house).

gobiensis: L. adj. *gobiensis* meaning from Gobi desert, Mongolia (*-ensis*, geographic, occurrence in (location, toponym) (L.)).

Central Asia, China, and South Asia. Two subspecies are known from the region.

Subspecies

- *kashgaricus*: L. adj. *kashgaricus* belonging to Kashgar, Xinjiang, China (*-icus* belonging to, pertaining to (L./Gr.)).

  Kashgar Serotine. Afghanistan (Pamir Mountains) and India (Jammu and Kashmir).
- *centrasiaticus*: L. adj. *centrasiaticus* belonging to Central Asia; type locality is located in northwest Tibet (*-icus* belonging to, pertaining to (L./Gr.)).

  Central Asian Serotine. Possibly in the Himalayas in Nepal.

OEN Gobi Serotine

309. *Eptesicus tatei* Ellerman and Morrison-Scott, 1951
Sombre Bat

Eptesicus: Mod. L. *eptesicus* 'house-flyer' (Gr. *éptin* from *pétomai* to fly; Gr. *oíkos* house).
tatei: Eponym after George Henry Hamilton Tate (b. 1894–d. 1953), an English-born American zoologist, and botanist who worked in Southeast and East Asia, subsequently curator of American Museum of Natural History (*-i*, commemorating (dedication, eponym) (L.)).

Endemic to India (Darjeeling, West Bengal).

OEN Tate's Serotine (nom. form), Sombre Bat (S. Syn. *atratus*)

310. *Hesperoptenus tickelli* (Blyth, 1851)
Tickell's Bat

Hesperoptenus: Mod. L. *hesperoptenus* a crepuscular winged creature, a bat (Gr. *ésperos* evening, west; Gr. *ptènós* feathered or winged).
tickelli: Eponym after Colonel Samuel Richard Tickell (b. 1811–d. 1875), an English ornithologist in India and Burma (*-i*, commemorating (dedication, eponym) (L.)).

Southeast Asia and South Asia. Bangladesh, India, Nepal and Sri Lanka.

OEN Golden Bat (nom. form), Tickell's False Serotine (nom. form)
VN Sinh. *Podi Wawula*; Tam. (SL) *Sinna Vaval*

311. *Rhyneptesicus nasutus* (Dobson, 1877)
Sindh Bat

Rhyneptesicus: Mod. L. *rhyneptesicus* combination of *rhyn* 'cape' and *eptesicus* 'house-flyer' (Gr. *éptin* from *pétomai* to fly; Gr. *oíkos* house). No explanation was given.

nasutus: L. *nasutus* large-nosed (L. *nasus* nose).

Arabian Peninsula, Middle East, and South Asia. Afghanistan and Pakistan.

OEN Sind Bat (nom. form), Sindh Serotine Bat (nom. form)

Tribe Nycticeiini Gervais, 1855

312. *Scotoecus pallidus* (Dobson, 1876)

Desert Yellow Lesser House Bat

Scotoecus: Mod. L. *scotoecus* 'dweller of darkness' (Gr. *skōtos* darkness; Gr. *ōiko* to dwell); in allusion to its crepuscular habit.

pallidus: L. *pallidus* pallid, pale.

South Asia. Endemic to India and Pakistan.

OEN Pakistan Yellow Bat (nom. form), Yellow Desert Bat (nom. form)

313. *Scotomanes ornatus* (Blyth, 1851)

Harlequin Bat

Scotomanes: Gr. *skōtománes* slave of darkness (Gr. *skōtos* darkness, *mánes* slave); in allusion to its crepuscular habit.

ornatus: L. *ornatus* ornate, adorned (*ornare* to adorn).

Southeast Asia, China, and South Asia. Besides the nominate subspecies, *S. o. ornatus* Harlequin Bat, another subspecies is known from the region.

Subspecies

- *imbrensis*: L. adj. *imbrensis* meaning from 'the hills' (L. *imber*, *imbris* rain, cloud); in allusion to the type locality being in the hills experiencing heavy rainfall (*-ensis*, geographic, occurrence in (location, toponym) (L.)).

  Jantia Hills Harlequin Bat. India (Meghalaya).

OEN Alpine Bat (J. Syn. *nivicolus*), Large-eared Yellow Bat (J. Syn. *emarginatus*)

314. *Scotophilus kuhlii* Leach, 1821

Lesser Asiatic Yellow House Bat

Scotophilus: Mod. L. *scotophilus* 'lover of darkness' (Gr. *skōtos* darkness; Gr. *philos* loving); in allusion to its crepuscular habit.

kuhlii: Eponym after Heinrich Kuhl (b. 1797–d. 1821), a German naturalist and zoologist, collected and described species from South and Southeast Asia (*-ii*, commemorating (dedication, eponym) (L.)).

Southeast Asia and South Asia. Bangladesh, India, Pakistan, and Sri Lanka.

OEN Khuli's Yellow House Bat (nom. form), Asiatic Lesser Yellow House Bat (nom. form), Lesser Asian House Bat (nom. form), Lesser Asiatic Yellow Bat (nom. form), Lesser Yellow Bat (J. Syn. *temminckii*), Wroughton's Bat (J. Syn. *wroughtoni*)

VN Sinh. *Podi Wawula, Heen Kaha-vavula*; Tam. (SL) *Sinna Vaval*

315. *Scotophilus heathii* (Horsfield, 1831)
Greater Asiatic Yellow House Bat

Scotophilus: Mod. L. *scotophilus* 'lover of darkness' (Gr. *skōtos* darkness; Gr. *philos* loving); in allusion to its crepuscular habit.
heathii: Eponym after Josiah Marshall Heath (d. 1851), an English metallurgist, businessman, and ornithologist, worked at Porto Novo in the Madras Presidency (*-ii*, commemorating (dedication, eponym) (L.)).

Southeast Asia and South Asia. Afghanistan, Bangladesh, Bhutan, India, Nepal, Pakistan, and Sri Lanka.

OEN Asiatic Greater Yellow House Bat (nom. form), Asiatic Greater Yellow House Bat (nom. form), Common Yellow-bellied Bat (nom. form), Large Yellow bat (nom. form), Greater Asiatic Yellow Bat (nom. form), Greater Yellow Bat (Subsp. *heathii*), Bengal Yellow Bat (J. Syn. *luteus*)

VN Sinh. *Wawula, Maha Kaha-vavula*; Tam. (SL) *Vava, Vaval*

Tribe Pipistrellini Tate, 1942

316. *Nyctalus noctula* (Schreber, 1774)
Noctule

Nyctalus: Mod. L. *nyctalus* (>Gr. *niktalós* drowsy), in allusion to its crepuscular habits.
noctula: New L. *noctula* a bat, >Fr. *noctule* common name for a bat (>L. *noct-* or *nox* night).
Europe, Middle East, Central Asia, South Asia (in Himalayas), China, and Southeast Asia. One subspecies occurs in the region.

Subspecies

- *labiata*: L. adj. *labiata* meaning two-lipped (L. *labia* lip).
  Thick-lipped Noctule. India, Nepal, and Pakistan.

OEN Common Noctule (nom. form), Common Noctule Bat (nom. form), Indian Noctule Bat (Subsp. *labiata*)

317. *Nyctalus leisleri* (Kuhl, 1817)
Leisler's Noctule

Nyctalus: Mod. L. *nyctalus* (>Gr. *niktalós* drowsy), in allusion to its crepuscular habits.

leisleri: Eponym after Johann Philipp Achilles Leisler (b. 1772–d. 1813), a German physician and naturalist (*-i*, commemorating (dedication, eponym) (L.)).

Europe, Mediterranean region, Middle East, South Asia (in Himalayas). Afghanistan, India, and Pakistan.

OEN Hairy-armed Bat (nom. form), Leisler's Bat (nom. form), Lesser Noctule Bat (nom. form), Lesser Noctule (nom. form)

318. *Nyctalus montanus* (Barret-Hamilton, 1906)
Mountain Noctule

Nyctalus: Mod. L. *nyctalus* (>Gr. *niktalós* drowsy), in allusion to its crepuscular habits.

montanus: L. *montanus* of the mountains, mountain- (L. *mons*, *montis* mountain).

South Asia. Endemic to Afghanistan, India, and Nepal.

319. *Pipistrellus pipistrellus* (Schreber, 1774)
Common Pipistrelle

Pipistrellus: Mod. L. *pipistrellus* pipistrelle bat (>Ital. *pipistrello*, *vispitrello* dim. of L. *vespertilio* a bat), *-ellus*, diminutive (comparison), somewhat (adj.) (L.).

pipistrellus: Mod. L. *pipistrellus* pipistrelle bat (>Ital. *pipistrello*, *vispitrello* dim. of L. *vespertilio* a bat), *-ellus*, diminutive (comparison), somewhat (adj.) (L.).

Europe, Mediterranean region, Middle East, China and South Asia. One subspecies occurs in the region.

Subspecies

- *aladdin*: No explanation was given, perhaps after Aladdin (a protagonist of one of the Middle Eastern folktales Aladdin).

  Aladdin's Pipistrelle. Afghanistan, India, and Pakistan.

320. *Pipistrellus kuhlii* (Kuhl, 1817)
Kuhl's Pipistrelle

Pipistrellus: Mod. L. *pipistrellus* pipistrelle bat (>Ital. *pipistrello*, *vispitrello* dim. of L. *vespertilio* a bat), *-ellus*, diminutive (comparison), somewhat (adj.) (L.).

kuhlii: Eponym after Heinrich Kuhl (b. 1797–d. 1821), a German naturalist and zoologist, collected and described species from South and Southeast Asia, (*-ii*, commemorating (dedication, eponym) (L.)).

Europe, Mediterranean region, Middle East, Arabian Peninsula, Madagascar, and South Asia. One subspecies occurs in the region.

Subspecies

- *lepidus*: L. *lepidus* charming, elegant.

  Blyth's Pipistrelle. Afghanistan, India, and Pakistan.

OEN Great Pipistrelle (nom. form), Kuhl's Pipistrelle Bat (nom. form), Kuhl's Bat (nom. form), White-bordered Bat (nom. form), Hoary Bat (J. Syn. *canus*); Lobe-eared Bat (J. Syn. *lobatus*), White-eared Bat (J. Syn. *leucotis*)

321. *Pipistrellus coromandra* (Gray, 1838)
Indian Pipistrelle

Pipistrellus: Mod. L. *pipistrellus* pipistrelle bat (>Ital. *pipistrello*, *vispitrello* dim. of L. *vespertilio* a bat), *-ellus*, diminutive (comparison), somewhat (adj.) (L.).

coromandra: Mod. L. *coromandra* from the Coromandel (Cholamandalam) Coast, India.

South Asia and Southeast Asia. Afghanistan, Bangladesh, Bhutan, India, Nepal, Pakistan, and Sri Lanka.

OEN Coromandel Pipistrelle (nom. form), Coromandel bat (nom. form), Little Indian Bat (nom. form)

VN Sinh. *Kosetta Wawula*, *Indu Koseta-vavula*, Tam. (SL) *Sinna Vaval*

322. *Pipistrellus javanicus* (Gray, 1838)
Javan Pipistrelle

Pipistrellus: Mod. L. *pipistrellus* pipistrelle bat (>Ital. *pipistrello*, *vispitrello* dim. of L. *vespertilio* a bat), *-ellus*, diminutive (comparison), somewhat (adj.) (L.).

javanicus: L. adj. *javanicus* Javan (*-icus* belonging to, pertaining to (L./Gr.)).

Southeast Asia and South Asia. Three subspecies are known from the region.

Subspecies

- *camortae*: Belonging to Camorta Islands, Nicobar Archipelago, India (*-ae*, commemorating (dedication); geographic (location, toponym) (L.)).
  Camorta Pipistrelle. Endemic to India; only from Nicobar Islands.
- *babu*: No explanation was given.
  Thomas's Pipistrelle. Afghanistan, Bangladesh, India, Nepal, and Pakistan.
- *peguensis*: L. adj. *peguensis* meaning from Pegu (now Bago), Burma (*-ensis*, geographic, occurrence in (location, toponym) (L.)).
  Pegu Pipistrelle. India (West Bengal).

OEN Yellow-headed Pipistrelle (nom. form), Babu Pipistrelle (Subsp. *babu*), Himalayan Pipistrelle (Subsp. *babu*)

323. *Pipistrellus abramus* (Temminck, 1840)
Japanese Pipistrelle

Pipistrellus: Mod. L. *pipistrellus* pipistrelle bat (>Ital. *pipistrello*, *vispitrello* dim. of L. *vespertilio* a bat), *-ellus*, diminutive (comparison), somewhat (adj.) (L.).

abramus: L. *abramus* beautiful-mouse (L. *abrus* beautiful, graceful; Gr. *mȳs* mouse).
China, Japan, Southeast Asia and South Asia (India).

324. *Pipistrellus tenuis* (Temminck, 1840)
Least Pipistrelle

Pipistrellus: Mod. L. *pipistrellus* pipistrelle bat (>Ital. *pipistrello*, *vispitrello* dim. of L. *vespertilio* a bat), *-ellus*, diminutive (comparison), somewhat (adj.) (L.).
tenuis: L. *tenuis*, *tenue* weak, slight; in allusion to its size.
South Asia and Southeast Asia. One subspecies occurs in the region.

Subspecies

- *mimus*: L. *mimus* mimic.
  Wroughton's Least Pipistrelle. Afghanistan, Bangladesh, India, Nepal, Pakistan, and Sri Lanka.

OEN Indian Pygmy Bat (Subsp. *mimus*), Southern Dwarf Pipistrelle (Subsp. *mimus*), Pygmy Bat (Subsp. *mimus*)
VN Sinh. *Kosetta Wawula*, *Heen Koseta-vavula*; Tam. (SL) *Sinna Vaval*

325. *Pipistrellus ceylonicus* (Kelaart, 1852)
Kelaart's Pipistrelle

Pipistrellus: Mod. L. *pipistrellus* pipistrelle bat (>Ital. *pipistrello*, *vispitrello* dim. of L. *vespertilio* a bat), *-ellus*, diminutive (comparison), somewhat (adj.) (L.).
ceylonicus: L. adj. *ceylonicus* Sri Lankan (*-icus* belonging to, pertaining to (L./Gr.)).
South Asia and Southeast Asia. Besides the nominate subspecies, *P. c. ceylonicus* Kelaart's Pipistrelle. Endemic to Sri Lanka, another subspecies is known from the region.

Subspecies

- *indicus*: L. adj. *indicus* Indian, belonging to India (*-icus* belonging to, pertaining to (L./Gr.)).
  Indian Pipistrelle. Bangladesh, India, and Pakistan.

OEN Kelaart's Bat (Subsp. *ceylonicus*), Small Bat (Subsp. *ceylonicus*), Southern Kelaart's Pipistrelle (Subsp. *indicus*), Golden-haired Pipistrelle (J. Syn. *chrysothrix*), Northern Kelaart's Pipistrelle (J. Syn. *subcanus*), Gujarat Pipistrelle (J. Syn. *subcanus*)
VN Sinh. *Kiri Wavula*, *Rathbora Koseta-vavula*; Tam (SL) *Vaval*

326. *Pipistrellus paterculus* Thomas, 1915
Mount Popa Pipistrelle

Pipistrellus: Mod. L. *pipistrellus* pipistrelle bat (>Ital. *pipistrello*, *vispitrello* dim. of L. *vespertilio* a bat), *-ellus*, diminutive (comparison), somewhat (adj.) (L.).

paterculus: L. *paterculus* paternal (L. *paterculus* dim. from Mod. L. *pater* father).
Southeast Asia and South Asia. India and Nepal.

OEN Indochina Pipistrelle (nom. form), Paternal Pipistrelle (nom. form)

327. *Scotozous dormeri* Dobson, 1875
Dormer's Pipistrelle

Scotozous: Mod. L. *scotozous* 'one who lives in darkness' (Gr. *skōtos* darkness; *zòós* living); in allusion to its crepuscular habit.
dormeri: Eponym after Major General James Charlemagne Dormer (b. 1834–d. 1893), a British officer serving in the Madras Presidency (-*i*, commemorating (dedication, eponym) (L.)).
South Asia. Endemic to Bangladesh, India, and Pakistan.

OEN Dormer's Bat

Tribe Plecotinini Gray, 1866

328. *Barbastella darjelingensis* (Hodgson, 1855)
Large Barbastelle

Barbastella: L. *barbastella* (>Fr. *barbastelle*, from L. *barba* beard); in allusion to the hairy chin.
darjelingensis: L. adj. *darjelingensis* meaning from Darjeeling, West Bengal, India (-*ensis*, geographic, occurrence in (location, toponym) (L.)).
China, South Asia, and Middle East. Afghanistan, Bhutan, India, Nepal, and Pakistan.

OEN Darjiling Bat

329. *Otonycteris hemprichii* Peters, 1859
Hemprich's Desert Bat

Otonycteris: Mod. L. *otonycteris* eared-bat (Gr. *otus* ear; Gr. *nikterís* bat); in allusion to its long ears.
hemprichii: Eponym after Wilhelm Friedrich Hemprich (b. 1796–d. 1825), a German naturalist and explorer, collected in the Middle-East (-*ii*, commemorating (dedication, eponym) (L.)).
Middle East, South Asia, Arabian Peninsula, and North Africa. Afghanistan, India, and Pakistan.

OEN Hemprich's Arrow-eared Bat (nom. form), Hemprich's Long-eared Bat (nom. form), Hemprich's Desert Long-eared Bat (nom. form), Hemprich's Long-eared Desert Bat (nom. form), Long-eared Desert Bat (nom. form), Desert Long-eared Bat (nom. form), Hemprich's Big-eared Bat (nom. form)

330. *Otonycteris leucophaea* (Severcov, 1873)
Turkestani Long-eared Bat

Otonycteris: Mod. L. *otonycteris* eared-bat (Gr. *otus* ear; Gr. *nikterís* bat); in allusion to its long ears.
leucophaea: Mod. L. *leucophaea* whitish (Gr. *leukos* white, *pháos* light).
Middle East, South Asia, and Central Asia. Afghanistan.

331. *Plecotus homochrous* Hodgson, 1847
Hodgson's Long-eared Bat

Plecotus: Mod. L. *plecotus* twisted-ears (Gr. *plékò* to twine, to twist; Gr. *otus* ear).
homochrous: Gr. *homokhrous* uniform, of one colour (Gr. *homos* common; Gr. *khroa* colour).
Endemic to South Asia; India, Nepal, and Pakistan.

OEN Darjeeling Long-eared Bat

332. *Plecotus wardi* Thomas, 1911
Ward's Long-eared Bat

Plecotus: Gr. *plecotus* twisted-ears (Gr. *plékò* to twine, to twist; Gr. *otus* ear).
wardi: Eponym after Colonel A.E. Ward (dates not known), an amateur zoologist, sportsman, and collector, active in the Kashmir Himalayas (-*i*, commemorating (dedication, eponym) (L.)).
Endemic to South Asia; India, Nepal, and Pakistan.

333. *Plecotus strelkovi* Spitzenberger, 2006
Strelkov's Long-eared Bat

Plecotus: Gr. *plecotus* twisted-ears (Gr. *plékò* to twine, to twist; Gr. *otus* ear).
strelkovi: Eponym after Petr Petrovich Strelkov (b. 1931–d. 2012), a Russian biologist known for his work on *Plecotus* taxonomy (-*i*, commemorating (dedication, eponym) (L.)).
China, Middle East, and South Asia. Afghanistan.

Tribe Vespertilionini Gray, 1821

334. *Falsistrellus affinis* (Dobson, 1871)
Chocolate Pipistrelle

Falsistrellus: Mod. L. *falsistrellus* false pipistrelle (Ital. *falsi* fake; L. *pipistrello*, *vispitrello*, dim. of L. vespertilio, a bat).
affinis: L. *affinis* related, allied.
South Asia and Southeast Asia. India, Nepal, and Sri Lanka.

OEN Chocolate Bat (nom. form), Grizzled Pipistrelle (reported as *Pipistrellus mordax*), Highland Pipistrelle (reported as *Pipistrellus mordax*).
VN Sinh. *Kiri Wawula, Bora Koseta-vavula*; Tam. (SL) *Vaval*

335. *Hypsugo savii* (Bonaparte, 1837)
Savi's Pipistrelle

Hypsugo: Mod. L. *hypsugo* high-flying bat (Gr. *ÿpsi* on high, aloft; ending *-ugo*; formed in analogy of L. *vesperugo* a bat).
savii: Eponym after Pablo Savi (b. 1798–d. 1871), a Professor of Zoology at University of Pisa, Italy, (*-ii*, commemorating (dedication, eponym) (L.)).

Europe, Mediterranean region, Middle East, and South Asia. Two subspecies are known from the region.

Subspecies

- *austenianus*: No explanation was given; most possibly an eponym after Lieutenant-Colonel Henry Haversham Godwin-Austen (b. 1834–d. 1923), an English topographer, naturalist, geologist, and surveyor (*-anus*, belonging to, pertaining to (L.)).
  Dobson's Pipistrelle. Bangladesh and India (in the Himalayas).
- *caucasicus*: L. adj. *caucasicus* belonging to Caucasus mountains (*-icus* belonging to, pertaining to (L./Gr.)).
  Caucasian Pipistrelle. Afghanistan and India (Jammu and Kashmir).

OEN Savi's Pipistrelle Bat (nom. form), Caucasian Savi's Pipistrelle (Subsp. *caucasicus*)

336. *Hypsugo cadornae* (Thomas, 1916)
Cadorna's Pipistrelle

Hypsugo: Mod. L. *hypsugo* high-flying bat (Gr. *ÿpsi* on high, aloft, ending *–ugo*; formed in analogy of L. *vesperugo* a bat).
cadornae: Eponym after Italian Field Marshall Luigi Cadorna (b. 1850–d. 1928).

Southeast Asia and South Asia. India (Sikkim and West Bengal).

OEN Thomas' Pipistrelle

337. *Ia io* Thomas, 1902
Great Evening Bat

Ia: L. *Ia* a young woman of classical times, as perceived of many women of those times, a bat is essentially flighty.
io: L. *Io* a nymph who was seduced by Zeus.

China, Southeast Asia, and South Asia. India and Nepal (in the Himalayas).

OEN Ia Bat

338. *Philetor brachypterus* (Temminck, 1840)
Rohu's Bat

Philetor: Gr. *philêtòr* lover.
brachypterus: Mod. L. *brachypterus* short-winged (Gr. *brakhus* short; L. *-pterus* -winged >Gr. *pteron* wing).
Melanesia, Southeast Asia, and South Asia. India and Nepal (in the Himalayas).

OEN Short-winged Pipistrelle (nom. form), Narrow-winged Brown Bat (nom. form), Short-winged Brown Bat (nom. form)

339. *Tylonycteris fulvida* (Blyth, 1850)
Blyth's Lesser Bamboo Bat

Tylonycteris: Mod. L. *tylonycteris* padded bat (Gr. *týlos* knob, knot; Gr. *nikterís* bat); in allusion to the expanded fleshy pads on the under surface of the base of the thumb.
fulvidus: Mod. L. *fulvidus* (L. *fulvus* fulvous, tawny).
Southeast Asia, China, and South Asia. Bangladesh and India.

OEN Pale Club-footed Bat

340. *Tylonycteris aurex* (Thomas, 1915)
Thomas' Lesser Bamboo Bat

Tylonycteris: Mod. L. *tylonycteris* padded bat (Gr. *týlos* knob, knot; Gr. *nikterís* bat); in allusion to the expanded fleshy pads on the under surface of the base of the thumb.
aurex: L. *aurex* from *aureus* golden (L. *aurum* gold).
Endemic to India.

OEN Golden Club-footed Bat

341. *Tylonycteris malayana* Chasen, 1940
Malayan Greater Bamboo Bat

Tylonycteris: Mod. L. *tylonycteris* padded bat (Gr. *týlos* knob, knot; Gr. *nikterís* bat); in allusion to the expanded fleshy pads on the under surface of the base of the thumb.
malayana: L. *malayana* belonging to Malay or Malaysia (*-ana*, belonging to, pertaining to geographic (location, toponym) (L.)).
Southeast Asia, China, and South Asia. Two subspecies are known from the region.

Subspecies

- *malayana*: L. *malayana* belonging to Malay or Malaysia (*-ana*, belonging to, pertaining to geographic (location, toponym) (L.)).
  Malayan Greater Bamboo Bat. India (Mizoram).

- *eremtaga*: L. noun *eremtaga* derived from the Aka-Kora dialect of the Great Andamanese language, meaning 'forest-dweller'.
  Andaman Greater Bamboo Bat. India (Andaman Islands).

OEN Greater Flat-headed Bat (nom. form)

342. *Vespertilio murinus* Linnaeus, 1758
Particoloured Bat

Vespertilio: L. *vespertilio* a bat, so-called due to its flying habit in the evening – probably from the L. *vespertinus* meaning of the evening.
murinus: L. *murinus* of mice, mouse- >*mus*, *muris* mouse; in allusion to its mouse-like appearance.
Europe, Central Asia, China, Mongolia, Russia, and South Asia. Afghanistan and India (Jammu and Kashmir).

Subfamily Myotinae Tate, 1942

343. *Myotis emarginatus* (E. Geoffroy, 1806)
Geoffroy's Myotis

Myotis: Mod. L. *myotis* mouse-eared bat (Gr. *mȳs* mouse; Gr. *otus*, *ōtos* ear), in allusion to its mouse-like ears.
emarginatus: L. *emarginatus* emarginated (L. *emarginare* to provide with a margin; -*atus*, provided with (L.)).
Europe, Mediterranean region, Middle East, and South Asia. One subspecies occurs in the region.

Subspecies

- *desertorum*: L. *desertorum* of the deserts (L. *desertum* desert).
  Desert Myotis. Afghanistan.

OEN Geoffroy's Bat (nom. form), Notch-eared bat (nom. form), Marginate Vespertilion (nom. form), Notch-eared Bat (nom. form)

344. *Myotis formosus* (Hodgson, 1835)
Hodgson's Myotis

Myotis: Mod. L. *myotis* mouse-eared bat (Gr. *mȳs* mouse; Gr. *otus*, *ōtos* ear), in allusion to its mouse-like ears.
formosus: L. *formosus* beautiful (L. *forma* beauty).
China, Southeast Asia, and South Asia. Afghanistan, Bangladesh, India, and Nepal.

OEN Hodgson's Bat (nom. form), Golden-winged Myotis (nom. form), Copper-winged Bat (nom. form), Beautiful Bat (nom. form), Bartel's Myotis (nom. form)

345. ***Myotis hasseltii*** **(Temminck, 1840)**
Lesser Large-footed Myotis

Myotis: Mod. L. *myotis* mouse-eared bat (Gr. *mȳs* mouse; Gr. *otus*, *ōtos* ear), in allusion to its mouse-like ears.

hasseltii: Eponym after Johan Coenraad van Hasselt (b. 1797–d. 1823), a Dutch physician, zoologist, and botanist, worked and collected in Java (-*ii*, commemorating (dedication, eponym) (L.)).

Southeast Asia and South Asia. India (West Bengal) and Sri Lanka.

OEN Van Hasselt's Bat (nom. form), Large-footed Bat (nom. form), Hasselt's Large-footed Myotis (nom. form), Lesser Large-tooth Bat (nom. form), Brown Bat (nom. form), Small Brown Bat (nom. form)

VN Sinh. *Podi Wawula, Bora-vavula*; Tam. (SL) *Sinna Vaval*

346. ***Myotis horsfieldii*** **(Temminck, 1840)**
Horsfield's Myotis

Myotis: Mod. L. *myotis* mouse-eared bat (Gr. *mȳs* mouse; Gr. *otus*, *ōtos* ear), in allusion to its mouse-like ears.

horsfieldii: Eponym after Dr. Thomas Horsfield (b. 1773–d. 1859), an American physician and naturalist, later curator of the East India Company Museum in London (-*ii*, commemorating (dedication, eponym) (L.)).

Southeast India and South Asia. Two subspecies are known from the region.

Subspecies

- *dryas*: Gr. Myth. *Dryas*, tree-nymph, dryad.
  Andaman Myotis. Endemic to Andaman Islands, India.
- *peshwa*: No explanation was given; *peshwa* (>Arab. *pēshwā*, foremost, leader) is the titular equivalent of a modern Prime Minister, possibly in allusion to its type locality Poona (= Pune), Maharashtra, the capital of Mahrata Kingdom.
  Wroughton's Myotis. Endemic to India.

OEN Horsfield's Bat (nom. form), Lesser Large-tooth Bat (nom. form), Common Asiatic Myotis (nom. form)

347. ***Myotis muricola*** **(Gray, 1846)**
Hairy-faced Myotis

Myotis: Mod. L. *myotis* mouse-eared bat (Gr. *mȳs* mouse; Gr. *otus*, *ōtos* ear), in allusion to its mouse-like ears.

muricola: Mod. L. *muricola* wall-dweller (L. *murus* wall; L. -*cola* inhabitant >L. *colere* to dwell).

Southeast Asia and South Asia. Eastern Himalayas in Bangladesh, India, and Nepal.

OEN Nepalese Whiskered Myotis (nom. form), Nepalese Whiskered Bat (nom. form), Whiskered Myotis (nom. form), Wall-roosting Mouse-eared Bat (nom. form), Wall Bat (nom. form)

348. *Myotis siligorensis* (Horsfield, 1855)
Himalayan Whiskered Myotis

Myotis: Mod. L. *myotis* mouse-eared bat (Gr. *mŷs* mouse; Gr. *otus*, *ōtos* ear), in allusion to its mouse-like ears.
siligorensis: L. adj. *siligorensis* meaning from Siliguri, West Bengal, India (-*ensis*, geographic, occurrence in (location, toponym) (L.)).
Southeast Asia and South Asia. Bhutan, India, and Nepal.

OEN Himalayan Whiskered Myotis (nom. form), Siliguri Bat (nom. form), Small-toothed Myotis (nom. form), Terai Bat (nom. form), Darjeeling Bat (J. Syn. *darjilingensis*)

349. *Myotis blythii* (Tomes, 1857)
Lesser Myotis

Myotis: Mod. L. *myotis* mouse-eared bat (Gr. *mŷs* mouse; Gr. *otus*, *ōtos* ear), in allusion to its mouse-like ears.
blythii: Eponym after Edward Blyth (b. 1810–d. 1873), an English zoologist and pharmacist, curator of Museum at the Royal Asiatic Society of Bengal (-*ii*, commemorating (dedication, eponym) (L.)).
Mediterranean region, Middle East, China, Central Asia, and South Asia. Afghanistan, India, Nepal, and Pakistan.

OEN Lesser Mouse-eared Bat (nom. form), Lesser Mouse-eared Myotis (nom. form), Blyth's Vespertilion (nom. form), Blyth's Bat (nom. form), Kashmir Lesser Mouse-eared Bat (nom. form), Small Mouse-like Bat (J. Syn. *murinoides*), Mouse-like Bat (J. Syn. *murinus*)

350. *Myotis davidii* (Peters, 1869)
David's Myotis

Myotis: Mod. L. *myotis* mouse-eared bat (Gr. *mŷs* mouse; Gr. *otus*, *ōtos* ear), in allusion to its mouse-like ears.
davidii: Eponym after Father Armand 'Pere' David (b. 1826–d. 1900), missionary and priest and naturalist; collected in China (-*ii*, commemorating (dedication, eponym) (L.)).
China and South Asia. Afghanistan.

351. ***Myotis annectans*** **(Dobson, 1871)**
Hairy-faced Myotis

Myotis: Mod. L. *myotis* mouse-eared bat (Gr. *mÿs* mouse; Gr. *otus*, *ōtos* ear), in allusion to its mouse-like ears.
annectans: L. *annectans* connecting (L. *annectere* to connect).
Southeast Asia and South Asia. India (Nagaland and West Bengal).

OEN Hairy-faced Bat (nom. form), Intermediate Bat (nom. form), Thomas' Myotis (J. Syn. *primula*)

352. ***Myotis laniger*** **(Peters, 1871)**
Chinese Water Myotis

Myotis: Mod. L. *myotis* mouse-eared bat (Gr. *mÿs* mouse; Gr. *otus*, *ōtos* ear), in allusion to its mouse-like ears.
laniger: L. *laniger* woolly.
China and South Asia. India (Meghalaya).

353. ***Myotis nipalensis*** **(Dobson, 1871)**
Nepal Whiskered Myotis

Myotis: Mod. L. *myotis* mouse-eared bat (Gr. *mÿs* mouse; Gr. *otus*, *ōtos* ear), in allusion to its mouse-like ears.
nipalensis: L. adj. *nipalensis* meaning from Nepal (*ni* holy, *pal* land), (*-ensis*, geographic, occurrence in (location, toponym) (L.)).
Middle East, Central Asia, China, and South Asia. Afghanistan, Bhutan, India, Nepal, and Pakistan.

OEN Nepal Myotis (nom. form), Meinertzhagen's Bat (J. Syn. *meinertzhageni*)

354. ***Myotis longipes*** **(Dobson, 1873)**
Kashmir Cave Myotis

Myotis: Mod. L. *myotis* mouse-eared bat (Gr. *mÿs* mouse; Gr. *otus*, *ōtos* ear), in allusion to its mouse-like ears.
longipes: L. *longipes* long-footed (L. *longus* long, L. *pes* foot, from Gr. *pous* foot).
China and South Asia. Afghanistan, India, and Nepal.

OEN Kashmir Myotis (nom. form), Large-footed Kashmir Myotis (J. Syn. *macropus*), Theobald's Mouse-bat (J. Syn. *theobaldi*)

355. ***Myotis montivagus*** **(Dobson, 1874)**
Burmese Whiskered Myotis

Myotis: Mod. L. *myotis* mouse-eared bat (Gr. *mÿs* mouse; Gr. *otus*, *ōtos* ear), in allusion to its mouse-like ears.
montivagus: L. *montivagus* mountain roaming (L. *mons*, *montis* mountain; L. *vagari* to wander).

China, Southeast Asia, and South Asia. India (Mizoram).

OEN Indochina Myotis, Burmese Whiskered Bat, Large Brown Myotis

356. *Myotis peytoni* Wroughton and Ryley, 1913
Peyton's Whiskered Myotis

Myotis: Mod. L. *myotis* mouse-eared bat (Gr. *mȳs* mouse; Gr. *otus*, *ōtos* ear), in allusion to its mouse-like ears.
peytoni: Eponym after General Walter Peyton (dates not known), an English soldier, naturalist and Conservator of Forests stationed at Kanara (*-i*, commemorating (dedication, eponym) (L.)).
Endemic to India.

OEN Indian Whiskered Bat

357. *Myotis sicarius* Thomas, 1915
Mendelli's Mouse-eared Myotis

Myotis: Mod. L. *myotis* mouse-eared bat (Gr. *mȳs* mouse; Gr. *otus*, *ōtos* ear), in allusion to its mouse-like ears.
sicarius: L. *sicarius* assassin, murderer.
Endemic to South Asia; India and Nepal (in the Himalayas).

OEN Mandelli's Mouse-eared Bat (nom. form), Sikkim Myotis (nom. form)

358. *Myotis csorbai* Topál, 1997
Csorba's Myotis

Myotis: Mod. L. *myotis* mouse-eared bat (Gr. *mȳs* mouse; Gr. *otus*, *ōtos* ear), in allusion to its mouse-like ears.
csorbai: Eponym after Dr. Gábor Csorba (b. 1961), curator of Mammals at Hungarian Museum of Natural History, collected bats in the Himalayas (*-i*, commemorating (dedication, eponym) (L.)).
Endemic to Nepal.

359. *Submyotodon caliginosus* (Tomes, 1859)
Tomes' Myotis

Submyotodon: Mod. L. *submyotodon* near or having close affinity to myotodon (L. *sub* under, close to; Gr. *mȳs* mouse; Gr. *ous*, ear; Gr. *ódôn* tooth), in allusion to its nearly myotodont development of the postcristid in the lower molars.
caliginosus: L. *caliginosus* dark, obscure, misty (L. *caligo*, *caliginis* mist, fog).
Southeast Asia and South Asia. Afghanistan, India, and Pakistan.

OEN Mustachioed Bat, Tomes Whiskered Bat

Subfamily Murininae Miller, 1907

360. *Harpiocephalus harpia* (Temminck, 1840)

Lesser Hairy-winged Bat

Harpiocephalus: Mod. L. *harpiocephalus* bearing head of a harpy (Gr. Myth. *harpya* harpy a mythological winged monster, ravenous and filthy, with the head of a woman and the wings of a bird of prey; Gr. *kephalus* head).

harpia: Gr. Myth. *harpya* harpy a mythological winged monster, ravenous and filthy, with the head of a woman and the wings of a bird of prey.

Southeast Asia, China, South Asia. Two subspecies are known from the region.

Subspecies

- *lasyurus*: Mod. L. *lasyurus* hairy-tailed (Gr. *lásios* hairy; Gr. *oúrá* tail); the entire interfemoral membrane is hairy on the dorsal side.
  Hodgson's Hairy-winged Bat. India (in the Himalayas).
- *madrassius*: Mod. L. *madrassius* pertaining to the Madras Presidency in British India (*-ius* commemorating or dedication to (L./Gr.)).
  South Indian Hairy-winged Bat. Endemic to India; Tamil Nadu (in the Western Ghats).

OEN Hairy-winged Bat (nom. form), Harpy-headed Bat (nom. form), Myanmar Hairy-winged Bat (Sub. J. Syn. *mordax*), Greater Hairy-winged Bat (Sub. J. Syn. *mordax*)

361. *Murina aurata* Milne-Edwards, 1872

Little Tube-nosed Bat

Murina: L. *murina* mouse-like (L. *muris* mouse); in allusion to the shape of its ear and head.

aurata: L. *auratus* gilded, ornamented with gold (*aurum* gold).

China, Southeast Asia, and South Asia. India and Nepal (in the Himalayas).

OEN Tibetan Tube-nosed Bat (nom. form)

362. *Murina cyclotis* Dobson, 1872

Round-eared Tube-nosed Bat

Murina: L. *murina* mouse-like (L. *muris* mouse); in allusion to the shape of its ear and head.

cyclotis: L. *cyclotis* round-eared (Gr. *kuklos* circle; L. *otus, ōtos* ear).

Southeast Asia, China, and South Asia. Besides the nominate subspecies, *M. c. cyclotis* Round-eared Tube-nosed Bat from India and Nepal, another subspecies is known from the region.

Subspecies

- *eileenae*: Eponym after Ms. Eileen Wynell-Mayow (b. 1927–b. 2012), daughter of W.W.A. Phillips, a naturalist in Sri Lanka, author of *Manual of the Mammals of Sri Lanka* (*-ae*, commemorating (dedication, eponym) (L.)).

  Sri Lankan Round-eared Tube-nosed Bat. Endemic to Sri Lanka.

OEN Orange Tube-nosed Bat (Subsp. *cyclotis*), Ceylon Tube-nosed Bat (Subsp. *eileenae*)

VN Sinh. *Podi Wawula*, *Nalanehe-vavula*; Tam. (SL) *Sinna Vaval*

363. *Murina huttoni* (Peters, 1872)
Hutton's Tube-nosed Bat

Murina: L. *murina* mouse-like (L. *muris* mouse); in allusion to the shape of its ear and head.

huttoni: Eponym after Captain Thomas Hutton (b. 1806–d. 1875), an English officer in the Bengal Native Infantry, served in Afghan War, (*-i*, commemorating (dedication, eponym) (L.)).

South Asia, China, Southeast Asia. Himalayas in India, Nepal, and Pakistan.

364. *Murina leucogaster* Milne-Edwards, 1872
Greater Tube-nosed Bat

Murina: L. *murina* mouse-like (L. *muris* mouse); in allusion to the shape of its ear and head.

leucogaster: Mod. L. *leucogaster* white-bellied (Gr. *leukogastēr* white-belly, from Gr. *leukos* white; Gr. *gastēr* belly).

China, South Asia, and Southeast Asia. One subspecies is known from the region.

Subspecies

- *rubex*: Mod. L. *rubex* red (L. *ruber* red).

  Thomas' Greater Tube-nosed Bat. India and Nepal (in the Himalayas).

OEN White-bellied Tube-nosed Bat (nom. form), Rufous Tube-nosed Bat (Subsp. *rubex*)

365. *Murina tubinaris* (Scully, 1881)
Scully's Tube-nosed Bat

Murina: L. *murina* mouse-like (L. *muris* mouse); in allusion to the shape of its ear and head.

tubinaris: L. *tubinaris* tube-nosed (L. *tubis* tube; L. *naris* nose).

Southeast Asia and South Asia. India and Pakistan (in the Himalayas).

366. *Murina feae* (Thomas, 1891)
Fea's Tube-nosed Bat

Murina: L. *murina* mouse-like (L. *muris* mouse); in allusion to the shape of its ear and head.
feae: Eponym Leonardo Fea (b. 1852–d. 1903), Italian naturalist, collected in Burma in 1885 (*-e*, commemorating (dedication, eponym) (L.)) (*Murina*).
Southeast Asia and South Asia. India (in the Himalayas).

367. *Murina pluvialis* Ruedi, Biswas and Csorba, 2012
Rainforest Tube-nosed Bat

Murina: L. *murina* mouse-like (L. *muris* mouse); in allusion to the shape of its ear and head.
pluvialis: L. *pluvialis* relating to rain (*pluvia* rain).
South Asia (Himalayas in Meghalaya, India and possibly in Bangladesh).

368. *Murina jaintiana* Ruedi, Biswas and Csorba, 2012
Jaintia Tube-nosed Bat

Murina: L. *murina* mouse-like (L. *muris* mouse); in allusion to the shape of its ear and head.
jaintiana: Mod. L. *jaintiana* from Jaintia Hills, Meghalaya, India (*-ana*, belonging to, pertaining to (L.) geographic (location, toponym)).
South Asia (Himalayas in Meghalaya, India) and Myanmar.

369. *Murina guilleni* Soisook et al., 2013
Guillen's Tube-nosed Bat

Murina: L. *murina* mouse-like (L. *muris* mouse); in allusion to the shape of its ear and head.
guilleni: Eponym after Antonio Guillen-Servent (b. 1964), a Spanish zoologist, now at Institute of Ecology, Veracruz, Mexico, who worked on bats in southeast Asia (*-i*, commemorating (dedication, eponym) (L.)).
South Asia (Great Nicobar Island, Andaman and Nicobar Islands, India) and Southeast Asia (Thailand).

Subspecies

- *nicobarensis*: L. adj. *nicobarensis* Nicobarese or belonging to Nicobar Islands, India (*-ensis* belonging to, pertaining to (L./Gr.)). Nicobar Guillen's Tube-nosed Bat. Endemic to Nicobar Islands.

370. *Harpiola grisea* (Peters, 1872)
Peter's Tube-nosed Bat

Harpiola: Mod. L. *harpiola* somewhat like *Harpiocephalus* (see *Harpiocephalus*), (*-ola*, diminutive (comparison); somewhat (adj.) (L.)).

grisea – Med. L. *griseus* grey (>Old Fr. *gris* grey).
Endemic to India; Himalayas in Mizoram and Uttarakhand.

OEN Grey Tube-nosed Bat (nom. form)

Subfamily Kerivoulinae Miller, 1907

371. *Kerivoula picta* (Pallas, 1767)
Painted Woolly Bat

Kerivoula: Mod. L. *kerivoula* from the Sinhala *kehelvoulha* (*kehe wawula*) or *kiri wawula*, a name for this bat.
picta: L. *pictus* painted (L. *pingere* to paint).

Southeast Asia and South Asia. Bangladesh, India, Nepal, and Sri Lanka.

OEN Painted Bat (nom. form)
VN Sinh. *Kiri Wawula, Kehe Wawula, Visithuru Kehel-vavula*; Tam. (SL) *Sinna Vaval*

372. *Kerivoula hardwickii* (Horsfield, 1824)
Hardwicke's Woolly Bat

Kerivoula: Mod. L. *kerivoula* from the Sinhala *kehelvoulha* (*kehe wawula*) or *kiri wawula*, a name for this bat.
hardwickii: Eponym after Major-General Thomas Hardwicke (b. 1756–d. 1835), an English soldier and naturalist, travelled extensively throughout the Indian subcontinent, (*-i*, commemorating (dedication, eponym) (L.)).

Southeast Asia and South Asia. Bangladesh, India, Pakistan, and Sri Lanka.

OEN Hardwicke's Forest Bat (nom. form), Lesser Painted Bat (nom. form), Common Woolly Bat (nom. form), Trumpet-eared Bat (nom. form), Malpas's Bat (J. Syn. *malpasi*), Small Brown Bat (J. Syn. *malpasi*)
VN Sinh. *Podi Wawula*; Tam. (SL) *Sinna Vaval, Rathbora Kehel-vavula*

373. *Kerivoula lenis* Thomas, 1916
Lenis Woolly Bat

Kerivoula: Mod. L. *kerivoula* from the Sinhala *kehelvoulha* (*kehe wawula*) or *kiri wawula*, a name for this bat.
lenis: L. *lenis* soft, smooth.
Southeast Asia and South Asia. India (Tamil Nadu and West Bengal).

Family Miniopteridae Miller, 1907

374. *Miniopterus fuliginosus* (Hodgson, 1835)
Eastern Long-fingered Bat

Miniopterus: Gr. *miniopterus* small-winged (Gr. *miniós* small; Gr. *pterón* wing); from the very short phalanx of the third or longest finger.
fuliginosus: Mod. L. *fuliginosus* sooty (>L. *fuligo, fuliginis* soot).

Southeast Asia, China, South Asia, and Middle East. Afghanistan, India, Nepal, and Sri Lanka.

OEN Smoky Bat (nom. form), Hodgson's Long-winged Bat (nom. form)
VN Sinh. *Kiri Wawula, Dickpiya-vavula*; Tam. (SL) *Sinna Vaval*

375. *Miniopterus pusillus* Dobson, 1876
Small Long-fingered Bat

Miniopterus: Gr. *miniopterus* small-winged (Gr. *miniós* small; Gr. *pterón* wing); from the very short phalanx of the third or longest finger.
pusillus: L. *pusillus* tiny, very small.

Southeast Asia and South Asia. India and Nepal.

OEN Small Bent-winged Bat (nom. form), Nicobar Long-fingered Bat (nom. form)

376. *Miniopterus pallidus* (Thomas, 1905)
Pallid Long-fingered Bat

Miniopterus: Gr. *miniopterus* small-winged (Gr. *miniós* small; Gr. *pterón* wing); from the very short phalanx of the third or longest finger.
pallidus: L. *pallidus* pallid, pale.
Turkey, Middle East and South Asia (marginally). Afghanistan.

377. *Miniopterus magnater* Sanborn, 1931
Western Long-fingered Bat

Miniopterus: Gr. *miniopterus* small-winged (Gr. *miniós* small; Gr. *pterón* wing); from the very short phalanx of the third or longest finger.
magnater: Mod. L. *magnater* large (L. *magnus* great); in allusion to its large body and cranial size in comparison to *M. schreibersii*.
Southeast Asia and South Asia (marginally). One subspecies is known from the region.

Subspecies

- *macrodens*: Mod. L. *macrodens* (Gr. *makrodens*) large-toothed (Gr. *makros* long; L. *dens* tooth).

  Large-toothed Long-fingered Bat. India (Arunachal Pradesh).

OEN Large Bent-winged Bat (nom. form), Large Long-fingered Bat (nom. form)

Order Pholidota Weber, 1904
Family Manidae Gray, 1821

378. *Manis crassicaudata* Gray, 1827
Indian Pangolin

Manis: L. *manis* ghost; in allusion to the animal's nocturnal habit.
crassicaudata: Mod. L. *crassicaudata* heavy-tailed (L. *crassus* thick, heavy, dense; L. *caudatus* tailed, having a (long/short) tail (L. *cauda* tail)).

Endemic to South Asia; Bangladesh, Bhutan, India, Nepal, Pakistan, and Sri Lanka.

OEN Scaly Anteater (nom. form), 'Armadillo' (nom. form)
VN Hind. *Sillu, Sal, Salu, Sakun-khor, Bajar-Kit, Bajra Kapta*; Sans. *Vajra Kapta, Bajr-Kit*; Marw. *Shalma*; Beng. *Keyot-Mach, Kat-pohu*; Dakh. *Ban-Rohu*; Kol. *Armoi*; Mar. *Kaulimah, Kowli-manjra, Kassoli-manjur*; Mal. *Alangu, Enampechi*; Tam. *Alangu, Azhangu*; Tel. *Alawa, Polusu Pandi, Nela Chepa*; Kan. *Chippu Hundi*; Odi. *Bajrakapta*; Sinh. *Kaba-laya, Kaballewa*; Tam. (SL) *Alangu*

379. *Manis pentadactyla* Linnaeus, 1758
Chinese Pangolin

Manis: L. *manis* ghost; in allusion to the animal's nocturnal habit.
pentadactyla: Mod. L. *pentadactyla* five-fingered (L. *penta-* five- (in comp.); L. *dactylus* (>Gr. *daktulos*) finger or toe).

China, Southeast Asia, and South Asia. One subspecies is known from the region.

Subspecies

- *auritus*: L. *auritus* long-eared.
  Long-eared Chinese Pangolin. Bangladesh, Bhutan, India, and Nepal.

OEN Eared Pangolin (subsp. *auritus*), Sikkim Scaly Ant-eater (subsp. *auritus*)
VN Beng. *Keyot-Mach, Kat-pohu*

Order Carnivora Bowdich, 1821
Suborder Caniformia Kretzoi, 1938
Family Canidae Fischer, 1817

380. *Canis aureus* Linnaeus, 1758
Golden Jackal

Canis: L. *canis* Dog.
aureus: L. *aureus* golden (*aurum* gold; *-eus* having the quality of).
North Africa, Eastern Europe, Middle East, Arabian Peninsula, South Asia, and Southeast Asia. Two subspecies are known from the region.

Subspecies

- *indicus*: L. adj. *indicus* Indian, belonging to India (*-icus* belonging to, pertaining to (L./Gr.)).
  Indian Jackal. Afghanistan, Bangladesh, Bhutan, India, Nepal, and Pakistan.
- *naria*: From Sinh. *nariya*, a local name for the jackal.
  Wroughton's Jackal. Endemic to South Asia; India (southern) and Sri Lanka.

OEN Himalayan Golden Jackal (Subsp. *indicus*), Sri Lankan Jackal (Subsp. *naria*), Western India Jackal (J. Syn. *kola*), Ceylon Jackal (J. Syn. *lanka*)

VN Kash. *Gidah*, *Shial*, *Shal*; Pers. *Shigal*; Lepch. *Syal*; Pahari W. *Syal*; Mar. *Kola*, *Koila*; Hind. *Ghidar*, *Kola*, Dakh. *Kola*, *Shighal*; Kan. *Nari*; Tel. *Nakka*; Gond. *Nerka*; Khan. *Neru-Koela*; Wad. *Nakka*, *Tada Nakka*; Sinh. *Nariya*, *Hiwala*; Tam. (SL) *Narie*, *Kulla narie*

381. *Canis lupus* Linnaeus, 1758
Wolf

Canis: L. *canis* Dog.
lupus: L. *lupus* Wolf.
Most of North America, Europe, and Asia, south up to India and Central China. Two subspecies are known from the region.

Subspecies

- *pallipes*: Mod. L. *pallipes* pale-footed (L. *pallidus* pale; L. *pes* foot, >Gr. *pous* foot).
  Indian Wolf. Afghanistan, India, Nepal, and Pakistan.
- *chanco*: From Tib. *chanko*, a local name for the wolf.
  Tibetan Wolf. Himalayas in Afghanistan, India, and Pakistan.

OEN Red Wolf (Subsp. *chanco*), Gold Wolf (Subsp. *chanco*), Black Wolf (J. Syn. *niger*)

VN Kut. *Bagad*; Sind. *Bagyár*; Hind. *Bheriya*, *Gúng*, *Hondár*, *Nekra*, *Bighána*; Dakh. *Lángdá*; Gond. *Lángdá*; Kan. *Tola*; Tel. *Toralú*, *Todelu*; Kum. *Chankodi*; Lad. *Chanco*, *Chanku*; Tib. *Shanku*, *Changu*, *Hakpo-chanko*; Lep. *Sitim*; Bhot. *Phao*; Kash. *Ratnakin*

382. *Cuon alpinus* (Pallas, 1811)
Dhole

Cuon: Gr. *cyon* Dog.
alpinus: L. *alpinus* alpine, of the Alps.
China, Russia, Southeast Asia, Central Asia, and South Asia. Four subspecies are known from the region.

Subspecies

- *dukhunensis*: L. adj. *dukhunensis* meaning from Dukhun (=Deccan – Indian peninsular plateau), India, (*-ensis*, geographic, occurrence in (location, toponym) (L.)).
  Deccan Wild Dog. Endemic to India.
- *primaevus*: L. *primaevus* early in life, youthful (L. *primus* first; L. *aevum* an age).
  Himalayan Wild Dog. Himalayas in Bhutan, India, and Nepal.
- *laniger*: L. *laniger* woolly.
  Kashmir Wild Dog. India (Jammu and Kashmir).
- *adustus*: L. *adustus* burnt, fuliginous, soot-coloured.
  Burmese Wild Dog. Hill tracts in Bangladesh and India (northeastern).

OEN Asiatic Wild Dog (nom. form), Red Dog (nom. form), Indian Dhole (subsp. *dukhunensis*), Nilgiri Dhole (subsp. *dukhunensis*), Himalayan Dhole (Subsp. *primaevus*), Kashmir Dhole (Subsp. *laniger*), Burma Dhole (Subsp. *adustus*)

VN Dakh. *Jungli kutta*, *Sakki-sarai*; Hind. *Son-kutta*, *Rám-kutta*, *Jangli-kutta*, *Ban-kutta*, *Ban-kuka*; Mar. *Kolsun*, *Kolasna*, *Kolasra*, *Kolsa*; Gond. *Eram-naiko*; Ho *Tani*; Kol *Tani*; Tam. *Vatai-Karau*; Tel. *Reza-kútá*, *Adavi Kútá*, *Adavi Kuká*, *Rezu-kuká*; Kurg. *Ken-nai*, *Chen-nai*; Mal. *Shin-nai*, *Shen-nai*; Kash. *Rám-hun*, *Rán-hun*; Lad. *Siddaki*; Pahari W. *Bhaosa*, *Bhánsa*, *Bhuánsú*; Tib. *Hazí*, *Phará*, *Sidda-ki*; Bhut. *Paoho*; Lepch. *Suhu-tum*, *Sa-tum*

383. *Vulpes vulpes* (Linnaeus, 1758)

Red Fox

Vulpes: L. *vulpes* Fox.

vulpes: L. *vulpes* Fox.

Practically throughout the northern hemisphere. Three subspecies are known from the region.

Subspecies

- *griffithi*: Eponym after William Griffith (b. 1810–d. 1845), an English doctor, naturalist, and botanist, who made the first scientific collection of mammals in Afghanistan (*-i*, commemorating (dedication, eponym) (L.)).
  Griffith's Red Fox. Afghanistan and Pakistan.
- *montana*: L. *montana* of the mountains, mountain- (L. *mons*, *montis* mountain).
  Montane Red Fox. Himalayas in Bhutan, India, and Nepal.
- *pusilla*: L. *pusillus* tiny, very small.
  Little Red Fox. Semiarid regions of India and Pakistan.

OEN Hill Fox (Subsp. *montana*), Punjab Fox (Subsp. *pusilla*)
VN Hind. *Lomri*, *Lúmri*; Sind. *Lokri*; Balochi *Lombar*; Pers. *Rubah*; Kash. *Loh* (male), *Luh* (male), *Laash* (female); Nep. *Wamu*

384. *Vulpes corsac* (Linnaeus, 1768)
Corsac Fox

Vulpes: L. *vulpes* Fox.
corsac: Mod. L. *corsac* from Russ. *korsak* (>Kirghiz *karsak*), a native name for the species.

Central Asia, Middle East, China, Mongolia and South Asia (marginally); Afghanistan.

385. *Vulpes bengalensis* (Shaw, 1800)
Bengal Fox

Vulpes: L. *vulpes* Fox.
bengalensis: L. adj. *bengalensis* meaning from Bengal, India, (-*ensis*, geographic, occurrence in (location, toponym) (L.)).

Endemic to South Asia; Bangladesh, Bhutan, India, Nepal, and Pakistan.

OEN Indian Fox (nom. form), Indian Fox (J. Syn. *indicus*), Dukhun Fox (J. Syn. *kokree*), Fulvous-tailed Dog (J. Syn. *chrysurus*), Hodgson's Fox (J. Syn. *hodgsonii*)
VN Dakh. *Lomri*, *Nomri*; Hind. *Lúmri*, *Löm*, *Lokri*, *Lokeria*, *Kek-Siyal*; Bundel. *Kukhariya*; Bihar. *Khekar*, *Khikir*; Beng. *Khek-siyal*; Mar. *Kokri*; Gond. *Khekri*; Tel. *Konka-nakka*; *Gunta-nakka*, *Potti Nara*; Kan. *Konk*, *Kemp nari*, *Chandak-nari*

386. *Vulpes rueppelli* (Schinz, 1825)
Ruppell's Fox

Vulpes: L. *vulpes* Fox.
rueppelli: Eponym after Wilhelm Peter Edward Simon Rueppell (b. 1794–d. 1884), a German naturalist, worked in northeast Africa (-*i*, commemorating (dedication, eponym) (L.)).

Deserts of North Africa, Arabian Peninsula, Middle East, and South Asia. One subspecies is known from the region.

Subspecies

- *zarudnyi*: Eponym after Dr. N.A. Zarudny (b. 1859–d. 1919), curator of Tashkent Museum, Uzbekistan, and collected in Persia (-*i*, commemorating (dedication, eponym) (L.)).

  Zarudny's Fox. Afghanistan and Pakistan.

OEN Ruppell's Sand Fox (nom. form)

387. *Vulpes ferrilata* Hodgson, 1842
Tibetan Fox

Vulpes: L. *vulpes* Fox.

ferrilata: Mod. L. *ferrilata* iron-grey sided (L. *ferrum* iron; L. *latus* flank); in allusion to its colour on the lateral sides.

China and South Asia. India and Nepal (in the Himalayas).

OEN Tibetan Sand Fox (nom. form), Sand Fox (nom. form)
VN Balt. *Dē-dze*

388. *Vulpes cana* Blanford, 1877
Blanford's Fox

Vulpes: L. *vulpes* Fox.
cana: L. *cāna* hoary.

Arabian Peninsula, Middle East, and South Asia. Afghanistan and Pakistan.

VN Baloch. *Poh*; Pers. *Kúrba-shákál*

Family Mustelidae Fischer de Waldheim, 1817

389. *Aonyx cinerea* (Illiger, 1815)
Oriental Small-clawed Otter

Aonyx: Mod. L. *aonyx* without nail (Gr. *a* without; Gr. *ónyx* claw, nail); due to its very rudimentary claws.
cinerea: L. *cinereus* ash-grey, ash-coloured (L. *cinis*, *cineris* ashes).

Southeast Asia and South Asia. Two subspecies are known from the region.

Subspecies

- *concolor*: L. *concolor* uniform, similar in colour, plain.
  Rafinesque's Small-clawed Otter. Himalayas in Bangladesh, Bhutan, India, and Nepal.
- *nirnai*: From Kan. *nirnai*, a local name for the otter (*nir* water, *nai* dog).
  South Indian Small-clawed Otter. Endemic to India, Western Ghats of Karnataka, Kerala, and Tamil Nadu.

OEN Asian Small-clawed Otter (nom. form), Clawless Otter (nom. form), Himalayan Small-clawed Otter (J. Syn. *indigitatus*)
VN Bhut. *Chusam*; Lepch. *Suriam*

390. *Lutra lutra* (Linnaeus, 1758)
European Otter

Lutra: L. *lutra* otter.
lutra: L. *lutra* otter.

Europe, Mediterranean region, Middle East, Central Asia, China, Mongolia, Russia, South Asia and Southeast Asia. Four subspecies are known from the region.

Subspecies

- *nair*: From kan. *nair*, a local name for the otter (*nair* water dog).
  Indian Otter. Endemic to South Asia, India (southern states), and Sri Lanka.
- *aurobrunneus*: Mod. L. *aurobrunneus* gold-brown (L. *aurum* gold; Mod. L. *brunneus* brown); in allusion to its coat color – rich chestnut brown above and golden red below.
  Himalayan Otter. Endemic to South Asia; India (Himalayas in Uttarakhand and Himachal Pradesh) and Nepal (Himalayas in Far-Western Nepal).
- *monticola*: L. *monticola* mountain-dweller, mountaineer (L. *mons*, *montis* mountain; L. *-cola* inhabitant >L. *colere* to dwell).
  Hill Otter. Himalayas in Bangladesh, Bhutan, India, and Nepal.
- *kutab*: Mod. L. *kutab* possibly from Urdu *kutab*, local name for the pole-cat (Subsp. *Lutra lutra* Kashmir Otter).
  Kashmir Otter. Himalayas in Afghanistan, India, and Pakistan.

OEN Southern Indian Otter (Subsp. *nair*), Ceylon Otter (J. Syn. *ceylonica*), Northern Indian Otter (J. Syn. *indica*), Nepal Otter (J. Syn. *nepalensis*)

VN Kan. *Nir nai*; Dakh. *Pani Kutta*; Hind. *Ud*, *Hud*, *Pani kuta*, *Udbillao*, *Udni*, *Udbilli*; Mar. *Hud*, *Hada*, *Jalmanus*, *Jal-manjer*; Wad. Datwai Bekk; Sinh. *Diya-balla*; Tam. (SL) *Neer-nai*

391. *Lutrogale perspicillata* (I. Geoffroy Saint-Hilaire, 1826)
Smooth-coated Otter

Lutrogale: Mod. L. *lutrogale* otter-weasel, generic name combination (L. *lutra* otter; Gr. *gale* weasel).

perspicillata: Mod. L. *perspicillatus* spectacled (>L. *perspicillum* lens, spectacle >L. *perspicere* to see through).

Southeast Asia and South Asia. Besides the nominate subspecies, *L. p. perspicillata* Smooth-coated Otter from Bangladesh, Bhutan, India, and Nepal, another subspecies is known from the region.

Subspecies

- *sindica*: L. adj. *sindica* belonging to Sindh, Pakistan (*-ica* belonging to, pertaining to (L./Gr.)).
  Sindh Smooth-coated Otter. India and Pakistan.

OEN Terai Otter (J. Syn. *tarayensis*), Elliot's Otter (J. Syn. *ellioti*)

Subfamily Mustelinae Fischer, 1817

392. *Arctonyx collaris* F. Cuvier, 1825
Hog-Badger

Arctonyx: Mod. L. *arctonyx* having bearlike claws (Gr. *arktos* a bear; Gr. *ónyx* claw); owing to the long, slightly curved, blunt claws.

collaris: L. *collaris* of the neck, collar (L. *collum* neck).

China, Southeast Asia, and South Asia (marginally). Besides the nominate subspecies, *A. c. collaris* Hog-Badger from the Himalayas in Bhutan, India, and Nepal, another subspecies is known from the region.

Subspecies

- *consul*: L. *cōnsul* (from Old L. *consol*) annually elected highest-ranking official of the Roman empire; in allusion to its larger size in comparison to the nominate form.

  Large Hog-Badger. Bangladesh and India.

OEN Bear Pig (nom. form)
VN Hind. *Bhalu-sur*

393. *Martes foina* (Erxleben, 1777)
Beech Marten

Martes: L. *martes* marten.

foina: Mod. L. *foina* (Fr. *fouine*, marten from L. *fagina* pertaining to a beech tree) marten.

Europe, Middle East, Central Asia, China, Mongolia, and South Asia. Two subspecies are known from the region.

Subspecies

- *toufoeus*: Mod. L. *toufoeus* >Tib. *toufee*, a local name for this marten in Tibet, toufee pelts are highly prized in Tibet and China.
  Lhasa Beech Marten. India (Jammu and Kashmir).
- *intermedia*: L. *intermedius* intermediate (*cf.* Med. L. *intermediatus* intermediate).

  Himalayan Beech Marten. Himalayas in Afghanistan, India, Nepal, and Pakistan.

VN Tib. *Toufee*

394. *Martes flavigula* (Boddaert, 1785)
Yellow-throated Marten

Martes: L. *martes* marten.

flavigula: Mod. L. *flavigula* yellow-throated (L. *flavus* yellow; *gula* throat).

China, Southeast Asia, and South Asia. Himalayas in Afghanistan, Bhutan, India, Nepal, and Pakistan.

OEN Indian Marten (nom. form), Northern Indian Marten (nom. form), Golden-bellied Marten (J. Syn. *chyrsogaster*), Nepal Marten (J. Syn. *hardwickei*)
VN Bhut. *Huniah*, *Aniar*; Lepch. *Sakku*; Kum. *Tuturala*; Nep. *Mal Sampra*

395. *Martes gwatkinsii* Horsfield, 1851
Nilgiri Marten

Martes: L. *martes* marten.
gwatkinsii: Eponym after R. Gwatkins (dates not known), who collected in the Himalayas, but not in the Nilgiris (*-ii*, commemorating (dedication, eponym) (L.)). Wrong credit: the correct collector's name is Walter Elliot.
Endemic to India, Western Ghats of Karnataka, Kerala, and Tamil Nadu.

396. *Meles meles* (Linnaeus, 1758)
European Badger

Meles: L. *meles* badger.
meles: L. *meles* badger.
Europe, Middle East, and South Asia (marginally). Afghanistan.

397. *Mellivora capensis* (Schreber, 1776)
Honey Badger

Mellivora: L. *mellivora* honey-eating (L. *mel* honey; L. *voro* to devour), from the animal's favourite food.
capensis: L. adj. meaning from Cape of Good Hope, South Africa, (*-ensis*, geographic, occurrence in (location, toponym) (L.)).
Africa, Arabian Peninsula, Middle East, and South Asia. Two subspecies are known from the region.

Subspecies

- *indicus*: L. adj. *indicus* Indian, belonging to India (*-icus* belonging to, pertaining to (L./Gr.)).
  Indian Ratel. India and Pakistan.
- *inauritus*: L. *inauritus* without ears (L. *in-* not; L. *auritus* -eared).
  Nepal Ratel. India and Nepal (in the Himalayas).

OEN Indian Badger (subsp. *indicus*), Nepal Badger (subsp. *inauritus*)
VN Hind. *Biju*, *Bhajrubhal*; Tam. *Tavakaradi*; Tel. *Biyu-khawar*; Nep. *Oker* (Subsp. *inauritus*); Tib. *Tum-pha*

398. *Melogale moschata* (Gray, 1831)
Chinese Ferret-Badger

Melogale: Mod. L. *melogale* badger-weasel, generic name combination (L. *meles* badger; Gr. *gale* weasel).
moschata: Mod. L. *moschatus* musky (Gr. *moskhos* musk >Pers. *mušk*).
China, Southeast Asia, and South Asia. One subspecies is known from the region.

Subspecies

- *millsi*: Eponym after J. Phillip Mills (b. 1890–d. 1960), British Imperial Civil Service officer serving in Nagaland (*-i*, commemorating (dedication, eponym) (L.)).
  Nagaland Ferret-Badger. India (northeast).

399. ***Melogale personata*** **I. Geoffroy Saint-Hilaire, 1831**
Burmese Ferret-Badger

Melogale: Mod. L. *melogale* badger-weasel, generic name combination (L. *meles* badger; Gr. *gale* weasel).

personata: L. *personatus* masked (L. *persona* mask).

Southeast Asia and South Asia. One subspecies is known from the region.

Subspecies

- *nipalensis*: L. adj. *nipalensis* meaning from Nepal (*ni* holy, *pal* land), (-*ensis*, geographic, occurrence in (location, toponym) (L.)).
  Nepal Ferret-Badger. Himalayas of Bhutan, India, and Nepal.

400. ***Mustela erminea*** **Linnaeus, 1758**
Ermine

Mustela: L. *mustela* weasel.

erminea: L. *erminea* ermine.

Throughout the temperate and alpine regions of northern hemisphere. One subspecies is known from the region.

Subspecies

- *ferghanae*: After Ferghana, Kazakhstan, (-*ae*, geographic (location, toponym)).
  Ferghana Ermine. Himalayas of Afghanistan, India, and Pakistan.

401. ***Mustela nivalis*** **Linnaeus, 1758**
Least Weasel

Mustela: L. *mustela* weasel.

nivalis: L. *nivalis* snowy, snow-white (L. *nix*, *nivis* snow).

Throughout the temperate and alpine regions of the northern hemisphere. One subspecies is known from the region.

Subspecies

- *stoliczkana*: Eponym after Dr. Ferdinand Stoliczka (b. 1838–d. 1874), a British paleontologist who participated in Yarkand Missions.
  Stoliczka's Weasel. Afghanistan.

OEN Weasel (nom. form)

402. ***Mustela sibirica*** **Pallas, 1773**
Siberian Weasel

Mustela: L. *mustela* weasel.

sibirica: L. adj. *sibirica* belonging to Siberia, Asiatic Russia (-*ica* belonging to, pertaining to (L./Gr.)).

Russia, China, Southeast Asia (marginally), and South Asia (marginally). Three subspecies are known from the region.

Subspecies

- *subhemachalana*: Mod. L. *subhemachalana* sub-Himalayan (L. *sub* beneath, Hind. *himachal* in the lap of the Himalayas).

Himalayan Weasel. Himalayas in Bhutan, India, and Nepal.

- *hodgsoni*: Eponym after Brian Houghton Hodgson (b. 1801–d. 1894), a British civil servant, naturalist, and ethnologist stationed in Nepal (*-i*, commemorating (dedication, eponym) (L.)).
  Hodgson's Weasel. Himalayas in India.
- *canigula*: Mod. L. *canigula* hoary-necked, grey-necked (L. *canus* grey; L. *gula* throat).
  Hoary-necked Weasel. India and Nepal (in the Himalayas).

OEN Hodgson's Weasel (Subsp. *hodgsoni*), Tibetan Weasel (Subsp. *canigula*)
VN Bhut. *Zimiong*; Kash. *Kran*; Lepch. *Sang-king*

403. *Mustela altaica* Pallas, 1811
Mountain Weasel

Mustela: L. *mustela* weasel.
altaica: L. adj. *altaica* belonging to the Altai mountains, East Central Asia (*-ica* belonging to, pertaining to (L./Gr.)).
Russia, Mongolia, China, Central Asia, and South Asia. One subspecies is known from the region.

Subspecies

- *temon*: Mod. L. *temon* >Tib. *temon*, a local name for the weasel.
  Tibetan Mountain Weasel. Himalayas in Bhutan, India, Nepal, and Pakistan.

OEN Longstaff's Weasel (J. Syn. *longstaffi*)

404. *Mustela kathiah* Hodgson, 1835
Yellow-bellied Weasel

Mustela: L. *mustela* weasel.
kathiah: Mod. L. *kathiah* from native name *Káthiah nyúl* of the Yellow-bellied Weasel in Nepal and India in its range.
China, Southeast Asia, and South Asia. Besides the nominate subspecies, *M. k. kathiah* Yellow-billed Weasel, Himalayas in Bhutan, India, and Nepal, another subspecies is known from the region.

Subspecies

- *caporiaccoi*: Eponym after Ludovico di Caporiacco (b. 1901–d. 1951), an Italian arachnologist, (*-i*, commemorating (dedication, eponym) (L.)).
  Kashmir Yellow-bellied Weasel. Himalayas in India (Jammu and Kashmir) and Pakistan.

OEN Golden-bellied Weasel (J. Syn. *auriventer*)
VN Nep. *Káthiah nyúl, Kathiah nyal*

405. *Mustela strigidorsa* Gray, 1853
Back-striped Weasel

Mustela: L. *mustela* weasel.

strigidorsa: Mod. L. *strigidorsa* back-striped (L. *strigis* furrow, groove; L. *dorsum* back).

China (southern), Southeast Asia and South Asia. India and Nepal (in the Himalayas).

OEN Striped Weasel (nom. form)

406. *Vormela peregusna* (Guldenstaedt, 1770)
Marbled Polecat

Vormela: Mod. L. *vormela* >German *würmlein* meaning little worm, in allusion to its body shape.

peregusna: Mod. L. *peregusna* >Ukranian *pereguznya*, a name for the species.

Temperate regions of Europe (eastern), Middle East, Central Asia, China, and South Asia. Afghanistan and Pakistan.

Family Ailuridae Gray, 1843

407. *Ailurus fulgens* F. G. Cuvier, 1825
Red Panda

Ailurus: Gr. *aílouros* cat (later a weasel), so called due to its resemblance to a cat.

fulgens: L. *fulgens* glittering (L. *fulgere* to shine).

China, South Asia. Himalayas in Bhutan, India, and Nepal.

OEN Cat-bear (nom. form), Red Cat-bear (nom. form), Panda (nom. form), Nepalese Red Panda (J. Syn. *ochraceus*)
VN Bhut. *Wahdonka*; Lepch. *Sunnam*, *Suknam*; Nep. *Negalya-ponya*, *Wah*

Family Ursidae Fischer de Waldheim, 1817

408. *Helarctos malayanus* (Raffles, 1822)
Sun Bear

Helarctos: Mod. L. *helarctos* sun bear (Gr. *hel* the sun's heat; Gr. *arktos* a bear).

malayanus: L. *malayanus* belonging to Malay or Malaysia, (-*anus*, belonging to, pertaining to geographic (location, toponym) (L.)).

Southeast Asia and South Asia (marginally). Himalayas in India (northeast) and Bangladesh.

OEN Malayan Sun Bear (nom. form)

409. *Melursus ursinus* (Shaw, 1791)
Sloth Bear

Melursus: L. *melursus* honey-bear (L. *mel* honey; L. *ursus* a bear), from its fondness for honey.

ursinus: L. *ursinus* resembling a bear.

Endemic to South Asia. Besides the nominate subspecies, *M. u. ursinus* Indian Sloth Bear from Bangladesh, Bhutan, India, and Nepal, another subspecies is known from the region.

Subspecies

- *inornatus*: L. *inornatus* plain, unadorned.
  Sri Lankan Sloth Bear. Endemic to Sri Lanka.

OEN Indian Black Bear (Subsp. *ursinus*), Sri Lankan Black Bear (Subsp. *inornatus*), Ceylon Sloth Bear (Subsp. *inornatus*)

VN Hind. *Rínch, Rích, Adam-zád*; Beng. *Bhalúk*; Sans. *Riksha*; Mar. Aswal; Gond. *Yerid, Yedjal, Asal*; Kol *Banna, Bir Mendi*; Tel. *Elugu, Yelugu-bantu, Elagu-banti*; Tam. *Karadi*, Kan. *Kaddi, Karadi*; Mal. *Pani Karudi*; Sinh. *Walaha, Waelahinna*; Tam. (SL) *Karadee*

410. *Ursus arctos* Linnaeus, 1758

Brown Bear

Ursus: L. *ursus* a bear.

arctos: Gr. *arktos* a bear.

Temperate regions of the northern hemisphere, parts of Europe, China, and South Asia (marginally). One subspecies is known from the region.

Subspecies

- *isabellinus*: Mod. L. *isabellinus* fawn, greyish-yellow (>Fr. *Isabelle* or Spanish *Isabella*).
  Himalayan Brown Bear. Himalayas in Afghanistan, India, Nepal, and Pakistan.

OEN 'Snow Bear', 'Red Bear'

VN Hind. *Barf-ka-rinch, Barf-ka-reetch, Safed-bhalu, Siala-reech*; Kash. *Háput, Harput, Brabu*; Balt. *Drengmo*; Lad. *Drin-mor*; Nep. *Dúb*; Tib. *Tom Khaina*

411. *Ursus thibetanus* (G. Cuvier, 1823)

Asian Black Bear

Ursus: L. *ursus* bear.

thibetanus: Belonging to Tibet, toponym after Thibet (= Tibet), (*-anus*, belonging to, pertaining to (L.) geographic (location, toponym)).

China, Japan, Korea (North and South), Southeast Asia, South Asia, and Middle East (marginally). Besides the nominate subspecies, *U. t. thibetanus* Asian Black Bear from the Himalayas in Bangladesh, Bhutan, India (Eastern Himalayas), and Nepal, two more subspecies are known from the region.

Subspecies

- *gedrosianus*: Mod. L. *gedrosianus* belonging to Gedrosia (hellenized name of an area that corresponds to today's Balochistan), (*-anus*, belonging to, pertaining to (L.) geographic (location, toponym)).

Balochistan Black Bear. Pakistan and Iran.

- *laniger*: L. *laniger* woolly.
  Himalayan Black Bear. Endemic to South Asia, in the Himalayas of Afghanistan, India (Western Himalayas) and Pakistan.

OEN Asiatic Black Bear (nom. form), Burma Black Bear (nom. form), Pakistan Black Bear (Subsp. *gedrosianus*), Long-haired Black Bear (Subsp. *laniger*)

VN Hind. *Rínch, Rích, Reench, Reech, Bhalu, Kala Bhalu, Kala Reech*; Kash. *Haput*; Nep. *Sanár, Hingbong*; Beng. *Bhalak*; Bhut. *Dom, Thom*; Lepch. *Sona*; Limbú *Mágyen*; Daphla *Sulum*; Abor *Situm*; Garo *Mapol*; Kachari *Muphur, Musu-bhurma*; Kúki *Vumpi*; Mani. *Sawam*; Naga *Húghúm, Thágua, Thega, Chúp, Seván, Sápá*

Family Felidae Fischer de Waldheim, 1817

412. *Acinonyx jubatus* (Griffith, 1821)
Cheetah

Acinonyx: Mod. L. *acinonyx* pointed-claw Gr. *ákaina* thorn, prick (Gr. *ákis, ákidos* point (>*acantha* thorn or pick); Gr. *ónyx* claw); from the non-retractile, pointed claws.
jubatus: New L. *jubatus* crested, having a mane or crest (L. *iuba* mane, crest) (-*ātus* provided with or having (L.)).

Parts of Africa and Middle East. Extinct from South Asia, historically known from Afghanistan, India, and Pakistan. South Asian subspecies bears the name *venaticus* (from L. *venaticus* hunting-, of hunting (L. *venatus* hunting, the chase, or *venator* hunter). Indian Cheetah was often referred to as 'Hunting Leopard'.

OEN Hunting Leopard (Subsp. *venaticus*)

VN Beng. *Tendua-bāg*; Hind. *Yuz, Laggar, Cheeta*; Kan. *Chircha, Sivungi*; Dakh. *Chita*; Tel. *Chita puli*

413. *Caracal caracal* (Schreber, 1776)
Caracal

Caracal: Mod. L. *caracal* from Turkish *karakulak* (*kara* black, *kulak* ear), a native name of the species; in allusion to black-backed tufted ears.
caracal: Mod. L. *caracal* from Turkish *karakulak* (*kara* black, *kulak* ear), a native name of the species; in allusion to black-backed tufted ears.

Parts of Africa, Arabian Peninsula, Middle East, and South Asia (only in the arid and semiarid regions). One subspecies is known from the region.

Subspecies

- *schmitzi*: Eponym after Ernst Johann Schmitz (b. 1845–d. 1922), a German naturalist, ornithologist, and entomologist who collected the type specimens, (-*i*, commemorating (dedication, eponym) (L.)).

Schmitz's Caracal. Arid and semi-arid regions of Afghanistan, India, and Pakistan.

OEN Red Lynx (nom. form), Bengal Caracal (J. Syn. *bengalensis*)
VN Turk. *Karakal*, Pers. *Siyeh Gush*; Hind. *Siah Gosh*; Lad. *Ech*; Balt. *Tsogde*

414. *Felis chaus* Schreber, 1777
Jungle Cat

Felis: L. *felis* cat (probably from root *fe*, to produce, bear young).
chaus: Apparently from common name *chaus* in Hindi.

South Asia, Southeast Asia, Middle East, Egypt, and China. Five subspecies are known from the region.

Subspecies

- *affinis*: L. *affinis* related, allied.
  Himalayan Jungle Cat. Bhutan, India (in the Himalayas and northern parts) and Nepal.
- *kutas*: Mod. L. *kutas*, possibly from the local name for this animal.
  Bengal Jungle Cat. Endemic to South Asia; Bangladesh and India (central and western parts).
- *maimanah*: Toponym after Maimana, Afghanistan.
  Maimanah Jungle Cat. Endemic to Afghanistan.
- *prateri*: Eponym after Stanley Henry Prater (b. 1890–d. 1960), a British naturalist in India, who was the curator of mammal collections at Bombay Natural History Society and Prince of Wales Museum of Western India, Bombay (*-i*, commemorating (dedication, eponym) (L.)).
  Indian Jungle Cat. Endemic to South Asia; arid and semiarid regions of India and Pakistan.
- *kelaarti*: Eponym after Lieutenant Colonel Edward Frederick Kelaart (b. 1819–d. 1860), a Ceylonese-born physician and naturalist in Sri Lanka, author of *Prodromus fauna Zeylanica*, the first description of Ceylonese fauna (*-i*, commemorating (dedication, eponym) (L.)).
  Kelaart's Jungle Cat. Endemic to South Asia; India and Sri Lanka.

OEN Chinese Jungle Cat (Subsp. *affinis*), Ceylon Jungle Cat (Subsp. *kelaarti*), Sri Lankan Jungle Cat (Subsp. *kelaarti*), Red-eared Jungle Cat (J. Syn. *erythrotus*), Sooty Jungle Cat (J. Syn. *nigriscens*), Jacquemont's Jungle Cat (J. Syn. *jacquemonti*)
VN Beng. *Ban-beral*, *Katas*; Kut. *Jhang-Meno*; Malto *Berka*; Hind. *Khatas*, *Jangli Billi*, *Bhirka*; Kathi. *Mungra*; Kan. *Kadu-bekku*, *Mantbekku*, *Adaribekku*; Wad. *Kadabek*, *Karbek*, *Bellabek*; Mar. *Jungli Mámar*, *Baul*, *Bháoga*; Pard. *Burakatchki*; Kurg. *Kebbali*; Dakh. *Jungli-billi*; Tel. *Junka Pilli*, *Jinka Pilli*, *Jangam Pilli*, *Jangam-billi*; Mal. *Cherru puli*; Sinh. *Wal-balala*, *Handun Diviya*; Tam. (SL) *Kadu-poona*, *Kardup-poonai*

415. *Felis silvestris* Schreber, 1777
Wild Cat

Felis: L. *felis* cat (probably from root *fe*, to produce, bear young).
silvestris: Mod. L. *silvestris* of or pertaining to wood or forest, wild, living in forests (L. *silva* forest, wooded).

Africa, Mediterranean region, Europe (parts), Middle East, Arabian Peninsula (parts), China, and South Asia. One subspecies is known from the region.

Subspecies

- *ornata*; L. *ornatus* ornate, adorned (*ornare* to adorn).
  Asian Wild Cat. Afghanistan, Pakistan and India.

OEN Indian Desert wild Cat (Subsp. *ornata*), Spotted Wild Cat (J. Syn. *torquata*), Servaline Chaus (J. Syn. *servalina*)

416. *Felis margarita* Loche, 1858
Sand Cat

Felis: L. *felis* cat (probably from root *fe*, to produce, bear young).
margarita: Eponym after Jean Auguste Margueritte (b. 1823–d. 1870), a French general, the leader of the expedition during which Victor Loche (b. 1806–d. 1863) found this animal.

Africa, Middle East, Arabian Peninsula (parts) and South Asia. One subspecies is known from the region.

Subspecies

- *scheffeli*: Eponym after Walter Scheffel (dates not known), a fellow colleague of Prof. Helmut Hemmer, University of Mainz, Germany.
  Pakistan Sand Cat. Endemic to Pakistan.

417. *Lynx lynx* (Linnaeus, 1758)
European Lynx

Lynx: Gr. *lýgé* lynx, probably from its bright eyes (>Gr. root *lyk* eye or *lýkhnos* lamp).
lynx: Gr. *lýgé* lynx, probably from its bright eyes (>Gr. root *lyk* eye or *lýkhnos* lamp).

Temperate regions of Europe and Asia, China, Middle East (parts), Central Europe (parts), and South Asia (in the Himalayas). One subspecies is known from the region.

Subspecies

- *isabellinus*: Mod. L. *isabellinus* fawn, greyish-yellow (>Fr. *Isabelle* or Spanish *Isabella*).
  Tibetan Lynx. Himalayas in Afghanistan, India, Nepal, and Pakistan.

OEN Turkestan Lynx (Subsp. *isabellinus*)
VN Kash. *Patsalan*; Lah. *Phiauku*; Lad. *Eh*, *Ech*; Balt. *Tsogde*

418. ***Otocolobus manul*** **(Pallas, 1776)**
Pallas's Cat

Otocolobus: Mod. L. *otocolobus* mutilated-ears (Gr. *otus* ear; Gr. *kolobos* mutilated); in allusion to its short ears.
manul: Mod. L. *manul* > Mongol. *manuul*, the native name of the species.

Mongolia, Russia, China, South Asia and Iran. Two subspecies are known from the region.

Subspecies

- *ferruginea*: L. *ferruginea* rusty-coloured, ferruginous (L. *ferrugo*, *ferruginis* iron rust).

  Rufous Pallas's Cat. Afghanistan and Pakistan.
- *nigripecta*: Mod. L. *nigripecta* black-breasted (L. *nigri* (>L. *niger*) black; L. *pectus* breast).

  Himalayan Pallas's Cat. In the Himalayas of India (Western Himalayas) and Nepal.

OEN Black-chested Wild Cat (nom. form), Middle Asian Pallas's Cat (Subsp. *ferruginea*), Satunin's Cat (J. Syn. *satuni*)

419. ***Pardofelis temminckii*** **(Vigors and Horsfield, 1827)**
Asiatic Golden Cat

Pardofelis: Mod. L. *pardofelis* leopard-cat (L. *pardus* (Gr. *párdos*) leopard; L. *felis* cat), generic combination (*Pardus* and *Felis*).
temminckii: Eponym after Dutch ornithologist Coenraad Jacob Temminck (b. 1778–d. 1858), (*-ii*, commemorating (dedication, eponym) (L.)).

Southeast Asia, China, and South Asia. Himalayas in Bangladesh, Bhutan, India, and Nepal.

OEN Temminck's Golden Cat (nom. form), Eastern Golden Cat (nom. form), Moormi Cat (J. Syn. *moormensis*), Fire Cat (J. Syn. *aurata*)

420. ***Pardofelis marmorata*** **(Martin, 1837)**
Marbled Cat

Pardofelis: Mod. L. *pardofelis* leopard-cat (L. *pardus* (Gr. *párdos*) leopard; L. *felis* cat), generic combination (*Pardus* and *Felis*).
marmorata: L. *marmoratus* marbled (L. *marmor*, *marmoris* marble).

Southeast Asia, China, and South Asia. One subspecies is known from the region.

Subspecies

- *charltonii*: Eponym after Andrew Charlton (dates not known), of Liskard, Cheshire, who presented the type specimen to British Museum (Natural History), (*-ii*, commemorating (dedication, eponym) (L.)).

Nepalese Marbled Cat. Himalayas in Bhutan, India (northeast), and Nepal.

OEN Marbled Tiger Cat (nom. form), Sikkim Marbled Cat (J. Syn. *ogilbii*)
VN Lepch. *Dosal*; Bhut. *Sikmar*

421. *Prionailurus bengalensis* (Kerr, 1792)
Leopard Cat

Prionailurus: Mod. L. *prionailurus* cat bearing saw-tooth (Gr. *priòn* saw; Gr. *aílouros* cat (later a weasel)).
bengalensis: L. adj. *bengalensis* meaning from Bengal, India (*-ensis*, geographic, occurrence in (location, toponym) (L.)).

China, Southeast Asia, and South Asia. Besides the nominate subspecies, *P. b. bengalensis* Common Leopard Cat from Bangladesh and India, two more subspecies are known from the region.

Subspecies

- *horsfieldii*: Eponym after Dr. Thomas Horsfield (b. 1773–d. 1859), an American physician and naturalist, later curator of the East India Company Museum in London (*-ii*, commemorating (dedication, eponym) (L.)).

  Horsfield's Leopard Cat. Himalayas in Bhutan, India (northeast), and Nepal.
- *trevelyani*: Eponym after Col. W.R.F. Trevelyan (dates not known), a British naturalist and sportsman, who collected the type specimens (*-i*, commemorating (dedication, eponym) (L.)).

  Kashmir Leopard Cat. Himalayas in Afghanistan, India (western), and Pakistan.

OEN Lesser Leopard Cat (J. Syn. *jerdoni*)
VN Hind. *Chita Billi*, *Bandaris* (male), *Biralu* (female); Beng. *Ban Biral*; Kurg. *Borka*; Mar. *Wagati*; Kan. *Wagati*

422. *Prionailurus rubiginosus* (I. Geoffroy, 1831)
Rusty-spotted Cat

Prionailurus: Mod. L. *prionailurus* cat bearing saw-tooth (Gr. *priòn* saw; Gr. *aílouros* cat (later a weasel)).
rubiginosus: L. *rubiginosus* rusty, reddish (L. *rubigo*, *rubiginis* rust).

Endemic to South Asia. Besides the nominate subspecies, *P. r. rubiginosus* Indian Rusty-spotted Cat from India, another subspecies is known from the region.

Subspecies

- *phillipsi*: Eponym after Dr. William Watt Addison Phillips (b. 1892–d. 1981), a naturalist in Sri Lanka, author of *Manual of the Mammals of Sri Lanka* (*-i*, commemorating (dedication, eponym) (L.)).

  Sri Lankan Rusty-spotted Cat. Endemic to Sri Lanka.

OEN Ceylon Rusty-spotted Cat (Subsp. *phillipsi*), Small Jungle Cat (Subsp. *phillipsi*)

VN Tam. *Namali Pilli*; Wad. *Ark-philli*; Kan. *Kiraba-bekku*; Sinh. *Wal-balala, Handun Diviya, Kula Diya, Kola Diviya, Balal Diviya*; Tam. (SL) *Kadu-poona, Verewa puni, Kardup-poonai*

423. *Prionailurus viverrinus* (Bennett, 1833)
Fishing Cat

Prionailurus: Mod. L. *prionailurus* cat bearing saw-tooth (Gr. *priòn* saw; Gr. *aílouros* cat (later a weasel)).

viverrinus: Mod. L. *viverrinus* civet-like (L. *viverra* civet; *-inus* pertaining to (L.)).

Southeast Asia and South Asia. Bangladesh, India, Nepal, Pakistan, and Sri Lanka.

OEN Tiger-cat, Large Tiger-cat, Warwick's Cat (J. Syn. *himalayanus*), Hodgson's Fishing Cat (J. Syn. *viverriceps*), Bennett's Fishing Cat (J. Syn. *bennettii*)

VN Beng. *Bag-dasha*; Hind. *Mach-bagrul, Bag-dasha*; Sinh. *Handun Diviya, Kola Diviya*; Tam. (SL) *Koddy-puli*

424. *Neofelis nebulosa* (Griffith, 1821)
Clouded Leopard

Neofelis: Mod. L. *neofelis* new cat (Gr. *neos* new; L. *felis* cat; probably from root *fe*, to produce, bear young).

nebulosa: L. *nebulosus* misty, cloudy (L. *nebula* cloud, mist).

China, Southeast Asia, and South Asia. One subspecies is known from the region.

Subspecies

- *macrosceloides*: Probably after Gr. adj. *macroscelous* long-legged (*-oides* >Gr. *eidos*, apparent shape).

  Tibetan Clouded Leopard. Himalayas in Bangladesh, Bhutan, India (northeast), and Nepal.

VN Lepch. *Pungmar, Tungmar, Satchuk*; Limbú *Zik*; Bhut. *Kung, Zik*; Nep. *Lamchita*

425. *Panthera leo* (Linnaeus, 1758)
Lion

Panthera: L. *panthera* (Gr. *pánthēr*) panther.

leo: L. *leo* lion (>Gr. *léòn*).

Africa and South Asia. One subspecies is known from the region.

Subspecies

- *persica*: L. adj. *persica* Persian, belonging to Persia (*-ica* belonging to, pertaining to (L./Gr.)).

  Asiatic Lion. Endemic to India (Gir in Gujarat).

OEN Bengal Lion (J. Syn. *bengalensis*), Asiatic Lion (J. Syn. *asiaticus*), Gujerat Lion (J. Syn. *goojratensis*), Indian Lion (J. Syn. *goojratensis*, J. Syn. *indicus*), Maneless Lion (J. Syn. *goojratensis*)

VN Hind. *Sher, Babar-sher, Singh*; Sind. *Sher, Babar-sher*; Guj. *Untia-bágh*; Kathi. *Sawach*; Beng. *Shingal*; Kash. *Süh* or *Suh* (male), *Siming* (female); Brah. *Rastar*

426. *Panthera pardus* (Linnaeus, 1758)
Leopard

Panthera: L. *panthera* (Gr. *pánthēr*) panther.
pardus: L. *pardus* (Gr. *párdos*) leopard.

Africa, Arabian Peninsula, Middle East, China, Southeast Asia, and South Asia. Three subspecies are known from the region.

Subspecies

- *saxicolor*: Mod. L. *saxicolor* stone-coloured (>L. *saxum* stone, *color* colour).
  Persian Leopard. Afghanistan (Southwest).
- *fusca*: L. *fuscus* brown, dusky.
  Indian Leopard. Afghanistan, Bangladesh, Bhutan, India, Nepal, and Pakistan.
- *kotiya*: Mod. L. *kotiya* >Sinh. *kotiya* (meaning tiger) the name of the leopard in Sri Lanka.
  Sri Lankan Leopard. Endemic to Sri Lanka.

OEN North Persian Leopard (Subsp. *saxicolor*), Central Asian Leopard (Subsp. *saxicolor*), Caucasian Leopard (Subsp. *saxicolor*), West Asian Leopard (Subsp. *saxicolor*), Long-tailed Leopard (J. Syn. *longicaudata*), Tibetan Leopard (J. Syn. *perniger*), Ni Boer Leopard (J. Syn. *perniger*), Nepal Leopard (J. Syn. *perniger*), Sind Leopard (J. Syn. *sindica*), Baluchistan Leopard (J. Syn. *sindica*), Shun Tak Leopard (J. Syn. *sindica*), Kashmir Leopard (J. Syn. *millardi*)

VN Hind. *Tendwa, Chita, Sona-chita, Chita-bágh, Adnára*; Pers. *Palang*; Baloch. *Diho*; Kash. *Súh*; Bundel. *Tidua, Srighas*; Dakh. *Gorbacha, Borbacha, Tenduwa, Bibla*; Mar. *Karda, Asnea, Singhal, Bibia-bágh*; Kan. *Honiga, Kerkal*; Kol Teon-Kula; Malto Jerkos; Gond. Burkál, Gordág; Korku *Sonora*; Tam. *Chiru-thai*; tel. *Chinna-puli*; Mal. *Puli*; Pahari W. *Bai-hira, Tahir-hé, Goral- hé, Ghor- hé*; Tib. *Sik*; Lepch. *Syik, Siyak, Sejjak*; Mani. *Kajengla*; Kúki *Misi patrai, Kam-kei*; Naga *Hurrea Kon, Morrh, Rusa, Tekhu Khuia, Kekhi*; Sinh. *Kotiya, Diviya* (male), *Dividena* (female); Tam. (SL) *Puli, Sarrugu-puli, Sirruthai-pulie*

427. *Panthera tigris* (Linnaeus, 1758)
Tiger

Panthera: L. *panthera* (Gr. *pánthēr*) panther.
tigris: L. *tigris* (Gr. *tígris*) tiger.

Southeast Asia, South Asia, China (northeastern) and Siberia. Besides the nominate subspecies, *P. t. tigris* Bengal Tiger, from Bangladesh, Bhutan, India, and Nepal, another subspecies, *P. t. virgata* Caspian Tiger, now extinct, was historically reported from Afghanistan.

OEN Indian Tiger (nom. form)

VN Hind. *Bágh* (male), *Bághni* (female), *Sher* (male), *Sherni* (female); Hind. (in Central India) *Náhar*, *Sela-vágh*; Punj. *Babr*; Balochi *Mazar*; Sind. *Shinh*; Kash. *Padar-suh*; Mar. *Patayat-bágh*, *Waház*; Beng. *Go-vágh*; Malto *Tut*, *Sad*; Kol. *Garúmkúla*; Kurukh *Lákhra*; Kond. *Krodi*; Sant. *Kula*; Ho *Kula*; Korku *Kula*; Gond. *Púli*; Tam. *Púli-redda-púli*, *Peram-pilli*, *Púli*; Tel. *Pedda-púlli*, *Púli*; Mal. *Perian-púli*, *Kúdua*, *Púli*; Kan. *Huli*; Kurg. *Nári*; Dakh. *Bagh*, *Patayat Bag*; Toda *Pirri*, *Bürsh*; Tib. *Tag*, *Tagh*; Bhut. *Túkt*, *Túk*; Lepch. *Sathong*; Limbú *Keh-va*; Garo *Matsa*; Khasi *Kla*; Naga *Sa*, *Ragdi*, *Tekhu*, *Khudi*; Kúki *Humpi*; Adi *Sumyo*; Mani. *Kei*; Kachari *Misi*

428. *Panthera uncia* Schreber, 1775
Snow Leopard

Panthera: L. *panthera* (Gr. *pánthēr*) panther.
uncia: L. *uncia* >Fr. *once* (originally used for lynx, occasionally for other similar-sized cats).

Mongolia, China, Central Asia (parts), and South Asia. In the Himalayas of Afghanistan, Bhutan, India, Nepal, and Pakistan.

OEN Ounce

VN Bhut. *Ikar*, *Zig*, *Sachak*, *Sâh*; Lepch. *Pah-le*; Pahari W. *Bharal hé*, *Burhel Haye*; Kin. *Thurwagh*; Tib. *Iker*, *Stian*, *Safed Cheetah*

Family Prionodontidae Pocock, 1933

429. *Prionodon pardicolor* Hodgson, 1842
Spotted Linsang

Prionodon: Mod. L. *prionodon* saw-toothed (Gr. *príòn* saw; Gr. *ódôn* tooth).
pardicolor: Mod. L. *pardicolor* leopard-colored (L. *pardi*, *pardus* leopard; L. *color* colour).

China, Southeast Asia, and South Asia. In the Himalayas of Bhutan, India, and Nepal.

OEN Tiger-Civet (nom. form), Nepal Lingsang (nom. form)

VN Bhut. *Zik-chúm*; Lepch. *Súliyú*, *Silu*

Family Hyaenidae Gray, 1821

430. *Hyaena hyaena* (Linnaeus, 1758)
Striped Hyaena

Hyaena: Gr. *ÿaina* hyena.
hyaena: Gr. *ÿaina* hyena.

Africa (central and northern), Arabian Peninsula, Middle East, and South Asia. Afghanistan, India, Nepal, and Pakistan.

OEN Indian Striped Hyaena (J. Syn. *indica*)
VN Hind. *Lakar-bagha*, *Lakar-bágh*, *Lakra*, *Jhirak*, *Hondar*, *Harvágh*, *Taras*; Beng. *Harvágh*; Mar. *Taras*; Sind. *Cherak*, *Taras*; Balochi *Aptan*; Gond. *Renhra*; Dakh. *Taras*; Ho *Hebar Kula*; Kol *Hebar Kula*; Malto *Derko Tud*; Korku *Dhopre*; Kan. *Kirba*, *Kutt-Kirba*; Tel. *Dúmul-gúndu*, *Korna-gúndu*; Tam. *Kaluthai-Korachi*

Family Herpestidae Bonaparte, 1845

431. *Urva edwardsii* (E. Geoffroy Saint-Hilaire, 1818)
Indian Grey Mongoose

Urva: Mod. L. *urva* >Nep. *arva*, the name for the crab-eating mongoose.
edwardsii: Eponym after Henri Milne-Edwards (b. 1800–d. 1885), French zoologist (*-ii*, commemorating (dedication, eponym) (L.)).

South Asia, Iran, Saudi Arabia. Besides the nominate subspecies, *H. e. edwardsii* South Indian Grey Mongoose, endemic to India (southern parts), four subspecies are known from the region.

Subspecies

- *nyula*: Mod. L. *nyula* >Hind. *nyul* the name of the mongoose in Gangetic plains in India.
  North Indian Grey Mongoose. Endemic to South Asia; Bangladesh, Bhutan, India, and Nepal.
- *ferrugineus*: L. *ferrugineus* rusty-coloured, ferruginous (*ferrugo*, *ferruginis* iron rust).
  Desert Grey Mongoose. Endemic to South Asia; India and Pakistan.
  montanus: L. *montanus* of the mountains, mountain- (L. *mons*, *montis* mountain).
  Montane Grey Mongoose. Endemic to Pakistan.
- *lanka*: Mod. L. *lanka* meaning belonging to or from Sri Lanka.
  Sri Lankan Grey Mongoose. Endemic to Sri Lanka.

OEN Nyula (Subsp. *nyula*), Blanford's Indian Mongoose (Subsp. *ferrigineus*), Edward's Mongoose (nom. form), Silvery Mongoose (Subsp. *lanka*), Anderson's Grey Mongoose (J. Syn. *andersoni*), Carnatic Mongoose (J. Syn. *carnaticus*), Elliot's Mongoose (J. Syn. *ellioti*), Madras Mongoose (J. Syn. *griseus*), Pale Grey Mongoose (J. Syn. *pallidus*), Pondicherry Mongoose (J. Syn. *ponticeriana*)

VN Hind. *Newal, Newala, Nyul, Newar, Dhor, Rasu*; Beng. *Baji, Biji*; Mar. *Mangús*; Dakh. *Mangús*; Kan. *Mungali, Mungili*; Tel. *Yentawa, Mungisa*; Wad. *Mungsi, Antur*; Pard. *Mungsi, Antur*; Kurg. *Kera, Hon-Kera*; Kat. *Nurlia*; Gond. *Koral*; Ho *Binquidaro*; Kol *Sarambumbui*; Sinh. *Alu Mugatiya*; Tam. (SL) *Kirri, Kirri-pulle, Keerie*

432. *Urva vitticollis* Bennett, 1835
Striped-necked Mongoose

Urva: Mod. L. *urva* >Nep. *arva*, the name for the crab-eating mongoose. *vitticollis*: L. *vitticollis* striped-necked (L. *vitta* band, ribbon; Mod. L. *-collis* -throated (>L. *collum* neck)).

Endemic to South Asia. Besides the nominate subspecies, *H. v. vitticollis* Striped-necked Mongoose, from India (southern Western Ghats) and Sri Lanka, another subspecies is known from the region.

Subspecies

- *inornatus*: L. *inornatus* plain, unadorned.

  Pocock's Striped-necked Mongoose. Endemic to India; Karnataka (in the Western Ghats).

OEN Stripe-necked Mongoose (Subsp. *vitticollis*), Elliot's Mongoose (Subsp. *vitticollis*), Badger Mongoose (Subsp. *vitticollis*), Pocock's Striped-necked Mongoose (Subsp. *inornatus*), Rusty Striped-necked Mongoose (J. Syn. *rubiginosus*)

VN Kurg. *Quoki-Balu, Kati-Kera*; Sinh. *Maha-mugatiya, Gal-mugatiya, Locu-mugatiya*; Tam. (SL) *Malam-keerie*

433. *Urva auropunctatus* (Hodgson, 1836)
Small Indian Mongoose

Urva: Mod. L. *urva* >Nep. *arva*, the name for the crab-eating mongoose. *auropunctatus*: Mod. L. *auropunctatus* gold-spotted (L. *aurum* gold; Mod. L. *punctatus* spotted (>L. *punctum* spot)); 'freckled with golden-yellow' (Hodgson, 1836).

South Asia, Iran, China and Southeast Asia. Besides the nominate subspecies, *H. a. auropunctatus* Small Indian Mongoose, endemic to South Asia from Bangladesh, Bhutan, India (northern parts), and Nepal, one subspecies is known from the region.

Subspecies

- *pallipes*: Mod. L. *pallipes* pale-footed (L. *pallidus* pale, *pes* foot, >Gr. *pous* foot).

  Pale-footed Small Mongoose. Afghanistan, India, and Pakistan.

OEN Gold-spotted Mongoose (nom. form)
VN Kash. *Núl*; Pers. *Mush-i-Khourma*; Arab. *Jeraydee ma'l Nak-hala*, *Abu alarrais*

434. *Urva urva* (Hodgson, 1836)
Crab-eating Mongoose

Urva: Mod. L. *urva* >Nep. *arva*, the name for the crab-eating mongoose.
urva: Mod. L. *urva* >Nep. *arva*, the name for the crab-eating mongoose.

Southeast Asia, China (southern) and South Asia. Bangladesh, Bhutan, India, and Nepal.

OEN Brown Crab Mongoose (J. Syn. *fusca*), Crab Mongoose (J. Syn. *cancrivora*)
VN Nep. *Arva*

435. *Urva smithii* Gray, 1837
Ruddy Mongoose

Urva: Mod. L. *urva* >Nep. *arva*, the name for the crab-eating mongoose.
smithii: Eponym after Charles Hamilton Smith (b. 1776–d. 1859), an English naturalist and soldier (*-ii*, commemorating (dedication, eponym) (L.)).

Endemic to South Asia. Besides the nominate subspecies, *H. s. smithii* Indian Ruddy Mongoose, endemic to India, two subspecies are known from the region.

Subspecies

- *thysanurus*: Mod. L. *thysanurus* tufted-tail (Gr. *thusanos* tassel, fringe, tuft; Gr. *-ouros* -tailed (Gr. *oúrá* tail)).
  Kashmir Ruddy Mongoose. Endemic to India (Jammu and Kashmir).
- *zeylanius*: Mod. L. *zeylanius* belonging to Ceylon (= Sri Lanka), (*-ius* belonging to, pertaining to (L./Gr.)).
  Ceylon Ruddy Mongoose. Endemic to Sri Lanka.

OEN Smith's Mongoose (Subsp. *smithii*), Red Mongoose (Subsp. *zeylanius*)
VN Tel. *Konda Yentawa*; Sinh. *Hotamba*, *Rath Mugatiya*, *Hothambuwa*; Tam. (SL) *Keerie*, *Seng-keerie*, *Raja-keerie*

436. *Urva fuscus* Waterhouse, 1838
Indian Brown Mongoose

Urva: Mod. L. *urva* >Nep. *arva*, the name for the crab-eating mongoose.
fuscus: L. *fuscus* brown, dusky.

Endemic to South Asia. Besides the nominate Indian endemic subspecies, *H. f. fuscus* Indian Brown Mongoose, from south India in the Western Ghats, four subspecies are known from the region.

Subspecies

- *flavidens*: Mod. L. *flavidens* yellow-toothed (L. *flavus* yellow; *dens* tooth).
  Ceylon Wetzone Brown Mongoose. Endemic to Sri Lanka.

- *maccarthiae*: Eponym after Mrs. MacCarthy (dates not known), a naturalist and collector, wife of Sir Charles J. MacCarthy, a British Governor-General of Sri Lanka (*-ae*, commemorating (dedication, eponym) (L.)).
  Northern Ceylon Brown Mongoose. Endemic to Sri Lanka.
- *siccatus*: L. *sicca* dry place, dry land.
  Ceylon Dryzone Brown Mongoose. Endemic to Sri Lanka.
- *rubidior*: Mod. L. *rubidior* redder, ruddier (*cf.* L. *rubidus* red, ruddy).
  Western Ceylon Brown Mongoose. Endemic to Sri Lanka.

OEN South Indian Mongoose (subsp. *fuscus*), Nilgiri Brown Mongoose (subsp. *fuscus*), Highland Ceylon Brown Mongoose (Subsp. *flavidens*), Northern Brown Mongoose (Subsp. *maccarthiae*), Sandy Mongoose (Subsp. *siccatus*), Ceylon Brown Mongoose (J. Syn. *ceylanicus*), Central Ceylon Brown Mongoose (J. Syn. *phillipsi*)

VN Kurg. *Sendali-Kera*; Sinh. *Mugatiya*, *Ram-mugatiya*; Tam. (SL) *Keerie-pulle*, *Karrung-keerie*, *Poo-keerie*, *Pambu-keerie*

437. *Urva palustris* Ghose, 1965
Indian Marsh Mongoose

Urva: Mod. L. *urva* >Nep. *arva*, the name for the crab-eating mongoose.

palustris: L. *palustris* marshy (L. *palus*, *paludis* swamp, marsh).

Endemic to India (West Bengal).

OEN Bengal Mongoose (nom. form)

Family Viverridae Gray, 1821

438. *Arctictis binturong* (Raffles, 1821)
Binturong

Arctictis: Mod. L. *arctitis* bear-weasel (Gr. *arktos* a bear; Gr. *ictis* a weasel); owing to its appearance.

binturong: L. *binturong* from Malay *binturung*, a native name for the species.

Southeast Asia and South Asia. One subspecies is known from the region.

Subspecies

- *albifrons*: Mod. L. *albifrons* white-headed (L. *albus* white; L. *frons* forehead, front).
  Himalayan Binturong. In the Himalayas of Bangladesh, Bhutan, India (northeast), and Nepal.

OEN Bear-Cat (nom form), Asian Bear-Cat (nom form), Black Bear-Cat (J. Syn. *ater*), Indian Binturong (Subsp. *albifrons*)

VN Assam. *Young*

439. *Arctogalidia trivirgata* (Gray, 1832)
Small-toothed Palm Civet

Arctogalidia: Mod. L. *arctogalidia* small bear-weasel (Gr. *arktos* a bear; Gr. *galidia* (dim. of *gale*) a weasel), (*-idia* diminutive (comparison) (L. >Gr.)).
trivirgata: Mod. L. *trivirgata* three-striped (L. *tri* (*tres*) >Sans. *tri* three; L. adj. *virgatus* striped (L. *virga* stripe)).

Southeast Asia and South Asia. One subspecies is known from the region.

Subspecies

- *millsi*: Eponym after J. Phillip Mills (b. 1890–d. 1960), British Imperial Civil Service officer serving in Nagaland (*-i*, commemorating (dedication, eponym) (L.)).
  Indian Small-toothed Palm Civet. Bangladesh and India (northeast).

OEN Three-striped Palm Civet (nom. form), Three-streaked Paguma (nom. form)

440. *Paguma larvata* (Hamilton-Smith, 1827)
Masked Palm Civet

Paguma: Origin unknown, probably a coined word.
larvata: L. *larvatus* masked (L. *larva* mask).

China, Southeast Asia, and South Asia. Four subspecies are known from the region.

Subspecies

- *grayi*: Eponym after John Edward Gray (b. 1800–d. 1875), a British zoologist, Keeper of Zoology in the British Museum (Natural History) (*-i*, commemorating (dedication, eponym) (L.)).
  Gray's Masked Palm Civet. Himalayas in India (in Uttarakhand) and Nepal.
- *tytleri*: Eponym after Robert Christopher Tytler (b. 1818–d. 1872), a British soldier, naturalist, and photographer, who worked as the superintendent of the convict settlement on Port Blair, Andaman (*-i*, commemorating (dedication, eponym) (L.)).
  Tytler's Masked Palm Civet. Endemic to India; Andaman Islands.
- *wroughtoni*: Eponym after Robert Charles Wroughton (b. 1849–d. 1921), an English officer of the Indian Forest Service in the Bombay Presidency, member of the Bombay Natural History Society, who described many new taxa resulting from the Mammal Survey of British India (*-i*, commemorating (dedication, eponym) (L.)).
  Wroughton's Masked Palm Civet. Endemic to South Asia; India (Northwest Himalayas) and Pakistan (in the Himalayas).
- *neglecta*: L. *neglectus* ignored, overlooked, neglected, disregarded (L. *neglegere* to neglect).
  Naga Hills Masked Palm Civet. Bangladesh (in the Himalayas) and India (Northeast Himalayas).

OEN Himalayan Palm Civet (nom. form), Gem-faced Civet (nom. form), Hill Musang (Subsp. *grayi*), Hill Tree-cat (Subsp. *grayi*), Nepal Paguma (Subsp. *grayi*), Nepal Masked Palm Civet (J. Syn. *nipalensis*)

441. *Paradoxurus hermaphroditus* (Pallas, 1777)

Common Palm Civet

Paradoxurus: Mod. L. *paradoxurus* strange tail (Gr. *paradoxus* strange, marvelous; Gr. *oúrá* tail).

hermaphroditus: Gr. Myth. *hermaphroditus* (Gr. *hermaphróditos*, from *Hermês* and *Aphrodítē*), the son of Hermes and Aphrodite, who merged bodies with a naiad.

Southeast Asia and South Asia. Besides the nominate subspecies, *P. h. hermaphroditus* Common Palm Civet, from Bangladesh, India and Sri Lanka, five subspecies are known from the region.

Subspecies

- *bondar*: Mod. L. *bondar* >Beng. *bondar* the name of the civet.
  Terai Palm Civet. Endemic to South Asia; India (in Indo-Gangetic Plains) and Nepal.
- *pallasi*: Eponym after Peter Simon Pallas (b. 1741–d. 1811), a German zoologist and botanist who worked in Russia (*-i*, commemorating (dedication, eponym) (L.)).
  Pallas's Palm Civet. Himalayas in India (eastern) and Nepal.
- *nictitatans*: Mod. L. *nictitatans* bearing nictitating membrane, after well-developed nictitating membrane (L. *membrane nictitans*).
  Taylor's Palm Civet. Endemic to India (Odisha).
- *scindiae*: Eponym after Madho Rao Scindia (b. 1876–d. 1925), the fifth Maharaja of Gwalior belonging to the Scindian dynasty of the Marathas, a benefactor who helped start the Mammal Survey of British India (*-ae*, commemorating (dedication, eponym) (L.)).
  Central Indian Palm Civet. Endemic to India (Madhya Pradesh).
- *vellerosus*: Mod. L. *vellerosus* thick-furred (L. *vellera* fur, >*vellus* fleece, wool; *-osus* abundance, fullness, quality of (L.)).
  Kashmir Palm Civet. Endemic to South Asia; India and Pakistan (in the Himalayas).

OEN Toddy Cat (nom. form), Bondar (Subsp. *bondar*), Terai Musang (Subsp. *bondar*), Terai Tree-cat (Subsp. *bondar*), Mush-Cat (Subsp. *bondar*), Bondar Paradoxurus (Subsp. *bondar*), Black Palm-cat (Subsp. *hermaphroditus*), Pallas's Paradoxurus (Subsp. *pallasi*), Cross's Palm Civet (J. Syn. *crossi*), Cross's Paradoxurus (J. Syn. *crossi*), Cross's Paguma (J. Syn. *crossi*), Nepal Palm Civet (J. Syn. *hirsutus*), Indian Toddy Cat (J. Syn. *nigra*), Black-fronted Paradoxure (J. Syn. *nigrifrons*), Kuttaas (J. Syn. *pennantii*), Pennant's Palm Civet (J. Syn. *pennantii*), Pennant's Paradoxurus (J. Syn. *pennantii*), Prehensile Paradoxurus

(J. Syn. *prehensilis*), Luwack (J. Syn. *typus*), Golden Palm Civet (J. Syn. *zeylonensis*), Golden Palm Cat (J. Syn. *zeylonensis*), Ceylon Palm Civet (J. Syn. *zeylonensis*), Red Palm Civet (J. Syn. *zeylonensis*)

VN Hind. *Khotas, Menuri, Lakáti, Changar, Chingar, Jhar-ka-kuta*; Nep. *Malwa, Machabba*; Mar. *Ud*; Kurg. *Kullibekku*; Dakh. *Menuri*; Tel. *Manupilli, Punugupilli*; Beng. *Bhondar, Baghdankh, Bham*; Kan. *Punaginabekku, Kerabekku, Kerbek*; Wad. *Nulla Philli*; Pard. *Mahngutchi*; Sinh. *Ugguduwa, Kalawedda*; Tam. (SL) *Marram Nai, Mara-nai*

442. *Paradoxurus aureus* F. Cuvier, 1822
Golden Wet-zone Palm Civet

Paradoxurus: Mod. L. *paradoxurus* strange tail (Gr. *paradoxus* strange, marvelous; Gr. *oúrá* tail).
aureus: L. *aureus* golden (*aurum* gold, -*eus* having the quality of).
Endemic to Sri Lanka.

OEN Golden Palm Civet (S. Syn. *zeylonensis*)

443. *Paradoxurus montanus* Kelaart, 1852
Sri Lankan Brown Palm Civet

Paradoxurus: Mod. L. *paradoxurus* strange tail (Gr. *paradoxus* strange, marvelous; Gr. *oúrá* tail).
montanus: L. *montanus* of the mountains, mountain- (L. *mons, montis* mountain)).
Endemic to Sri Lanka.

444. *Paradoxurus jerdoni* Blanford, 1885
Jerdon's Palm Civet

Paradoxurus: Mod. L. *paradoxurus* strange tail (Gr. *paradoxus* strange, marvelous; Gr. *oúrá* tail).
jerdoni: Eponym after Thomas Caverhill Jerdon (b. 1811–d. 1872), a British physician, zoologist, and botanist, who worked extensively in the Madras Presidency (-*i*, commemorating (dedication, eponym) (L.)).
Endemic to India. Besides the nominate subspecies, *P. j. jerdoni* Jerdon's Palm Civet, from Kerala and Tamil Nadu, another subspecies is known from the region.

Subspecies

- *caniscus*: Mod. L. *caniscus* light grey (L. *canus* grey, -*iscus* diminutive (comparison) (L./Gr.)).
  Coorg Palm Civet. Endemic to India (in the Western Ghats of Kodagu District, Karnataka and Nilgiris District, Tamil Nadu).

OEN Brown Palm Civet (nom. form)

445. *Paradoxurus stenocephalus* Groves et al., 2009
Golden Dry-zone Palm Civet

Paradoxurus: Mod. L. *paradoxurus* strange tail (Gr. *paradoxus* strange, marvelous; Gr. *oúrá* tail).
stenocephalus: Mod. L. *stenocephalus* narrow-headed (Gr. *stēnos* narrow, thin; Gr. *kephalus* head).
Endemic to Sri Lanka.

446. *Viverra zibetha* Linnaeus, 1758
Large Indian Civet

Viverra: L. *viverra* civet.
zibetha: Med. L. *zibethum* or Ital. *zibetto*, both >Arab. *zabād*, civet perfume.
China, Southeast Asia, and South Asia. Bangladesh, India, and Nepal.

OEN Large Civet-Cat (nom. form), North Indian Large Civet (nom. form), Oriental Civet-Cat (J. Syn. *orientalis*)
VN Hind. *Khatás*; Beng. *Mach-Bondar, Bágdas, Pudo-ganda, Puda-gauda*; Nep. *Bhrán, Nit-birahi, Nit-biralu*; Bhut. *Kung*; Lepch. *Saphiong*; Pahari W. *Ningalichitua*

447. *Viverra civettina* Blyth, 1862
Malabar Large Spotted Civet

Viverra: L. *viverra* civet.
civettina: Diminutive of Fr. *civette* civet cat (*-ina* belonging to (L.)).
Endemic to India (in the Western Ghats of Karnataka and Kerala).

OEN Malabar Civet

448. *Viverricula indica* (E. Geoffroy Saint-Hilaire, 1803)
Small Indian Civet

Viverricula: L. *viverricula* small civet, dim. of *viverra* civet (*-ula* diminutive (comparison), somewhat (adj.) (L.)).
indica: L. adj. *indica* belonging to India (*-ica* belonging to, pertaining to (L./Gr.)).
China, Southeast Asia, and South Asia. Besides the nominate subspecies, *V. i. indica* Small South Indian Civet, from Bangladesh, India, and Pakistan, four subspecies are known from the region.

Subspecies

- *deserti*: L. *deserti* desert, waste, solitude.
  Small Rajasthan Civet. Endemic to South Asia; India (Rajasthan) and Pakistan (Sindh).
- *baptistae*: Eponym after N.A. Baptista (dates not known), taxidermist who collected in India for the Mammal Survey of British India (*-ae*, commemorating (dedication, eponym) (L.)).

Small North Indian Civet. Himalayas in Bhutan, India (eastern), and Nepal.

- *mayori*: Eponym after Major E.W. Mayor (dates not known), collected in Sri Lanka for the Mammal Survey of British India (*-i* commemorating (dedication, eponym) (L.)).
  Small Sri Lankan Civet. Endemic to Sri Lanka.
- *wellsi*: Eponym after H.W. Wells (dates not known), collected in India for the Mammal Survey of British India (*-i* commemorating (dedication, eponym) (L.)).
  Well's Small Civet. Endemic to India (Himachal Pradesh and Uttarakhand).

OEN Little Civet (nom. form), Small Ceylon Civet (Subsp. *mayori*), Ceylon Small Civet (Subsp. *mayori*), Civet-cat (Subsp. *mayori*), Ring-tailed Civet (Subsp. *mayori*), Bengal Small Civet (J. Syn. *bengalensis*)

VN Hind. *Mashk-billa, Katus*; Beng. *Gundha Gokal, Gando Gaula*; Kut. *Jabadio*; Kan. *Punagina-Bekku*; Kum. *Malpusa*; Nep. *Saiyar, Bagnyul*; Mar. *Fowádi-manjur*; Wad. *Puluk-philli*; Pard. *Punkassibekku*; Kurg. *Pulunguotay Punugu*; Dakh. *Mushak-billi*; Tel. *Punugu-pilli*; Sinh. *Urulaeva*; Tam. (SL) *Poolu poona, Veerugu poonai, Punugoo poonai*

Order Perissodactyla Owen, 1848
Family Equidae Gray, 1821

449. *Equus hemionus* Pallas, 1775
Asian Wild Ass

Equus: L. *equus* horse.

hemionus: L. *hemionus* half-ass (L. *hemi-* half- (>Gr. *hēmi-* half-); *onus* (Gr. *ónos*) ass).

India, Iran, and Turkmenistan. One subspecies is known from the region.

Subspecies

- *khur*: Mod. L. *khur* after Hind. name *ghor-khur* for the species.
  Indian Wild Ass. India (Rann of Kachchh, Gujarat), extinct from Pakistan.

OEN Asiatic Wild Ass (nom. form), Hemionus (nom. form), Baluchi Wild Ass (J. Syn. *blanfordi*)

VN Hind. *Ghor-khur*; Pers. *Ghor*

450. *Equus kiang* Moorcroft, 1841
Tibetan Wild Ass

Equus: L. *equus* horse.

kiang: Mod. L. *kiang* from Tib. *kyang*, a name for the species.

India and China (Tibetan Plateau). Besides the nominate subspecies, *E. k. kiang* Kiang, from India (Jammu and Kashmir), another subspecies is known from the region.

Subspecies

- *polyodon*: Gr. *poluodous* with many teeth (Gr. *polus* many; Gr. *odous, odontos* tooth), owing to two extra molars in the upper jaw.
  Hodgson's Kiang. Himalayas in India (Sikkim) and Nepal.

VN Tib. *Kyang, Kiang, Dzightai*

Family Rhinocerotidae Gray, 1821

451. *Dicerorhinus sumatrensis* (Fischer, 1814)
Sumatran Rhinoceros

Dicerorhinus: Mod. L. *dicerorhinus* two horns on the nose (Gr. *di* two, *kéras* horn; Gr. *rhis, rhinos* nose); in allusion to the transversely paired nasal horns.
sumatrensis: L. adj. *sumatrensis* meaning from Sumatra, Indonesia (*-ensis* geographic, occurrence in (location, toponym) (L.)).

Indonesia and Malaysia. One subspecies was historically known from the region; probably extinct in South Asia.

Subspecies

- *lasiotis*: Mod. L. *lasiotis* hairy-eared (Gr. *lásios* hairy; L. *otus, ōtos* ear); in allusion to its hairy ears.
  Northern Sumatran Rhinoceros. Extinct from Bangladesh, Bhutan, and India; historically reported from the Himalayas in these countries.

OEN Asian Two-horned Rhinoceros (nom. form)

452. *Rhinoceros unicornis* Linnaeus, 1758
Great One-horned Rhinoceros

Rhinoceros: Mod. L. *rhinoceros* (>Gr. *rhinōkéras*) horned-nose (Gr. *rhis, rhinos* nose*;* Gr. *kéras* horn).
unicornis: L. *unicornis* one-horned (L. *uni* one, L. *cornu* horn).

Terai regions in Gangetic Plains of Bhutan, India, and Nepal. Historically also from Afghanistan, Bangladesh, and Pakistan, where the species is extinct.

OEN Great Indian Rhinoceros (nom. form), Great Indian One-horned Rhinoceros (nom. form), Greater One-horned Rhino (nom. form), Indian Rhinoceros (nom. form, J. Syn. *indicus*)

VN Hind. *Genda, Gonda, Genra*

453. *Rhinoceros sondaicus* Desmarest, 1822
Lesser One-horned Rhinoceros

Rhinoceros: Mod. L. *rhinoceros* (>Gr. *rhinōkéras*) horned-nose (Gr. *rhis*, *rhinos* nose; Gr. *kéras* horn).
sondaicus: L. adj. *sondaicus* meaning from Sunda, Indonesia (*-icus*, belonging to, pertaining to (location, toponym) (L.)).

Indonesia and Vietnam. One subspecies was historically known from the region; probably extinct in South Asia.

Subspecies

- *inermis*: Mod. L. *inermis* unarmed, without weapons (L. *in* prefix for no, L. *arma* arm).

  Lesser Indian Rhinoceros. Extinct from Bangladesh, Bhutan, and India; historically reported from the Himalayas in these countries.

OEN Sunda Rhinoceros (nom. form), Javan Rhinoceros (nom. form)

Order Artiodactyla Owen, 1848
Family Suidae Gray, 1821

454. *Porcula salvania* (Hodgson, 1847)
Pygmy Hog

Porcula: Mod. L. *porcula* little piglike (L. *porca* a sow).
salvania: Origin not known, possibly either from L. *silvanus* of the woods (L. *silva* woodland) or L. *silvania* wood-nymph (L. *silva* wood, forest).

Terai regions in Gangetic Plains of Bhutan and India.

VN Hind. *Chota Sur*, *Chota Suwar*; Nep. *Sano-banei*

455. *Sus scrofa* (Linnaeus, 1758)
Wild Boar

Sus: L. *sus* (Gr. *hûs*) pig.
scrofa: L. *scrofa* sow.

Europe, Mediterranean region, Middle East, South Asia, Southeast Asia, China, Russia (southern), and Japan. Three subspecies are known from the region.

Subspecies

- *vittatus*: L. *vittatus* banded (*vitta* ribbon, band).
  Asian Wild Pig. Andaman and Nicobar Islands; the taxa *andamanensis* and *nicobarensis* are based on domestic or feral populations of this species.
- *cristatus*: L. *cristatus* crested (L. *crista* crest).

  Indian Wild Pig. Afghanistan, Bangldesh, Bhutan, India, Nepal, and Sri Lanka.

- *davidi*: Eponym after Reuben David (1912–1989), a founder and superintendent of the Ahmedabad Zoo, Gujarat, where the specimens of this taxon were first sighted by Colin Groves (*-i*, commemorating (dedication, eponym) (L.)).

  Pakistan Wild Pig. Endemic to South Asia; India (northwest) and Pakistan (Sindh).

OEN Wild Pig
VN Hind. *Sur*, *Suwar*, *Bura Janwar*, *Kis*; Kan. *Handi*, *Mikka*, *Jewadi*; Gond. *Paddi*; Mar. *Paddi*; Tel. *Pandi*; Tam. *Panni*; Sinh. *Wal Ura*; Tam. (SL) *Kadu-pandi*, *Kardup-pandri*

Family Tragulidae Milne-Edwards, 1864

456. *Moschiola meminna* (Erxleben, 1777)
White Spotted Chevrotain

Moschiola: Mod. L. *moschiola* moschus-like (Gr. *móskhos* musk >Pers. *mušk*); in allusion to the musk glands of the male (*-ola* dim. meaning in comparison, or somewhat like (adj.) (L.)).
meminna: Mod. L. *meminna* after Sinhalese name *meeminna* for the species.

Endemic to Sri Lanka.

OEN White Spotted Mouse-deer, Sri Lankan Mouse-deer
VN Sinh. *Meeminna*, *Capita-meeminna*, *Wal-miya*; Tam. (SL) *Sarrugumann*, *Ukkulam*

457. *Moschiola indica* (Gray, 1843)
Indian Spotted Chevrotain

Moschiola: Mod. L. *moschiola* moschus-like (Gr. *móskhos* musk >Pers. *mušk*); in allusion to the musk glands of the male (*-ola* dim. meaning in comparison, or somewhat like (adj.) (L.)).
indica: L. adj. *indica* belonging to India (*-ica* belonging to, pertaining to (L./Gr.)).

Endemic to South Asia; India and Nepal (only in the foothills).

OEN Indian Spotted Mouse-deer, Pissora, Pissay
VN Hind. *Pisuri*, *Pisora*, *Pisai*; Beng. *Jitri Harn*; Mar. *Mirgi*, *Pisuri*, *Pisora*, *Pisai*; Tel. *Kuru-Pandi*; Odi. *Gandwa*; Kol *Yar*

458. *Moschiola kathygre* Groves and Meijaard, 2005
Yellow Striped Chevrotain

Moschiola: Mod. L. *moschiola* moschus-like (Gr. *móskhos* musk >Pers. *mušk*); in allusion to the musk glands of the male (*-ola* dim. meaning in comparison, or somewhat like (adj.) (L.)).

kathygre: Mod. L. *kathygre* living in wet places (Gr. *kathygre, kathygros* >Gr. *kata* down from, on account of; Gr. *hygros* wetness); with reference to its wet zone habitat in Sri Lanka.

Endemic to Sri Lanka.

Family Moschidae Gray, 1821

459. *Moschus chrysogaster* Hodgson, 1839
Golden-bellied Musk Deer

Moschus: Med. L. *moschus* (Gr. *móskhos* musk >Pers. *mušk*) musk; in allusion to the musk glands of the male.
chrysogaster: Mod. L. *chrysogaster* golden-bellied (Gr. *khrusos* gold; Gr. *gastēr* belly).

Besides the nominate subspecies, *M. c. chrysogaster* Alpine Musk Deer, from the Himalayas in Bhutan, India, and Nepal, another subspecies is known from the region.

Subspecies

- *sifanicus*: L. adj. *sifanicus* meaning from Sifan, Gansu, China (*-icus*, belonging to, pertaining to (location, toponym) (L.)).

  Kansu Musk Deer. Himalayas in India (Sikkim) and Nepal.

OEN Golden-eyed Musk Deer (nom. form), Himalayn Forest Musk Deer (nom. form), Alpine Musk Deer (Subsp. *sifanicus*), West China Musk Deer (Subsp. *sifanicus*)

VN Hind. *Kastura, Kasturi*; Lad. *Rib-jo*; Tib. *La, Lawa*; Kash. *Kasturi, Roos, Rous*; Kin. *Bena*; Assam. *Gan Pahoo*

460. *Moschus leucogaster* Hodgson, 1839
White-bellied Musk Deer

Moschus: Med. L. *moschus* (Gr. *móskhos* musk >Pers. *mušk*) musk; in allusion to the musk glands of the male.
leucogaster: Mod. L. *leucogaster* white-bellied (Gr. *leukogastēr* white-belly, from Gr. *leukos* white; Gr. *gastēr* belly).

Himalayas of the Bhutan, India, and Nepal.

OEN Himalayan Musk Deer (nom. form)

VN Hind. *Kastura, Kasturi*; Lad. *Rib-jo*; Tib. *La, Lawa*; Kash. *Kasturi, Roos, Rous*; Kin. *Bena*; Assam. *Gan Pahoo*

461. *Moschus fuscus* Li, 1981
Dwarf Musk Deer

Moschus: Med. L. *moschus* (Gr. *móskhos* musk >Pers. *mušk*) musk; in allusion to the musk glands of the male.
fuscus: L. *fuscus* brown, dusky.

Himalayas of the Bhutan, India, and Nepal.

OEN Black Musk Deer (nom. form), Dusky Musk Deer (nom. form), Khumbu Musk Deer (nom. form)

462. *Moschus cupreus* Grubb, 1982
Kashmir Musk Deer

Moschus: Med. L. *moschus* (Gr. *móskhos* musk >Pers. *mušk*) musk; in allusion to the musk glands of the male.
cupreus: L. *cupreus* coppery (*cuprum* copper; *-eus* having the quality of (L.)).
Himalayas of the Afghanistan, India, and Pakistan.

OEN Rufous Musk Deer (nom. form)

Family Cervidae Goldfuss, 1820

463. *Axis axis* (Erxleben, 1777)
Spotted Deer

Axis: L. *axis* deer; perhaps of East Indian origin.
axis: L. *axis* deer; perhaps of East Indian origin.
Endemic to South Asia; Bangladesh, Bhutan, India, Nepal, and Sri Lanka.

OEN Ceylon Spotted Deer (J. Syn. *ceylonensis*)
VN Hind. *Chital*, *Chitra*, *Chitri*, *Jhank* (male), *Chatidah*, *Buriya*, *Harn*, *Harin*, *Hiran*; Tel. *Dupi*; Gond. *Lupi*; Sinh. *Tik-muwa*; Tam. (SL) *Pulli-mann*

464. *Axis porcinus* (Zimmermann, 1780)
Hog-Deer

Axis: L. *axis* deer; perhaps of East Indian origin.
porcinus: Mod. L. *porcinus* hoglike (L. *porca* a sow).
Endemic to South Asia; Himalayas in the Bangladesh, Bhutan, India, Nepal, and Pakistan (also along the Indus river basin). Introduced in to Goa, India, and Sri Lanka.

OEN Ceylon Hog-Deer (J. Syn. *oryzus*), Paddyfield Deer (J. Syn. *oryzus*)
VN Hind. *Para*; Beng. *Nuthrini Harn*; Nep. *Sugoria*, *Khar Laguna*; Sinh. *Gona-muwa*, *Willa-muwa*; Tam. (SL) *Mann*

465. *Cervus elaphus* Linnaeus, 1758
Red Deer

Cervus: L. *cervus* stag, deer.
elaphus: Med. L. *elaphus* (Gr. *élaphos*) deer.
Temperate regions of the Northern Hemisphere. Three subspecies are known from the region.

Subspecies

- *wallichi*: Eponym after Dr. Nathaniel Wallich (1786–1854), a Danish surgeon, botanist, and naturalist who worked in India, the first curator of Indian Museum and force behind the development of Calcutta Botanical Garden (*-i*, commemorating (dedication, eponym) (L.)).
  Sikkim Stag. Himalayas in India (Sikkim) and Nepal.
- *hanglu*: Mod. L. *hanglu* after Kashmiri name *honglu* for the species.
  Kashmir Stag. Endemic to India; Himalayas in Jammu and Kashmir and Himachal Pradesh.
- *yarkandensis*: L. adj. *yarkandensis* meaning from Yarkand, Xinjiang Uyghur Autonomous Region, China (*-ensis*, geographic, occurrence in (location, toponym) (L.)).
  Yarkand Stag. Himalayas in Afghanistan.

OEN Wallich's Deer (Subsp. *wallichi*), Shou (Subsp. *wallichi*), Hangul (Subsp. *hanglu*), Yarkand Deer (Subsp. *yarkandensis*)

VN Kash. *Honglu*, *Hangul*; Hind. *Bara-singha*; Pers. *Maral*; Tib. *Shou*, *Sia*

466. *Muntiacus vaginalis* (Boddaert, 1785) Northern Red Muntjak

Muntiacus: L. *muntiacus* derivative meaning pertaining to *muntjak*, a native name of this animal in Sunda language (*-acus*, pertaining to (L./Gr.)).

vaginalis: Mod. L. *vaginalis* sheathed (L. *vagina* sheath); in allusion to its short antler that is mostly enclosed by hair-covered pedicels.

South Asia, Southeast Asia, and China (southern). Besides the nominate subspecies, *M. v. vaginalis* Northern Red Muntjak, from forested tracts of Bangladesh, Bhutan, India, and Nepal, two more subspecies are known from the region.

Subspecies

- *aureus*: L. *aureus* golden (*aurum* gold, *-eus* having the quality of).
  Golden Indian Muntjak. Endemic to India.
- *malabaricus*: L. adj. *malabaricus* meaning from Malabar – a region in southern India lying between the Western Ghats and the Arabian Sea, corresponding to the Malabar District of the Madras Presidency in British India (presently includes areas under five northern districts of Kerala), (*-icus*, belonging to, pertaining to (location, toponym) (L.)).
  South Indian Muntjak. Endemic to South Asia; India (in the Western Ghats) and Sri Lanka.

OEN Northern Red Barking-deer (Subsp. *vaginalis*), Golden Indian Barking-deer (Subsp. *aureus*), South Indian Barking-deer (Subsp. *malabaricus*)

VN Hind. *Kakur, Kakkar*; Beng. *Maya*; Bhut. *Karsiar*; Lepch. *Siku, Sikku, Suku*; Assam. *Hoogeree*; Kan. *Kan-kuri*; Dekh. *Jangli-Bakra*; Mar. *Bekra, Bekur*; Nep. *Ratwa*; Gond. *Gutra, Gutri*; Tel. *Kuka-gore*; Sinh. *Olu-muwa, Welli-muwa, Welly, Hoola Morha*; Tam. (SL) *Mann, Sembli-mann, Pulatar-mann*

467. *Muntiacus putaoensis* Amato, Egan and Rabinowitz, 1999
Leaf Muntjak

Muntiacus: L. *muntiacus* derivative meaning pertaining to *muntjak*, a native name of this animal in Sunda language (*-acus*, pertaining to (L./Gr.)).
putaoensis: L. adj. *putaoensis* meaning from Putao, Kachin State, Myanmar (*-ensis*, geographic, occurrence in (location, toponym) (L.)).

South Asia (northeast India), Southeast Asia (Myanmar).

468. *Rucervus duvaucelii* (Cuvier, 1823)
Swamp Deer

Rucervus: Mod. L. *rucervus* given name; generic combination, of first two letters of *rusa*, Malay name for sambar and L. *cervus* stag, deer.
duvaucelii: Eponym after Alfred Duvaucel (1793–1824), a French naturalist and explorer, collected widely in South and Southeast Asia for Paris Museum of Natural History (*-ii*, commemorating (dedication, eponym) (L.)). Species description was based on drawings of the antlers sent by Alfred Duvaucel.

Endemic to South Asia. Besides the nominate subspecies, *R. d. duvaucelii* North Indian Swamp Deer, from forested tracts of India (Uttarakhand) and Nepal (Far Western), two more subspecies are known from the region.

Subspecies

- *branderi*: Eponym after A.A. Dunbar-Brander (dates not known), Conservator of Forests, stationed in Central Province, British India (*-i*, commemorating (dedication, eponym) (L.)).

  South Indian Swamp Deer. Endemic to India (Kanha National Park, Madhya Pradesh).
- *ranjitsinhi*: Eponym after M.K. Ranjitsinh (dates not known), Indian civil servant, naturalist, and author of many books (*-i*, commemorating (dedication, eponym) (L.)).

  Assam Swamp Deer. Endemic to India (in Brahmaputra plains in the northeast).

OEN Barasingha (nom. form), Wetland Barasingha (Subsp. *duvaucelii*), Hard-ground Barasingha (Subsp. *branderi*), Eastern Barasingha (Subsp. *ranjitsinhi*)

VN Hind. *Goen* (male), *Gaoni* (female), *Geonjak*, *Bara-singha*; Bihar. *Potiya-harn*; Nep. *Baraya*

469. *Rucervus eldii* (McClelland, 1842)
Brow-antlered Deer

Rucervus: Mod. L. *rucervus* given name; generic combination, of first two letters of *rusa*, Malay name for sambar and L. *cervus* stag, deer.
eldii: Eponym after Colonel Lionel Percy Denham Eld (1808–1863), an officer of Bengal Infantry, stationed in Assam (*-ii*, commemorating (dedication, eponym) (L.)).

Southeast Asia (Myanmar, Cambodia, Laos, Hainan, and Vietnam) and South Asia (Manipur, India).

OEN Thamin (nom. form), Eld's Deer (nom. form), Manipur Dancing Deer (nom. form)
VN Hind. *Sing-nai*; Mani. *Tha'min*

470. *Rusa unicolor* (Kerr, 1792)
Sambar

Rusa: L. *rusa* from Malay name for sambar.
unicolor: L. *unicolor* plain, uniform (L. *uni-* single-, L. *color* colour).

Southeast Asia, South Asia, and China (southern). Only the nominate subspecies, *R. u. unicolor* Sambar, occurs in the region.

OEN Ceylon Sambar (nom. form)
VN Hind. *Sambar*; Mar. *Sambar, Meru*; Tel. *Kadathi, Kannadi*; Kan. *Kadavi, Kadaba*; Gond. *Ma-ao*; Pahari W. *Jarai, Jerrao*; Beng. *Ghous, Gaoj* (male), *Bhalogni* (female); Nep. *Maha*; Sinh. *Gona*; Tam. (SL) *Marrei, Komboo-marrei* (stags)

Family Bovidae Gray, 1821
Subfamily Antilopinae Gray, 1821

471. *Antilope cervicapra* (Linnaeus, 1758)
Blackbuck

Antilope: Med. L. *antalopus* (Gr. *ánthólops*) a horned animal, an antelope.
cervicapra: Mod. L. *cervicapra* she-goat–like deer (L. *cervus* deer; L. *capra* she-goat).

Endemic to India. Besides the nominate subspecies, *A. c. cervicapra* Indian Blackbuck, another subspecies is known from the region.

Subspecies

- *rajputanae*: Mod. L. *rajputanae* after Rājputāna, a historical region that included present-day Indian state of Rajasthan and also parts of Madhya Pradesh and Gujarat.

  Rajputana Blackbuck. Endemic to arid and semi-arid regions of northwestern India.

OEN Southeastern Blackbuck (Subsp. *cervicapra*), Indian Antelope (J. Syn. *bezoartica*), Sasin (J. Syn. *bezoartica*), Two-lined Antelope (J. Syn. *bilineata*), Central Indian Antelope (J. Syn. *centralis*), Hagenbeck's Antelope (J. Syn. *hagenbecki*)

VN Hind. *Harn* (male), *Harna* (male), *Kalwit* (male), *Harin* (female), *Mriga*, *Bureta* (near Bhagalpur, Bihar), *Guria* or *Goria* (near Tirhut, Bihar); Sans. *Mriga*; Kan. *Chigri*; Marw. *Alali* (male), *Gandoli* (female); Bihar. *Kalsar* (male), *Baoti* (female); Tel. *Krishna Jinka* (male), *Irri* (male), *Ledi* (female); Mar. *Phandayat*

472. *Gazella subgutturosa* (Guldenstaedt, 1780)
Goitered Gazelle

Gazella: Mod. L. *gazelle* >Fr. *gazelle* gazelle (from Arab.*ghazal* wild goat, gazelle).

subgutturosa: Mod. L. *subgutturosa* somewhat goitered (L. *sub* slightly, somewhat; L. *gutturosus* goitred (*guttur* throat, gullet)).

China, Middle East, South Asia, and Arabian Peninsula. Afghanistan and Pakistan (Balochistan).

OEN Arabian Sand Gazelle (nom. form), Persian Gazelle (nom. form)

VN Pers. *Ahu*

473. *Gazella bennettii* (Sykes, 1831)
Indian Gazelle

Gazella: Mod. L. *gazelle* >Fr. *gazelle* gazelle (>Arab. *ghazal* wild goat, gazelle).

bennettii – Eponym after Edward Truner Bennett (1797–1836), an English zoologist and writer, instrumental in establishing of Entomological Society that eventually led to the establishment of Zoological Society of London in 1826 (*-ii*, commemorating (dedication, eponym) (L.)).

South Asia and Iran. Besides the nominate endemic to India subspecies, *G. b. bennettii* Deccan Chinkara from central parts of India, three more subspecies are known from the region.

Subspecies

- *christyi*: Eponym after Dr Alexander Turnbull Christie (dates not known), a medical officer of the Madras Establishment, who collected in India (*-i*, commemorating (dedication, eponym) (L.)).
  Gujarat Chinkara. Endemic to Gujarat, India.
- *fuscifrons*: Mod. L. *fuscifrons* brown-foreheaded (L. *fuscus* dusky, brown; L. *frons* forehead, front).
  Baluchistan Chinkara. Endemic to South Asia; Afghanistan, India (Rajasthan), and Pakistan.
- *salinarum*: L. *salinarum* (L. *salina* salt pans), of salt pans.
  Salt Works Chinkara. Endemic to South Asia, India (Haryana and Punjab), and Pakistan (Punjab).

OEN Chinkara (nom. form), Chinkara Gazelle (nom. form), Ravine-Deer (nom. form), Goat-Antelope (nom. form), Jebeer Gazelle (Subsp. *fuscifrons*)

VN Hind. *Chinkara*, *Chikara*, *Kal-punch*; Marw. *Porsya* (male), *Chari*; Punj. *Hurnee*; Mar. *Kal-Sipi*; Balochi *Ast*; Kan. *Tiska*, *Budari*, *Mudari*; Tel. *Burudu-jinka*, *Gadi Jinka*

474. *Procapra picticaudata* Hodgson, 1846
Tibetan Gazelle

Procapra: Mod. L. *procapra* before-goat (L. *pro* before; L. *capra* she-goat).

picticaudata: Mod. L. *picticaudata* painted-tailed (L. *pictus* painted (L. *pingere* to paint); Mod. L. *caudatus* -tailed (L. *cauda* tail)).

China and South Asia. Himalayas in Jammu and Kashmir and Sikkim.

VN Tib. *Goa*, *Ra-goa*

Subfamily Bovinae Gray, 1821

475. *Bos gaurus* H. Smith, 1827
Indian Bison

Bos: L. *bos* ox.

gaurus: Mod. L. *gaurus* from Hind. *gaur* (Sans. *gaura-mriga*, name of the buffalo) the name of this species.

Southeast Asia and South Asia. Besides the nominate subspecies, *B. g. gaurus* Indian Gaur, another subspecies is known from the region.

Subspecies

- *laosiensis*: L. adj. *laosiensis* meaning from Laos (*-ensis*, geographic, occurrence in (location, toponym) (L.)).

Southeast Asian Gaur. Bangladesh (Chittagong), Bhutan, and India (Assom).

OEN Indian Gaur (nom. form), Wild Bull (J. Syn. *gour*), Asseel Gayal (J. Syn. *asseel*), Gauri Bull (J. Syn. *subhemachalus*, *cavifrons*)

VN Hind. *Gour*, *Gouri-gai*, *Bod* (near Seoni); Beng. *Ban-Gou*, *Vana-go*; Dakh. *Jangli Khulga*; Tam. *Katu-Yeni*; Tel. *Adavi Dhunna*; *Adavi Barre*; Gond. *Peru-mau*; Kan. *Kar-kona*

476. *Bos mutus* (Przewalski, 1883)
Wild Yak

Bos: L. *bos* ox.

mutus: L. *mutus* silent, dumb; in allusion to its silent nature in comparison to the domestic yak (*B. grunniens*, L. *grunniens* grunting).

China (Tibetan Plateau) and South Asia (Ladakh in Jammu and Kashmir, India).

OEN Yak (nom. form)

VN Tib. *G.yag* (male), *Dri* (female), *Nak* (female)

477. *Boselaphus tragocamelus* (Pallas, 1766)
Nilgai

Boselaphus: Mod. L. *Boselaphus* (generic combination of *bos* (L. *bos* ox) and *elaphus* (Gr. *élaphos* deer)).
tragocamelus: Mod. L. *tragocamelus* goat-camel (name combination of Mod. L. *tragos* (Gr. *trágos*) goat and L. *camelus* camel).
Endemic to South Asia; India, Nepal, and Pakistan.

OEN Blue Bull (nom. form), White-footed Antelope (J. Syn. *albipes*), Nilagau (J. Syn. *picta, hippelaphus*)
VN Hind. *Nil, Lil, Roz, Rojh, Rui*; Mar. *Roz, Rojh, Rui*; Gond. *Gurayi, Guriya*; Kan. *Maravi*; Tel. *Manu-potu*

478. *Bubalus arnee* (Kerr, 1792)
Wild Buffalo

Bubalus: L. *bubalus*, wild ox, from *bubalis* – a name given to an African antelope, meaning buffalo.
arnee: Mod. L. *arnee* from Hind. *arna* (fem. *arni*), a name for Indian wild buffalo.
Southeast Asia (Myanmar, Thailand, and Cambodia) and South Asia (Bhutan, India, and Nepal). Besides the nominate subspecies, *B. a. arnee* Indian Wild Buffalo from Chhattisgarh and Nepal (Kosi Tapu), another subspecies is known from the region.

Subspecies

- *fulvus*: L. *fulvus* tawny.
  Assam Wild Buffalo. Bangladesh (Chittagong), Bhutan, and India (Assom).

OEN Indian Buffalo (nom. form), Arna (nom. form)
VN Hind. *Jangli Bhains, Arna* (male), *Arni* (female), *Mung* (near Bhagalpur); Mar. *Arna* (male), *Arni* (female); Gond. *Gera Erumi*

479. *Tetracerus quadricornis* (de Blainville, 1816)
Four-horned Antelope

Tetracerus: Mod. L. *tetracerus* four-horned (Gr. *tetra* four, *kéras* horn).
quadricornis: Mod. L. *quadricornis* four-horned (L. *quadri* four, L. *cornu* horn).
Endemic to South Asia. Besides the nominate species endemic to India, *T. q. quadricornis* Four-horned Antelope, another subspecies is known from the region.

Subspecies

- *iodes*: From Gr. *iṑdēs* 'violet-coloured' or 'rusty-red'; in allusion to its pelage.
  Himalayan Four-horned Antelope. Himalayas in India (Uttarakhand) and Nepal.

OEN Chousingha (nom. form), Chikara (J. Syn. *chickara*, *tetracornis*), Chouka (J. Syn. *tetracornis*), Full-horned Antelope (J. Syn. *paccerois*)
VN Hind. *Chousingha*, *Chouka*, *Bhikri* (near Saugor), *Bhirul* (among Bhils), *Kotri* (in Bastar); Mar. *Bekra*; Dakh. *Jangli Bakra*; Kan. *Kondguri*; Tel. *Konda Gore*; Gond. *Kurus*, *Bhir*, *Bhir-kuru*

Subfamily Caprinae Gray, 1821

480. *Budorcas taxicolor* Hodgson, 1850
Takin

Budorcas: Mod. L. *budorcas* ox-like gazelle (L. *bu* from *bos* ox; Gr. *dorkás* gazelle); in allusion to its ox-like bulky body.
taxicolor: Mod. L. *taxicolor* badger-coloured (Mod. L. *taxus* badger; L. *color* colour).

Endemic to South Asia. Besides the nominate subspecies, *B. t. taxicolor* Mishmi Takin, another subspecies is known from the region.

Subspecies

- *whitei*: Eponym after John Claude White (1853–1918), British Commissioner in Sikkim, who presented two pairs of horns based on which R. Lydekker described this species (*-i*, commemorating (dedication, eponym) (L.)).

  Bhutan Takin. Himalayas in Bhutan and India (northeast).

VN Mish. *Tákin*; Khám. *Kin*

481. *Capra sibirica* (Pallas, 1776)
Siberian Ibex

Capra: L. *capra* she-goat.
sibirica: L. adj. *sibirica* belonging to Siberia, Asiatic Russia (*-ica* belonging to, pertaining to (L./Gr.)).

Siberia, Mongolia, Central Asia, and South Asia. Himalayas in Afghanistan, India (northwest), and Pakistan.

OEN Himalayan Ibex (nom. form), Baltistan Ibex (J. Syn. *wardi*)
VN Hind. *Skin*, *Skyin*, *Sakin*, *Iskin*; Tib. *Skin* (male), *Skyin* (male), *Sakin* (male), *Iskin* (male), *l'Danma* (female), *Dan-mo* (female); Lad. *Ski* (male), *Sakin* (male), *Dabino* (female), *Dajimo* (female); Kash. *A'f*; Kin. *Tangrol*, *Buz*; Balt. *Skin*

482. *Capra aegagrus* Erxleben, 1777
Wild Goat

Capra: L. *capra* she-goat.
aegagrus: Mod. L. *aegagrus* wild (Gr. *aigagros*, *aig-*, *aix* goat; Gr. *agrios* wild).

Middle East and South Asia. Afghanistan and Pakistan.
Two subspecies are known from the region.

Subspecies

- *blythi*: Eponym after Edward Blyth (b. 1810–d. 1873), an English zoologist and pharmacist, curator of the Museum at the Royal Asiatic Society of Bengal (*-i*, commemorating (dedication, eponym) (L.)).
  Sind Wild Goat. Balochistan and Sind Provinces in Pakistan.
- *chialtanensis*: L. adj. *chialtanensis* meaning from Chialtan (*-ensis* geographic, occurrence in (location, toponym) (L.)).
  Chiltan Wild Goat. Endemic to Pakistan (Balochistan).

OEN Bezoar Ibex (nom. form), Persian Wild Goat (nom. form), Sind Ibex (Subsp. *blythi*), Chiltan Markhor (Subsp. *chialtanensis*)

VN Pers. *Pasang* (male), *Boz* (female); Balochi *Surrah*

483. *Capra falconeri* (Wagner, 1839)
Markhor

Capra: L. *capra* she-goat.

falconeri: Eponym after Dr. Hugh Falconer (1808–1865), a Scottish geologist, botanist, paleontologist, and paleoanthropologist, who worked in the Himalayas and Burma (*-i*, commemorating (dedication, eponym) (L.)).

Tazikistan, Uzbekistan and South Asia. Besides the nominate species endemic to India, *C. f. falconeri* Kashmir Markhor, two more subspecies are known from the region.

Subspecies

- *megaceros*: Mod. L. *megaceros* big-horned (Gr. *megas* great, Gr. *kéras* horn), in allusion to the big horns.
  Sulaiman Markhor. Endemic to South Asia; Afghanistan and Pakistan.
- *heptneri*: Eponym after Dr. V.G. Heptner (dates not known), curator of Zoological Collections at Moscow State University (*-i*, commemorating (dedication, eponym) (L.)).
  Bukhara Markhor. Afghanistan.

OEN Flare-horned Markhor (nom. form), Astor (nom. form), Balistan Astor (nom. form), Astor Markhor (nom. form), Snake-eater (nom. form), Straight-horned Sulaiman Markhor (subsp. *megaceros*), Kabul Markhor (subsp. *megaceros*), Kabal Markhor (subsp. *megaceros*), Cakul Markhor (subsp. *megaceros*), Tajik Markhor (subsp. *heptneri*), Turkomen Markhor (subsp. *heptneri*), Bukhara Screw-horned Goat (subsp. *heptneri*), Pir Panjal Markhor (J. Syn. *cashmiriensis*), Kashmir Markhor (J. Syn. *cashmiriensis*), Chitral Markhor (J. Syn. *chitralensis*), Gilgit Markhor (J. Syn. *gilgitensis*), Hazara Markhor (J. Syn. *gilgitensis*), Suleiman Markhor (J. Syn. *jerdoni*), Straight-horned Markhor (J. Syn. *jerdoni*), Trans-Indus Markhor (J. Syn. *jerdoni*)

VN Lad. *Ra-pho che* (male), *Ra-che* (male), *Ra-moche* (female)

484. ***Capricornis thar*** **(Hodgson, 1831)**
Himalayan Serow

Capricornis: L. *capricornis* with goatlike horns (L. *capra* she-goat; *cornu* horn).
thar: Mod. L. *thar* a local name for the species in Nepal.
South Asia and Myanmar. Himalayas in Bangladesh, Bhutan, India, and Nepal.

OEN The Thar (J. Syn. *bubalina*), Forest Goat (J. Syn. *bubalina*), Hume's Forest Goat (J. Syn. *humei*), Chamba Serow (J. Syn. *rodoni*), Mainland Serow (J. Syn. *jamrachi*), Jamrach's Serow (J. Syn. *jamrachi*)

485. ***Hemitragus jemlahicus*** **(H. Smith, 1826)**
Himalayan Tahr

Hemitragus: Mod. L. *hemitragus* half-goat (Gr. *hēmi-* half-, Gr. *trágos* goat).
jemlahicus: L. adj. *jemlahicus* meaning from Jemla Hills, Nepal (*-icus*, belonging to, pertaining to (location, toponym) (L.)).
China (Tibet) and South Asia. Himalayas in Bhutan, India, and Nepal.

OEN The Tahr (nom. form)
VN Hind. *Tare*, *Tehr*, *Tahir*; Kash. *Kras*, *Jagla*; Kin. *Jhula*, *Kart*, *Thar* (male), *Tharni* (female); Nep. *Jharal*

486. ***Naemorhedus goral*** **(Hardwicke, 1825)**
Himalayan Goral

Naemorhedus: Mod. L. *naemorhedus* young goat of the woods (L. *nemus*, *nemoris* wood; L. *hedus* a young goat); in allusion to its montane woody habitat.
goral: Mod. L. *goral* from Nep. *ghural* the name of the species.
China (Tibet) and South Asia. Himalayas in Bhutan, India, Nepal, and Pakistan. Besides the nominate subspecies, *N. g. goral* Eastern Himalayan Goral from Bhutan and India, another subspecies is known from the region.

Subspecies

- *bedfordi*: Eponym after the 12th Duke of Bedford, Hastings William Sackville Russell (b. 1888–d. 1953), a keen naturalist (*-i*, commemorating (dedication, eponym) (L.)).
  Western Himalayan Goral. India, Nepal, and Pakistan.

OEN Himalayan Chamois (nom. form), Goral (nom. form), Duvaucell's Goral (J. Syn. *duvaucelii*), Bedford's Goral (Subsp. *bedfordi*), Nepal Goral (J. Syn. *hodgsoni*), Hodgson's Goral (J. Syn. *hodgsoni*)
VN Bhut. *Ra-giyu*; Lepch. *Shu-ging*; Kash. *Pijur*, *Pijur Rein*, *Pijur Rom*, *Ramu*; Pahari W. *Gural*, *Serao*, *Sarao*, *Bund-Bakree*; Lepch. *Suhging*

487. *Naemorhedus griseus* Milne-Edwards, 1872
Chinese Goral

Naemorhedus: Mod. L. *naemorhedus* young goat of the woods (L. *nemus*, *nemoris* wood; L. *hedus* a young goat); in allusion to its montane woody habitat.
griseus: Med. L. *griseus* grey (>Old French *gris* grey).
China, Southeast Asia, and South Asia. India (northeast).

OEN Grey Long-tailed Goral (nom. form), Gray Mountain Goral (nom. form)

488. *Naemorhedus baileyi* Pocock, 1914
Red Goral

Naemorhedus: Mod. L. *naemorhedus* young goat of the woods (L. *nemus*, *nemoris* wood; *hedus* a young goat); in allusion to its montane woody habitat.
baileyi: Eponym after Frederick Marshman Bailey (1882–1967), an Indian-born British Intelligence Officer, naturalist, and collector in Tibe (-*i*, commemorating (dedication, eponym) (L.)).
China and South Asia. India (Arunachal Pradesh).

OEN Tibetan Red Goral (nom. form)

489. *Nilgiritragus hylocrius* (Ogilby, 1838)
Nilgiri Tahr

Nilgiritragus: Mod. L. *nilgiritragus* Nilgiri-goat (Nilgiri (Sans. *nil*, *neel* blue; Sans. *giri* mountain) Hills; Gr. *trágos* goat).
hylocrius: Mod. L. *hylocrius* forest-dweller (>Gr. *ýlè* forest, wood; Gr. *ákrios* inhabitant).
Endemic to India; Kerala and Tamil Nadu (in the Western Ghats).

OEN Nilgiri Ibex (nom. form)
VN Tam. *Warri-ātu*, *Warra-aadu*

490. *Ovis ammon* (Linnaeus, 1758)
Argali

Ovis: L. *ovis* sheep.
ammon: Med. L. *ammon* from Hebrew *Amon* >Egypt *amun* or *amen* he who is hidden or concealed; the name of an Egyptian deity, worshipped in Africa under the form of a ram.
China, Central Asia, and South Asia. Two subspecies are known from the region.

Subspecies

- *hodgsoni*: Eponym after Brian Houghton Hodgson (1801–1894), a British civil servant, naturalist, and ethnologist stationed in Nepal (-*i*, commemorating (dedication, eponym) (L.)).

Tibetan Argali. Himalayas in India (Himachal Pradesh, Jammu and Kashmir and Sikkim) and Nepal.

- *polii*: Eponym after Marco Polo (1254–1324), a Venetian merchant traveler who first mentioned this sheep (-*i*, commemorating (dedication, eponym) (L.)).

  Pamir Argali. Afghanistan (Pamir mountains) and India (Himalayas in Jammu and Kashmir).

VN Tib. *Gnow, Niar, Nyund, Nyan, Hyan, Nuan*

491. *Ovis orientalis* Gmelin, 1774
Urial

Ovis: L. *ovis* sheep.
orientalis: L. *orientalis* eastern (*oriens*, *orientis* east), (-*alis*, pertaining to (L.)).

Middle East and South Asia. Three subspecies are known from the region.

Subspecies

- *vignei*: Eponym after Godfrey Thomas Vigne (1801–1863), a British traveler and naturalist, who perhaps was the firt Englsihman to visit Afghanistan (-*i*, commemorating (dedication, eponym) (L.)).

  Ladhak Urial. Himalayas in India (Jammu and Kashmir) and Pakistan.
- *cycloceros*: Mod. L. *cycloceros* spiral-horned (Gr. *kuklos* circle, Gr. *kéras* horn).

  Afghan Urial. Afghanistan, India (Himalayas in Jammu and Kashmir) and Pakistan.
- *punjabiensis*: L. adj. *punjabiensis* meaning from Punjab (-*ensis*, geographic, occurrence in (location, toponym) (L.)).

  Punjab Urial. Endemic to Pakistan (Salt Range, Punjab).

VN Hind. *Urial, Ooorial*; Pash. *Koch, Kuch*; Tib. *Sha-pao*; Lad. *Sha*

492. *Pantholops hodgsoni* (Abel, 1826)
Tibetan Antelope

Pantholops: Mod. L. *panthalops* all-antelope (Gr. *pan* all; Gr. *ánthólops* antelope).
hodgsoni: Eponym after Brian Houghton Hodgson (1801–1894), a British civil servant, naturalist, and ethnologist stationed in Nepal (-*i*, commemorating (dedication, eponym) (L.)).

China and South Asia (Ladhak, India).

OEN Chiru Antelope (nom. form)
VN Tib. *Chiru, Isoors, Choors*

493. *Pseudois nayaur* (Hodgson, 1833)
Blue Sheep

Pseudois: Mod. L. *pseudois* false-sheep (Gr. *pseudos* false; Gr. *ois* from L. *ovis* sheep); this species lacks facial gland and has a goatlike tail.
nayaur: Mod. L. *nayaur* from Nep. *nahoor* or *naur*, a name for this species.
China (Tibet) and South Asia. Himalayas in Bhutan, India, and Nepal.

OEN Blue Wild Sheep, Burhel
VN Hind. *Bharal*, *Menda* (male), *Bharur*, *Wa*, *War*; Lad. *Na*, *Sna*; Tib. *Na*, *Sna*; Nep. *Nervati*

Order Cetacea Brisson, 1762
Suborder Mysticeti Flower, 1864
Family Balaenidae Gray, 1821

494. *Eubalaena australis* (Desmoulins, 1822)
Southern Right Whale

Eubalaena: Mod. L. *eubalaena* true whale (Gr. *eÿ* well, typical; L. *balaena* whale).
australis: L. *australis* southern (L. *auster* south), (*-alis*, pertaining to (L.)).
Oceans in the southern Hmisphere. Historic records of *Eubalaena* sp. were reported from the coasts of Gujarat, India.

Family Balaenopteridae Gray, 1864

495. *Balaenoptera musculus* (Linnaeus, 1758)
Blue Whale

Balaenoptera: Mod. L. *balaenoptera* winged whale (L. *balaena* whale; Gr. *-pteros* -winged (>Gr. *pteron* wing)); in allusion to the strong dorsal fin.
musculus: Mod. L. *musculus* small mouse (Gr. *mÿs* mouse, *-ulus,* diminutive (comparison), somewhat (adj.) (L.)).
Cosmopolitan, all oceans except the Arctic. Two subspecies are known from the region.

Subspecies

- *brevicauda*: Mod. L. *brevicauda* short-tailed (L. *brevis* short; L. *cauda* tail).
  Pygmy Blue Whale. Arabian Sea, Bay of Bengal, and Indian Ocean south of Sri Lanka.
- *indica*: L. adj. *indica* belonging to India (*-ica* belonging to, pertaining to (L./Gr.)).
  Indian Ocean Blue Whale. Arabian Sea and Bay of Bengal.

OEN Sulphur-bottom Whale (nom. form), Northern Pygmy Blue Whale (nom. form), Indian Fin-Whale (Subsp. *indica*)

496. ***Balaenoptera physalus*** **(Linnaeus, 1758)**
Fin Whale

Balaenoptera: Mod. L. *balaenoptera* winged whale (L. *balaena* whale; Gr. *-pteros* -winged (>Gr. *pteron* wing)); in allusion to the strong dorsal fin.
physalus: Mod. L. *physalus* (Gr. *phýsalos*) whale.

Cosmopolitan, all oceans except the Arctic. Andaman Sea in Indian Ocean.

OEN Common Finback Whale (nom. form), Common Rorqual (nom. form)

497. ***Balaenoptera acutorostrata*** **Lacepede, 1804**
Common Minke Whale

Balaenoptera: Mod. L. *balaenoptera* winged whale (L. *balaena* whale; Gr. *-pteros* -winged (>Gr. *pteron* wing)); in allusion to the strong dorsal fin.
acutorostrata: Mod. L. *acutorostrata* pointed-snout (L. *acutus* sharp-pointed; *-rostratus* -billed (*rostrum* bill)).

Cosmopolitan, all oceans except the Arctic. Bay of Bengal and Indian Ocean south of Sri Lanka.

OEN Dwarf Minke Whale (nom. form), Little Piked Whale (nom. form), Pike Whale (nom. form), Least Rorqual (nom. form), Lesser Rorqual (nom. form)

498. ***Balaenoptera edeni*** **Anderson, 1879**
Bryde's Whale

Balaenoptera: Mod. L. *balaenoptera* winged whale (L. *balaena* whale; Gr. *-pteros* -winged (>Gr. *pteron* wing)); in allusion to the strong dorsal fin.
edeni: Eponym after Ashley Eden (1831–1887), a British diplomat stationed in Burma (*-i*, commemorating (dedication, eponym) (L.)).

Pacific, Indian, and Atlantic Oceans between about 40°N and 40°S. Arabian Sea, Bay of Bengal, and Indian Ocean south of Sri Lanka.

OEN Eden's Whale (nom. form)

499. ***Megaptera novaeangliae*** **(Borowski, 1781)**
Humpback Whale

Megaptera: Mod. L. *megaptera* great winged (Gr. *megas* great; Gr. *ptera* >Gr. *pteron* wing), in allusion to the strong dorsal fin.
novaeangliae: Mod. L. *novaeangliae* New Englander (L. *novae* new; L. *angliae* England), (*-ae*, geographic (location, toponym) (L.)).

Cosmopolitan. Arabian Sea, Bay of Bengal, and Indian Ocean south of Sri Lanka.

OEN Hump-backed Whale (nom. form)

Suborder Odontoceti Flower, 1867
Family Delphinidae Gray, 1821

500. *Delphinus capensis* Gray, 1828
Long-beaked Common Dolphin

Delphinus: Gr. *delphinus* dolphin.
capensis: L. adj. meaning from Cape of Good Hope, South Africa (*-ensis*, geographic, occurrence in (location, toponym) (L.)).

Coastal waters (up to 180 km into the ocean) in tropical waters. One subspecies is known from the region.

Subspecies

- *tropicalis*: Mod. L. *tropicalis* (>L. *tropicus*) tropical.
  Arabian Common Dolphin. Arabian Sea, Bay of Bengal and Indian Ocean south of Sri Lanka.

OEN Cape dolphin (nom. form), Arabian Dolphin (subsp. *tropicalis*), Arabian Saddle-backed Dolphin (subsp. *tropicalis*), Malabar Common Dolphin (subsp. *tropicalis*), Malabar Saddleback Dolphin (subsp. *tropicalis*), Saddleback Dolphin (subsp. *tropicalis*)
VN Sinh. *Mulla*; Tam. (SL) *Ongil*

501. *Feresa attenuata* Gray, 1875
Pygmy Killer Whale

Feresa: Mod. L. *feresa* from Fr. *feres*, a name for this dolphin.
attenuata: L. *attenuatus* shortened, abbreviated, thin (*attenuare* to attenuate).

Pacific, Indian and Atlantic oceans between about 40°N and 35°S. Arabian Sea, Bay of Bengal, and Indian Ocean south of Sri Lanka.

502. *Globicephala macrorhynchus* Gray, 1846
Short-finned Pilot Whale

Globicephala: Mod. L. *globicephala* round-headed (L. *globus* ball; Gr. *kephalus* head), in allusion to the globular shape of the head.
macrorhynchus: Mod. L. *macrorhynchus* (>Gr. *makrorhunkhos*) (Gr. *makros* long; Gr. *rhunkhos* bill) long-billed.

Pacific, Indian, and Atlantic Oceans between about 50°N and 40°S. Arabian Sea, Bay of Bengal, and Indian Ocean south of Sri Lanka.

503. *Grampus griseus* (G. Cuvier, 1812)
Risso's Dolphin

Grampus: Mod. L. *grampus* possibly a corruption of the Fr. *grand poisson*, 'great fish'.
griseus: Med. L. *griseus* grey (>Old French *gris* grey).

Pacific, Indian, and Atlantic Oceans between about 50°N and 40°S. Arabian Sea, Bay of Bengal, and Indian Ocean south of Sri Lanka.

504. *Lagenodelphis hosei* Fraser, 1956
Fraser's Dolphin

Lagenodelphis: Mod. L. *lagenodelphis* bottle-nosed dolphin (Gr. *lágè-nos* flagon, bottle; Gr. *delphis* dolphin); this nomen is a combination of two generic nomen, *Lagenorhynchus* and *Delphinus*, with which it shares some characters.
hosei: Eponym after Dr. Charles Hose (1863–1929), a British naturalist stationed at Sarawak, Borneo, who collected the skull in 1895 (*-i*, commemorating (dedication, eponym) (L.)).

Pacific, Indian, and Atlantic Oceans between about 35°N and 35°S. Arabian Sea, Bay of Bengal, and Indian Ocean south of Sri Lanka.

The common name is after Dr. Francis Fraser (dates not known), the curator of Marine Mammals at British Museum (Natural History).

505. *Orcaella brevirostris* (Owen, 1866)
Irrawady Dolphin

Orcaella: Mod. L. *orcaella* dim. of L. *orca* a kind of whale (*-ella*, diminutive (comparison); somewhat (adj.) (L.)).
brevirostris: Mod. L. *brevirostris* short-billed (L. *brevis* short; *-rostris* -billed (*rostrum* bill)).

Tropical and subtropical Indo-Pacific (Southeast Asia and South Asia), almost exclusively in estuarine and fresh waters. Bay of Bengal.

OEN Fresh-water Round-headed Dolphin (J. Syn. *fluminalis*)

506. *Orcinus orca* Linnaeus, 1758
Killer Whale

Orcinus: Mod. L. *orcinus* (>L. *orca*) a kind of whale.
orca: L. *orca* a kind of whale.

Cosmopolitan. Arabian Sea, Bay of Bengal, and Indian Ocean south of Sri Lanka.

507. *Peponocephala electra* (Gray, 1846)
Melon-headed Dolphin

Peponocephala: Mod. L. *peponocephala* melon-headed whale (Mod. L. *pepon* >Spanish *pepón* melon; Gr. *kephalus* head).
electra: L. *electra* (Gr. *ēlektra*) shining one.

Pacific, Indian, and Atlantic Oceans between about 35°N and 32°S. Arabian Sea, Bay of Bengal, and Indian Ocean south of Sri Lanka.

OEN Spindle-shaped Dolphin (J. Syn. *fusiformis*)

508. *Pseudorca crassidens* (Owen, 1846)
False Killer Whale

Pseudorca: Mod. L. *pseudorca* false whale (Gr. *pseudos* false; L. *orca* a kind of whale).

crassidens: Mod. L. *crassidens* heavy-toothed (L. *crassus* thick, heavy, dense; L. *dens* tooth).

Pacific, Indian and Atlantic Oceans between about 60°N and 60°S. Arabian Sea, Bay of Bengal, and Indian Ocean south of Sri Lanka.

OEN Large-toothed Killer Whale (nom. form)
VN Sinh. *Komaduva*

509. *Sousa chinensis* (Osbeck, 1765)
Indopacific Humpback Dolphin

Sousa: Mod. L. *sousa* from Bengali name *súsúk* or *sishúk*, the name of the Gangetic Dolphin.
chinensis: L. adj. *chinensis* meaning from China (*-ensis* geographic, occurrence in (location, toponym) (L.)).

Coastal waters (up to 180 km into the ocean) in tropical waters along the eastern coast of Africa, Arabian Peninsula, Indian Peninsula, Southeast Asia, and northern Australia. Arabian Sea, Bay of Bengal, and Indian Ocean south of Sri Lanka.

OEN Freckled Dolphin (J. Syn. *lentiginosa*), Speckled Dolphin (J. Syn. *lentiginosa*), Spotted Dolphin (J. Syn. *lentiginosa*), Dark Humpbacked Dolphin (J. Syn. *plumbea*), Plumbeous Dolphin (J. Syn. *plumbea*), Indian Humpback Dolphin (J. Syn. *plumbea*), Plumbeous Humpback Dolphin (J. Syn. *plumbea*), Spot-bellied Dolphin (J. Syn. *maculiventer*), Ferguson's Dolphin (J. Syn. *fergusoni*)
VN Sinh. *Kabara Mulla*

510. *Stenella longirostris* (Gray, 1828)
Spinner Dolphin

Stenella: Dim. of *steno*, genus *nomen* in honour of Dr. Nikolaus Steno (b. 1638–d. 1687), a celebrated Danish anatomist (*-ella*, diminutive (comparison), somewhat (adj.) (L.)).
longirostris: Mod. L. *longirostris* long-billed (L. *longus* long; *-rostris* -billed (rostrum bill)).

Pacific, Indian, and Atlantic Oceans between about 35°N and 32°S. Arabian Sea, Bay of Bengal, and Indian Ocean south of Sri Lanka.

OEN Gray's Spinner Dolphin (nom. form), Long-nosed Dolphin (nom. form)

511. *Stenella coeruleoalba* (Mayen, 1833)
Striped Dolphin

Stenella: Dim. of *steno*, genus *nomen* in honour of Dr. Nikolaus Steno (b. 1638–d. 1687), a celebrated Danish anatomist (*-ella*, diminutive (comparison), somewhat (adj.) (L.)).

coeruleoalba: Mod. L. *coeruleoalba* dark blue and white (L. *caeruleus* dark blue; L. *albus* white); in allusion to the prominent dark blue and white stripes on its body.

Pacific, Indian, and Atlantic Oceans between about 60°N and 50°S. Arabian Sea, Bay of Bengal, and Indian Ocean south of Sri Lanka.

512. *Stenella attenuata* (Gray, 1846)
Pantropical Spotted Dolphin

Stenella: Mod. L. *stenella* meaning Dim. of *steno*, genus *nomen* in honour of Dr. Nikolaus Steno (b. 1638–d. 1687), a celebrated Danish anatomist (*-ella*, diminutive (comparison), somewhat (adj.) (L.)).
attenuata: L. *attenuatus* shortened, abbreviated, thin (*attenuare* to attenuate).

Pacific, Indian, and Atlantic Oceans between about 45°N and 35°S. Arabian Sea, Bay of Bengal, and Indian Ocean south of Sri Lanka.

OEN Malay Dolphin (J. Syn. *malayana*)

513. *Steno bredanensis* (Lesson, 1828)
Rough-toothed Dolphin

Steno: Genus *nomen* in honour of Dr. Nikolaus Steno (b. 1638–d. 1687), a celebrated Danish anatomist.
bredanensis: L. adj. meaning from Breda, Netherlands (*-ensis*, geographic, occurrence in (location, toponym) (L.)). However, it is to be noted that the species is in fact named after an artist, 'van Breda', who first noticed the animal in Cuvier's collection.

Pacific, Indian, and Atlantic Oceans between about 47°N and 35°S. Arabian Sea, Bay of Bengal, and Indian Ocean south of Sri Lanka.

514. *Tursiops truncatus* (Montagu, 1821)
Bottle-nosed Dolphin

Tursiops: Mod. L. *tursiops* dolphin-like (L. tursio a kind of fish resembling a dolphin; Gr. *óps* aspect).
truncatus: L. *truncatus* mutilated, cut short (L. *truncare* to mutilate).

Pacific, Indian, and Atlantic Oceans between about 60°N and 50°S. Arabian Sea, Bay of Bengal, and Indian Ocean south of Sri Lanka.

OEN Bottlenose Dolphin (nom. form)
VN Sinh. *Mulla*; Tam. (SL) *Ongil*

515. *Tursiops aduncus* (Ehrenberg, 1833)
Indo-Pacific Bottlenose Dolphin

Tursiops: Mod. L. *tursiops* dolphin-like (L. *tursio* a kind of fish resembling a dolphin; Gr. *óps* aspect).
aduncus: Med. L. *aduncus* hooked (L. *ádunc* hook).
Coastal waters (up to 300 km into the ocean) in tropical waters along the eastern coast of Africa, Arabian Peninsula, Indian Peninsula, Southeast

Asia, and northern Australia. Arabian Sea, Bay of Bengal, and Indian Ocean south of Sri Lanka.

Common name Indo-Pacific Bottle-nosed Dolphin, Indo-Pacific owing to its distribution range and Bottle-nosed in allusion to its smooth bottle-like snout (or rostrum).

OEN Dawson's Dolphin (J. Syn. *dawsoni*), Black Dolphin (J. Syn. *perniger*)

Family Phocoenidae Gray, 1825

516. *Neophocaena phocaenoides* (G. Cuvier, 1829)
Finless Porpoise

Neophocaena: Gr. *neophocaena* new porpoise (Gr. *neos* new; Gr. *phôkaina* porpoise).

phocaenoides: Mod. L. *phocaenoides* porpoise-like (>Gr. *phôkaina* porpoise), (*–oides* (>Gr. *eidos*), apparent shape).

Coastal waters (up to 150 km into the ocean) in tropical waters along the Arabian Peninsula, Indian Peninsula, and Southeast Asia. Arabian Sea and Bay of Bengal.

Common name Finless Porpoise, due to it lacking a true dorsal fin. The fin is replaced by a low ridge covered in thick skin. The name Porpoise is from French *pourpois*, possibly from Med. L. *porcopiscis* (L. *porcus* pig, *piscis* fish).

OEN Little Indian Porpoise (nom. form), Karachi Finless Porpoise (J. Syn. *kurrachiensis*), Molagan Finless Porpoise (J. Syn. *molagan*)

VN Tam. *Ongil*, *Molagan*; Mal. *Minikutty*, *Eleyan Eedi*

Family Physeteridae Gray, 1821

517. *Physeter macrocephalus* Linnaeus, 1758
Sperm Whale

Physeter: Gr. *physètêr* blowpipe, a whale; from the single spiracle or blowpipe.

macrocephalus: Gr. *makrokephalos* (Gr. *makros* great, big; Gr. *kephalus* head) great-headed.

Cosmopolitan. Arabian Sea, Bay of Bengal, and Indian Ocean south of Sri Lanka.

The name Sperm Whale is shortened form of Spermaceti Whale, owing to its bearing spermaceti (wax-like susbstance stored in head cavities).

OEN Giant Sperm Whale (J. Syn. *catodon*)

Family Kogiidae Gill, 1871

518. *Kogia breviceps* (Blainville, 1838)
Pygmy Sperm Whale

Kogia: Possibly Latinized form of *codger*, a barbarous word, but it might be a tribute to Cogia Effendi, a Turk, who observed whales in the Mediterranean.

breviceps: Mod. L. *breviceps* short-headed (L. *brevis* short; *-ceps* -headed (L. *caput* head)).

Pacific, Indian, and Atlantic Oceans between about 60°N and 45°S. Arabian Sea, Bay of Bengal, and Indian Ocean south of Sri Lanka.

Common name Pygmy Sperm Whale, in allusion to its small size, grows up to 3.5 m. The name Sperm Whale is shortened form of Spermaceti Whale, owing to its bearing spermaceti (wax-like susbstance stored in head cavities).

519. *Kogia sima* (Owen, 1866)
Dwarf Sperm Whale

Kogia: Possibly Latinized form of *codger*, a barbarous word, but it might be a tribute to Cogia Effendi, a Turk, who observed whales in the mediterranean.

sima: L. *sima*, *simus* (Gr. *simos*) snubnosed.

Pacific, Indian, and Atlantic Oceans between about 45°N and 35°S. Arabian Sea, Bay of Bengal, and Indian Ocean south of Sri Lanka.

Common name Dwarf Sperm Whale, in allusion to its small size, grows up to 2.7 m. The name Sperm Whale is shortened form of Spermaceti Whale, owing to its bearing spermaceti (wax-like susbstance stored in head cavities).

Family Platanistidae Gray, 1846

520. *Platanista gangetica* (Roxburgh, 1801)
Gangetic Dolphin

Platanista: Mod. L. *platanista* given name for Gangetic Dolphin (Gr. *platanistês*, a fish of the Ganges, probably this species).

gangetica: L. adj. *gangetica* belonging to the Ganges river, India (*-ica* belonging to, pertaining to (L./Gr.)).

Endemic to South Asia. Beside the nominate subspecies, *P. g. gangetica* Ganges River Dolphin, endemic to river systems of Bangladesh (Meghna and its tributaries), Bhutan (in Manas and Puna Tsang Chu), India (Brahmaputra, Ganges, Son, Hooghly, Karnapuhli, and Bay of Bengal), and (Nepal tributaries of Ganges), another subspecies is known from the region.

Subspecies

- *minor*: L. *minor* smaller.

Indus River Dolphin. Endemic to Pakistan (Indus and its tributaries).

Common name Gangetic Dolphin, after its range in the Ganges river and its tributary.

OEN Gangetic Porpoise (nom. form)
VN Beng. *Shishuk*; Hind. *Susa*, *Sons*, *Susu*; Sans. *Sisumar*

Family Ziphiidae Gray, 1865

521. *Indopacetus pacificus* (Longman, 1926)
Tropical Bottlenose Whale

Indopacetus: Mod. L. *indopacetus* Indo-Pacific whale, generic nomen combination of *Indopa* from Indo-Pacific and L. *cetus* whale.
pacificus: L. adj. *pacificus* meaning from the Pacific Ocean (*-icus* belonging to, pertaining to (location, toponym) (L.)).

Tropical waters of Pacific and Indian Oceans. Arabian Sea, Bay of Bengal, and Indian Ocean south of Sri Lanka.

Common name Tropical Bottlenose Whale is in allusion to its range in tropical waters.

OEN Longman's Beaked Whale (nom. form)

522. *Mesoplodon densirostris* (Blainville, 1817)
Blainville's Beaked Whale

Mesoplodon: Mod. L. *mesoplodon* armed with a tooth in the middle (Gr. *mésos* middle; Gr. *ópla* arms; Gr. *ódoús*, *ódôn* tooth); in allusion to a prominent tooth in the lower jaw.
densirostris: Mod. L. *densirostris* thick-billed (L. *densus* thick; L. *-rostris* -billed (L. *rostrum* beak)).

Pacific, Indian, and Atlantic Oceans between about 66°N and 35°S. Arabian Sea, Bay of Bengal, and Indian Ocean south of Sri Lanka.

Common name Blainville's Beaked Whale, after Henri Marie Ducrotay de Blainville (1777–1850), a French zoologist and anatomist who first described this species.

OEN Dense-Beaked Whale (nom. form), Massive-toothed Whale (nom. form)

523. ***Mesoplodon hotaula*** **Deraniyagala, 1963**
Deraniyagala's Beaked Whale

Mesoplodon: Mod. L. *mesoplodon* armed with a tooth in the middle (Gr. *mésos* middle; Gr. *ópla* arms; Gr. *ódoús*, *ódôn* tooth); in allusion to a prominent tooth in the lower jaw.
hotaula: Mod. L. *hotaula* from Sinh. *Hota ul*, a common name for dolphins and whales.

Tropical waters of Pacific and Indian Oceans, conspicuously absent from the northern Arabian Sea. Arabian Sea, Bay of Bengal, and Indian Ocean south of Sri Lanka.

EN Indian Ocean Ginkgo-toothed Beaked Whale
VN Sinh. *Hota ul Thalmaha*

524. ***Ziphius cavirostris*** **G. Cuvier, 1823**
Goose-beaked Whale

Ziphius: Mod. L. *ziphius* swordfish (Gr. *xiphos* sword).
cavirostris: Mod. L. *cavirostris* hollow-billed (L. *cavus*, *cavi* hollow, hole (*cavus* excavated); L. *-rostris* -billed (*rostrum* beak)). Cosmopolitan. Arabian Sea, Bay of Bengal, and Indian Ocean south of Sri Lanka.

Common name Goose-beaked Whale, in allusion to the profile of its head bearing semblance to that of a goose.

OEN Cuvier's Beaked Whale (nom. form), Indian Ocean Cuvier's Beaked Whale (J. Syn. *indicus*)
VN Sinh. *Hota ul Thalmaha*

4 Dictionary of South Asian Mammalian Names

Aa

abramus L. *abramus* beautiful-mouse (L. *abrus* beautiful, graceful; Gr. *mŷs* mouse) (*Pipistrellus*).

achates Gr. Myth. *Achates*, companion and close friend of Aeneas, characters from Virgil's Latin epic poem, *Aeneid* (J. Syn. Satpura Langur, Western Hanuman Langur, see *Semnopithecus dussumieri*).

Acinonyx Mod. L. *acinonyx* pointed-claw Gr. *ákaina* thorn, prick (Gr. *ákis, ákidos* point (>*acantha* thorn or pick); Gr. *ónyx* claw); from the non-retractile, pointed claws (*A. jubatus*).

acrophilus Mod. L. *acrophilus* mountain-loving (Gr. *akron* mountain top, peak; Gr. *philos* -loving) (J. Syn. Abbott's Mountain Vole, see *Alticola stoliczkanus*).

Acomys Mod. L. *acomys* spiny mouse (Gr. *ákê* a sharp point (>*acantha* thorn or prick); Gr. *mŷs* mouse); owing to its spiny fur (*A. dimidiatus*).

acutorostrata Mod. L. *acutorostrata* pointed-snout (L. *acutus* sharp-pointed; *-rostratus* -billed (*rostrum* bill)) (*Balaenoptera*).

aduncus Med. L. *aduncus* hooked (L. *ádunc* hook) (*Tursiops*).

adustus L. *adustus* burnt, fuliginous, soot-coloured (Subsp. *Cuon alpinus* Burmese Wild Dog).

aegagrus Mod. L. *aegagrus* (Gr. *aigagros*, *aig-*, *aix* goat; Gr. *agrios* wild) (*Capra*).

aegyptiaca/aegyptiacus L. *aegyptiaca* or *aegyptiacus* Egyptian (*-aca*, *-acus*, belonging to, pertaining to (L. /Gr.)) (*Tadarida/Rousettus*).

aequicaudalis Mod. L. *aequicaudalis* equal-tailed (L. *aquus* equal, L. *cauda* tail) (J. Syn. Hodgson's Rat, see *Rattus rattus*).

affinis L. *affinis* related, allied (*Rhinolophus* and *Falsistrellus*) (Subsp. *Felis chaus* Himalayan Jungle Cat) (J. Syn. Indian Cynoptere, see *Rousettus leschenaultii*).

afghanus Mod. L. *afghanus* belonging to Afghan or Afghanistan (*-anus*, belonging to, pertaining to (L.) geographic (location, toponym)) (*Blanfordimys*/J. Syn. Afghan Jird, see *Meriones libycus*).

Ailurus Gr. *aílouros* cat (later, a weasel), so called due to its resemblance to a cat (*A. fulgens*).

aitchisoni Eponym after Dr. J.E.T. Aitchison (b. 1836–d. 1898), a Scottish surgeon and botanist (*-i*, commemorating (dedication, eponym) (L.)) (J. Syn. Aitchison's Vole, see *Hyperacrius fertilis*).

ajax Gr. Myth. *Ajax*, a hero of the Trojan War – a major character in Homer's *Iliad* (*Semnopithecus*).

aladdin No explanation was given, perhaps after Aladdin (a protagonist of one of the Middle Eastern folktales Aladdin) (*Pipistrellus*).

albicauda Mod. L. *albicauda* white-tailed (L. *albus* white, L. *cauda* tail) (*Alticola*).

albidiventris Mod. L. *albidiventris* white-bellied (L. *albidus* whitish, L. *ventris* belly) (J. Syn. White-bellied Field Mouse, see *Mus booduga*).

albifrons Mod. L. *albifrons* white-headed (L. *albus* white; L. *frons* forehead, front) (Subsp. *Arctictis binturong* Himalayan Binturong).

albipes Mod. L. *albipes* white-legged (L. *albus* white; L. *pes* foot, from Gr. *pous* foot) (J. Syn. White-legged Squirrel, see *Ratufa macroura*; White-footed Antelope, see *Boselaphus tragocamelus*).

albiventer Mod. L. *albiventer* bearing white belly (L. *albus* white, L. *venter* belly) (Subsp. *Petaurista petaurista* White-bellied Flying Squirrel).

alboniger Mod. L. *alboniger* white and black (L. adj. *albus* white, L. *niger* black) (*Hylopetes*).

alexandrinus L. *alexandrinus* of Alexandria, Egypt (J. Syn. Alexandria Black Rat or Egyptian House Rat or Egyptian Ship Rat or Alexandria Black Rat, see *Rattus rattus*).

Allactaga Mod. L. *allactaga* >Mongol *alak-daagha* (*alak* variegated, *daagha* colt); in allusion to its variegated coat of the type species of the genus (*A. elater, A. hotsoni,* and *A. williamsi*).

alophus Gr. *alophos* without a crest (J. Syn. Nepal Crestless Porcupine, see *Hystrix brachyura*).

alpinus L. *alpinus* alpine, of the Alps (*Cuon*).

altaica L. adj. *altaica* belonging to the Altai mountains, East Central Asia (*-ica* belonging to, pertaining to (L. /Gr.)) (*Mustela*).

Alticola/alticola Med. L. *alticola* inhabitant of high altitudes (L. *altus* high, *colo* to dwell, to inhabit), as the animal lives in elevated habitats (*A. albicauda, A. argentatus, A. montosa, A. roylei,* and *A. stoliczkanus*/*Cricetulus*).

amir Mod. L. *amir* from Arabic *amir* prince, or one who gives orders (J. Syn. Afghan Hedgehog, see *Paraechinus hypomelas*).

ammon Med. L. *ammon* from Hebrew *Amon* >Egypt *Amun* or *Amen* he who is hidden or concealed; the name of an Egyptian deity, worshipped in Africa under the form of a ram (*Ovis*).

Anathana Mod. L. *anathana* >Tamil *mungil anathaan* (for Bamboo Squirrel) (*A. ellioti*).

andamanensis L. adj. *andamanensis* meaning from Andaman Islands, India (*-ensis*, geographic, occurrence in (location, toponym) (L.)) (*Rattus* and *Crocidura*/Subsp. *Rhinolophus affinis* Andaman Horseshoe Bat/J. Syn. Andaman Short-nosed Fruit Bat, see *Cynopterus brachyotis*; Burmese Pig-tailed Macaque, see *Macaca leonina*).

angdawai L. adj. *angdawai* meaning 'of the Ang Dawa Sherpa of Nepal', (*-i*, commemorating (dedication) (L.)) (J. Syn. Ang Dawa Pika, see *Ochotona roylei*).

andersoni Eponym after Dr. John Anderson (b. 1833–d. 1900), a Scottish anatomist and zoologist, first curator of Indian Museum (*-i*, commemorating (dedication, eponym) (L.)). (J. Syn. Anderson's Grey Mongoose, see *Urva edwardsii*).

angulatus L. *angulatus* angular, cornered (L. *angulus* angle, corner) (Subsp. *Cynopterus sphinx* Miller's Short-nosed Fruit Bat).

annandalei Eponym after Thomas Nelson Annandale (b. 1876–d. 1924), a Scottish zoologist, herpetologist, and entomologist; Director of the Indian Museum in Calcutta and first Director of the Zoological Survey of India (*-i*, commemorating (dedication, eponym) (L.)). (J. Syn. Annandale's Jungle Palm Squirrel, see *Funambulus tristriatus*).

annectans L. *annectans* connecting (L. *annectere* to connect) (*Myotis*).

Anourosorex Mod. L. *anourosorex* shrew without a tail (Gr. *án* without; Gr. *oúrá* tail; L. *sorex* (>Gr. *ÿrax* shrew)); due to its very short tail (*A. assamensis, A. schmidi,* and *A. squamipes*).

Antilope Med. L. *antalopus* (Gr. *ánthólops*) a horned animal, an antelope (*A. cervicapra*).

Aonyx Mod. L. *aonyx* without nail (Gr. *a* without; Gr. *ónyx* claw, nail); due to its very rudimentary claws (*A. cinerea*).

apiculatus L. adj. *apiculatus* short pointed (>L. *apex*, *apicis* crown, point, extremity) (J. Syn. Hairy Leaf-nosed Bat or Pencilled Horseshoe Bat, see *Hipposideros speoris*).

Apodemus Mod. L. *apodemus* (Gr. *ápódèmos*, from *apodēmein* to go abroad, be abroad) abroad; meaning inhabiting fields (*A. draco, A. gurkha, A. latronum, A. rusiges,* and *A. pallipes*).

aquilo L. *aquilo* northwind; in allusion to northern range of the taxon (J. Syn. Anderson's Chestnut-bellied Squirrel, see *Callosciurus erythraeus*).

aquilus L. adj. *aquilus* dark-coloured (*Gerbillus*).

arabicus L. adj. *arabicus* Arabian, or from Arabia (*-icus*, belonging to, pertaining to (location, toponym) (L.)) (Subsp. *Rousettus aegyptiacus* Arabian Rousette).

arboreus L. *arboreus* arboreal, of a tree (L. *arbor, arboris* tree) (J. Syn. Arboreal Black Rat, see *Rattus rattus*).

Arctictis Mod. L. *arctitis* bear-weasel (Gr. *arktos* a bear; Gr. *ictis* a weasel); owing to its appearance (*A. binturong*).

Arctogalidia Mod. L. *arctitis* small bear-weasel (Gr. *arktos* a bear; Gr. *galidia* (dim. of *gale*) a weasel) (*-idia* diminutive (comparison) (L. >Gr.)) (*A. trivirgata*).

arctoides Mod. L. *arctoides* bearlike (Gr. *arktos* a bear; *-oides* (>Gr. *eidos*) apparent shape) (*Macaca*).

Arctonyx Mod. L. *arctonyx* having bearlike claws (Gr. *arktos* a bear; Gr. *ónyx* claw); owing to the long, slightly curved, blunt claws (*A. collaris*).

arctos Gr. *arktos* a bear (*Ursus*).

argentatus L. *argentatus* ornamented with silver (L. *argentum* silver) (*Alticola*/J. Syn. Silver-backed Capped Langur, see *Trachypithecus pileatus*).

argentescens L. *argentescens* silverish (L. *argenteus* of silver, silvery, *argentum* silver) (Subsp. *Funambulus pennantii* Pakistan Palm Squirrel, Sind Banyan Squirrel).

Arielulus Mod. L. *arielulus* somewhat like *ariel* (from Med. folklore *Ariel*, a spirit or sylph of the air) (*-ulus*, diminutive (comparison); somewhat (adj.) (L.)) (*A. circumdatus*).

armiger L. *armiger* bearing weapons, shield-bearer (*Hipposideros*).

arnee Mod. L. *arnee* from Hin. *arna* (fem. *arni*), a name for Indian wild buffalo (*Bubalus*).

aryabertensis L. adj. *aryabertensis* belonging to Aryavart (Sans. *Āryāvarta* abode of the Aryans, the tract between the Himalaya and the Vindhya ranges, from the Eastern (Bay of Bengal) to the Western Sea (Arabian Sea)) (*-ensis*, geographic, occurrence in (location, toponym) (L.)) (Subsp. *Lepus nigricollis* Nepal Hare).

asseel Mod. L. *asseel* from Hin. *asseel gayal* native name of the Indian Bison in Chittagong (from Hin. *Asseel* original, noble or untamed) (J. Syn. Asseel Gayal, see *Bos gaurus*).

asiaticus L. *Asiaticus* Asiatic (*-icus*, belonging to, pertaining to (location, toponym) (L.)) (J. Syn. Asiatic Elephant, see *Elephas maximus*; Asiatic Black Rat, see *Rattus rattus*; Asiatic Lion, see *Panthera leo*).

Asellia L. adj. *asellia* (used as n.) ass-like (>L. *asellus*, a little ass); in allusion to long, pointed ears (*A. tridens*).

assamensis L. adj. *assamensis* meaning from Assam, India (*-ensis*, geographic, occurrence in (location, toponym) (L.)) (*Macaca* and *Anourosorex*/J. Syn. Assam Tree Shrew, see *Tupaia belangeri*; Bengal Porcupine, see *Atherurus macrourus*; Assamese Pygmy Shrew, see *Suncus etruscus;* Assamese Flying Fox, see *Pteropus giganteus*).

ater L. *ater* black, dark, dull black (*Hipposideros*/J. Syn. Black Bear-Cat, see *Arctictis binturong*).

Atherurus Mod. L. *atherurus* brush-tailed (Gr. *ather* the beard of the ear of the corn; Gr. *oúrá* tail); in allusion to the bunch of flattened scaly bristles at the tip of the tail (*A. macrourus*).

atratus L. *atratus* clothed in mourning (L. *ater* black) (J. Syn. Black Pygmy Shrew, see *Suncus etruscus*; S. Syn. Sombre Bat, see *Eptesicus tatei*).

atridorsum Mod. L. *atridorsum* black-backed (L. *ater* black, *dorsum* back) (J. Syn. Barren Island Black Rat, see *Rattus rattus*).

attenuata L. *attenuatus* shortened, abbreviated, thin (L. *attenuare* to attenuate) (*Crocidura*, *Feresa* and *Stenella*).

aurata L. *auratus* gilded, ornamented with gold (L. *aurum* gold) (*Murina*) (J. Syn. Fire Cat, see *Pardofelis temminckii*).

aurea/aureus L. *aureus* golden (*aurum* gold) (*-ea*, *-eus* having the quality of) (*Canis* and *Paradoxurus*/Subsp. *Marmota caudata* Golden Marmot/Subsp. *Macaca fascicularis* Burmese Long-tailed Macaque or Golden Macaque, and *Muntiacus vaginalis* Golden Indian Muntjak).

aurex L. *aurex* from *aureus* golden (L. *aurum* gold) (*Tylonycteris*).

aurifrons Mod. L. *aurifrons* gold-fronted (L. *aurum*, *auri* gold; L. *frons* forehead, front) (Subsp. *Macaca sinica* Wet Zone Toque Macaque).

auritus L. *auritus* long-eared (*Hemiechinus*) (Subsp. *Manis pentadactyla* Long-eared Chinese Pangolin; *Ochotona macrotis* Long-eared Pika).

auriventer Mod. L. *auriventer* golden-bellied (L. *aurum* gold; L. *venter* belly) (J. Syn. Golden-bellied Weasel, see *Mustela kataiah*).

aurobrunneus Mod. L. *aurobrunneus* gold-brown (L. *aurum* gold; Mod. L. *brunneus* brown); in allusion to its coat color – rich chestnut brown above and golden red below (Subsp. *Lutra lutra* Himalayan Otter).

auropunctatus Mod. L. *auropunctatus* gold-spotted (L. *aurum* gold; Mod. L. *punctatus* spotted (>L. *punctum* spot)); 'freckled with golden-yellow' (Hodgson, 1836) (*Urva*).

austenianus No explanation was given; most possibly an eponym after Lieutenant-Colonel Henry Haversham Godwin-Austen (b. 1834–d. 1923), an English topographer, naturalist, geologist, and surveyor (*-anus*, belonging to, pertaining to (L.)) (Subsp. *Hypsugo savii* Dobson's Pipistrelle).

australis L. *australis* southern (L. *auster* south) (*-alis*, pertaining to (L.)) (*Eubalaena*/J. Syn. Mysore Rock-rat, see *Cremnomys cutchicus*).

Axis/axis L. *axis* deer; perhaps of East Indian origin (*A. axis*, *A. porcinus*/*Axis*).

Bb

baberi Eponym after Zahir-ud-din Muhammad Babur (also spelt as Baber) (b. 1483–d. 1530), a conqueror from central Asia who laid the foundation to Mughal Dynasty, a naturalist and a keen chronicler, his work *Baburnama* includes accounts of numerous animals, one such being this species which was confused with (*-i*, commemorating (dedication, eponym) (L.)) (Subsp. *Eoglaucomys fimbriatus* Small Afghan Flying Sqiurrel).

babu No explanation was given (Subsp. *Pipistrellus javanicus* Thomas's Pipistrelle).

bactriana/bactrianus L. adj. *bactriana/bactrianus* Bactrian, belonging to Bactria – an ancient country of central Asia, roughly corresponding to modern Turkistan and Balkh, northern Afghanistan) (*-ana*, *-anus*, geographic, belonging to, pertaining to (L.) geographic (location, toponym)) (J. Syn. Bactrian Small Five-toed Jerboa, see *Allactaga elater*/Subsp. *Mus musculus* Asiatic House Mouse; *Spermophilopsis leptodactylus* Bactrian Long-clawed Ground Squirrel).

badius L. *badius* chestnut-coloured (*Cannomys*).

bahadur From Hind. *bahadur* warrior (J. Syn. Karwar Spiny Mouse, see *Mus platythrix*).

baileyi Eponym after Captain Frederick Marshman Bailey (b. 1882–d. 1967), a British army officer, naturalist and a collector (*-i*, commemorating (dedication, eponym) (L.)) (*Episoriculus* and *Naemorhedus*).

Balaenoptera Mod. L. *balaenoptera* winged whale (L. *balaena* whale; Gr. *-pteros* -winged (>Gr. *pteron* wing)); in allusion to the strong dorsal fin (*B. acutorostrata*, *B. edeni*, *B. musculus* and *B. physalus*).

baluchi Mod. L. *baluchi* from Balochistan (*Calomyscus*).

Bandicota L. *bandicota* >*bandicoot*, from Tel. *pandi-kokku*, meaning pig-rat (*B. bengalensis* and *B. indica*).

baptistae Eponym after N.A. Baptista (dates not known), taxidermist and collected in India for the Mammal Survey of British India (*-ae*, commemorating (dedication, eponym) (L.)) (Subsp. *Meriones persicus* Baptista's Persian Jird; *Viverricula indica* North Indian Small Civet).

Barbastella L. *barbastella* (>Fr. *barbastelle*, from L. *barba* beard); in allusion to the hairy chin (*B. darjelingensis*).

barclayanus Eponym after Dr. Arthur Barclay (B. not known–d. 1893) of the Bengal Medical Services, who collected the type (*-anus* belonging to, pertaining to (L./Gr.)) (J. Syn. Barclay's Field-rat, see *Bandicota bengalensis*).

beaba R. C. Wroughton provided no explanation, but possibly from Austronesian *beáb* meaning a rat or rodent (J. Syn. Sindh Desert Mole Rat, see *Nesokia indica*).

beavanii Eponym after Captain A. C. Beavan (dates not known), a British officer who collected the type (*-ii*, commemorating (dedication, eponym) (L.)) (J. Syn. Beavan's Field Mouse, see *Mus terricolor*).

beddomei Eponym after CoL. Richard Henry Beddome (b. 1830–d. 1911), British military officer and naturalist (*-i*, commemorating (dedication, eponym) (L.)) (*Rhinolophus*).

bedfordi Eponym after the 12th Duke of Bedford, Hastings William Sackville Russell (b. 1888–d. 1953), a keen naturalist (*-i*, commemorating (dedication, eponym) (L.)) (Subsp. *Naemorhedus goral* Western Himalayan Goral or Bedford's Goral).

bedfordiae Eponym after the Mary, the Duchess of Bedford (b. 1865–d. 1937), an ornithologist (*-ae*, commemorating (dedication), feminine (L.)) (*Sorex*).

belangeri Eponym after Charles Belanger (b. 1805–d. 1881), a French traveler (*-i*, commemorating (dedication, eponym) (L.)) (*Tupaia*).

bellaricus L. adj. *bellaricus* meaning from Bellary, Karnataka (*-icus*, belonging to, pertaining to (location, toponym) (L.)) (Subsp. *Funambulus palmarum* Bellary Palm Squirrel).

Belomys Mod. L. *belomys* beautiful mouse (L. *bellus* beautiful; Gr. *mӱs* mouse) (*B. pearsonii*).

bengalensis L. adj. *bengalensis* meaning from Bengal, India (*-ensis*, geographic, occurrence in (location, toponym) (L.)) (*Bandicota*, *Nycticebus*, *Prionailurus* and *Vulpes*/Subsp. *Bandicota bengalensis* Lesser Bandicoot Rat; *Hystrix brachyura* Bengal Porcupine; *Prionailurus bengalensis* Common Leopard Cat/J. Syn. Bengal Elephant, see *Elephas maximus*; Bengal Palm Squirrel, see *Funambulus palmarum*; Bengal Caracal, see *Caracal caracal*; Bengal Lion, see *Panthera leo*; Bengal Small Civet, see *Viverricula indica*).

bennettii Eponym after Edward Truner Bennett (b. 1797–d. 1836), an English zoologist and writer, instrumental in establishing of Entomological Society that eventually led to the establishment of Zoological Society of London in 1826 (*-ii*, commemorating (dedication, eponym) (L.) (*Gazella*) J. Syn. Bennett's Fishing Cat, see *Prionailurus viverrinus*).

Berylmys Mod. L. *berylmys* crystalline mouse (L. *beryllus* beryl; Gr. *mӱs* mouse) (*B. bowersi*, *B. mackenziei* and *B. manipulus*).

bezoartica Mod. L. *bezoartica* pertaining to bezoar goat (after Persian *padzahr* or Arabic *bazahr*, a small stony concretion formed in stomachs of ruminants used as antidote (*-ica*, pertaining to (L.)) (J. Syn. Indian Antelope or Sasin, see *Antilope cervicapra*).

bhatnagari Eponym after R.K. Bhatnagar (dates not known), a zoologist (*-i*, commemorating (dedication, eponym) (L.)) (J. Syn. Bhatnagar's Mountain Vole, see *Alticola stoliczkanus*).

bhotia L. adj. *bhotia* meaning from Bhutan (*-ia*, commemorating (dedication) (L./Gr.)) (J. Syn. Long-snouted Bhutan Squirrel, see *Dremnomys lokriah*).

bhutanensis L. adj. *bhutanensis* meaning from Bhutan (*-ensis*, geographic, occurrence in (location, toponym) (L.)) (J. Syn. Northern Gee's Golden Langur, or Northern Golden Leaf Monkey, see *Trachypithecus geei*; Bhutan Red-bellied Squirrel, see *Callosciurus erythraeus*).

bicolor L. adj. *bicolor* having two colours (L. *bi* two; L. *color* colour) (*Ratufa*/J. Syn. Bicolored Tomb Bat, see *Taphozous melanopogon*; Bicoloured Spiny Rat or Bi-coloured Coelomys, see *Mus mayori*).

biddulphi Eponym after Captain John Biddulph (b. 1840–d. 1921), a British soldier, author and a naturalist (*-i*, commemorating (dedication, eponym) (L.)) (J. Syn. Biddulph's Hare, see *Lepus tibetanus*).

bidiana Eponym after surgeon Dr. G. Bidie (dates not known), a British Medical Officer who served as the Superintendent of the Madras Central Museum from 1872 to 1884 (*-ana*, commemorating (dedication, eponym) (L.)) (J. Syn. Bidie's Shrew, see *Suncus stoliczkanus*).

bilineata Mod. L. *bilineata* doublestriped (L. *bi-* two-; L. *lineatus* lined >*linea* line) (J. Syn. Two-lined Antelope, see *Antilope cervicapra*).

binturong L. *binturong* from Malay *binturung*, a native name for the species (*Arctictis*).

Biswamoyopterus Mod. L. *biswamoyopterus* 'winged-Biswamoy', the generic name is an eponym after Dr. Biswamoy Biswas (b. 1923–d. 1994), renowned ornithologist and former Director, Zoological Survey of India in conjunction with L. *-pterus* winged (Gr. *ptero* wing) (*B. biswasi*).

biswasi Eponym after S. Biswas (dates not known), a scientist at Zoological Survey of India (*-i* commemorating (dedication, eponym) (L.)) (*Biswamoyopterus*).

birrelli Eponym after Major Birrell, R.A.M.C. (dates not known), a British army officer and a collector (*-i*, commemorating (dedication, eponym) (L.)) (J. Syn. Birrel's Flying Squirrel, see *Petaurista petaurista*).

blanfordi Eponym after William Thomas Blanford (b. 1832–d. 1905), an English geologist and naturalist, collected in many parts of India, also instrumental in establishing Indian Museum (*-ii*, commemorating (dedication, eponym) (L.)) (*Jaculus*, *Madromys* and *Sphaerias*/Subsp. *Paraechinus hypomelas* Blanford's Hedgehog/J. Syn. Gilgit Vole, see *Alticola argentatus*; Baluchi Wild Ass, see *Equus hemionus*).

Blanfordimys Mod. L. *blanfordimys* Blanford's mouse (Blanford – William Thomas Blanford (b. 1832–1905), an English geologist and naturalist, collected in many parts of India, also instrumental in establishing Indian Museum and Gr. *mŷs* mouse) (*B. afghanus* and *B. bucharensis*).

blasii Eponym after Johann Heinrich Blasius (b. 1809–d. 1870), Professor of Natural Sciences and Director of Natural History Museum, Brunswick, Germany (*-ii*, commemorating (dedication, eponym) (L.)) (*Rhinolophus*).

blythi/blythii/blythianus Eponym after Edward Blyth (b. 1810–d. 1873), an English zoologist and pharmacist, curator of Museum at the Royal Asiatic Society of Bengal (*-i*, *-ii*, commemorating (dedication, eponym (L.)); *-anus* belonging to, pertaining to (L. /Gr.)) (*Myotis*/Subsp. *Callosciurus pygerythrus* Blyth's Squirrel and *Rhinolophus pusillus* Himalayan Least Horseshoe Bat/J. Syn. *blythii* Blyth's Pig-tailed Macaque, see *Macaca leonina*; Blanford's Vole, see *Phaiomys leucurus*; Bengal Field-Rat, see *Bandicota bengalensis*).

bocharicus L. adj. *bocharicus* meaning from Bukhara, Uzbekistan (*-icus* belonging to, pertaining to (L./Gr.)) (*Rhinolophus*).

bombayus Mod. L. *bombayus* meaning from Bombay Presidency, India (J. Syn. Bombay Red Squirrel, see *Ratufa indica*).

bombax L. *bombax* cotton, but perhaps to mean from Bombay (now Mumbai), Presidency, India (J. Syn. Bombay Bush-rat, see *Golunda ellioti*).

bondar Mod. L. *bondar* >Beng. *bondar* the name of the civet (Subsp. *Paradoxurus hermaphroditus* Terai Palm Civet).

booduga Mod. L. *booduga* from *buduga* native name of the species in Madras Presidency (*Mus*).

Bos L. *bos* ox (*B. gaurus* and *B. mutus*).

Boselaphus Mod. L. *boselaphus* (generic combination of *bos* (L. *bos* ox) and *elaphus* (Gr. *élaphos* deer)) (*B. tragocamelus*).

bowersi Eponym after Alexander Bowers (d. 1887), a British naval captain and merchantman stationed at Burma, collected from northern Burma and Yunnan (*-i*, commemorating (dedication, eponym) (L.)) (*Berylmys*).

brachelix Mod. L. *brachelix* short-eared (>Gr. *brakhus* short; Gr. *helix* incurved rim of the external ear) (J. Syn. Short-eared Vole, see *Hyperacrius fertilis*).

brachyotis/brachyotus Mod. L. *brachyotis* short-eared (Gr. *brakhus* short; Gr. *-otis* -eared (*ōtos* ear)) (*Cynopterus*/Subsp. *Hipposideros galeritus* Short-eared Leaf-nosed Bat).

brachypterus Mod. L. *brachypterus* short-winged (Gr. *brakhus* short; L. *-pterus* -winged >Gr. *pteron* wing) (*Philetor*).

brachysoma Mod. L. *brachysoma* short-bodied (Gr. *brakhusoma*, >Gr. *brakhus* short; Gr. *sōma* body) (Subsp. *Cynopterus brachyotis* Andaman Short-nosed Fruit Bat).

brachyura Mod. L. *brachyura* short-tailed (Gr. *brakhus* short; Gr. *-ouros* -tailed (Gr. *oúrá* tail)) (*Hystrix*).

brahma Mod. L. *brahma* for Brahma, a supreme Hindu deity (*Niviventer*/Subsp. *Trachypithecus pileatus* Buff-bellied Capped Langur, also White-bellied Capped Langur, Buff-bellied Capped Leaf Monkey, Buff-bellied Langur).

branderi Eponym after A. A. Dunbar-Brander (dates not known), Conservator of Forests, stationed in Central Province, British India (*-i*, commemorating (dedication, eponym) (L.)) (Subsp. *Rucervus duvaucelii* South Indian Swamp Deer).

bredanensis L. adj. meaning from Breda, Netherlands (*-ensis*, geographic, occurrence in (location, toponym) (L.)) However, it is to be noted that the species is in fact named after an artist, 'van Breda', who first noticed the animal in Cuvier's collection (*Steno*).

brevicauda/brevicaudus Mod. L. *brevicauda, brevicaudus* short-tailed (L. *brevis* short; L. *cauda* tail) (Subsp. *Balaenoptera musculus* Pygmy Blue Whale/J. Syn. Short-tailed Long-winged Bat, see *Taphozous longimanus*).

breviceps Mod. L. *breviceps* short-headed (L. *brevis* short; *-ceps* -headed (L. *caput* head)) (*Kogia*).

brevirostris Mod. L. *brevirostris* short-billed (L. *brevis* short; *-rostris* -billed (*rostrum* bill)) (*Orcaella*).

brodiei Eponym after A.O. Brodie (dates not known), a British officer in Ceylon and naturalist (*-i*, commemorating (dedication, eponym) (L.)) (Subsp. *Funambulus palmarum* Northern Ceylon Palm Squirrel).

brunneus Mod. L. *brunneus* brown (J. Syn. Brown Tree Rat, see *Rattus rattus*; Common Nepal House Rat, see *Rattus tanezumi*).

brunneusculus Mod. L. *brunneusculus* somewhat brown (Mod. L. *brunneus* brown, *-ulus*, diminutive (comparison); somewhat (adj.) (L.)) (J. Syn. Lesser Brown Rat, see *Rattus tanezumi*).

bubalina Mod. L. *bubalina*, like wild ox (>*bubalis* – a name given to an African antelope, meaning buffalo) (*-ina*, pertaining to (L.) J. Syn. The Thar or Forest Goat, see *Capricornis thar*).

Bubalus L. *bubalus*, wild ox, from *bubalis* – a name given to an African antelope, meaning buffalo (*B. arnee*).

bucharensis L. adj. *bucharensis* meaning from Bukhara, Uzbekistan (*-ensis*, geographic, occurrence in (location, toponym) (L.)) (*Blanfordimys*).

Budorcas Mod. L. *budorcas* ox-like gazelle (L. *bu* from *bos* ox, Gr. *dorkás* gazelle); in allusion to its ox-like bulky body (*B. taxicolor*).

burrus L. *burrus* old form of the personal name *Pyrrhus* (>Gr. *purrhos* red) (*Rattus*).

burrulus L. *burrulus* somewhat like *burrus* (old form of the personal name *Pyrrhus* >Gr. *purrhos* red) (*-ulus*, diminutive (comparison), somewhat (adj.) (L.)) (J. Syn. Andaman Brown Rat, see *Rattus andamanensis*).

Cc

cadornae Eponym after Italian Field Marshall Luigi Cadorna (b. 1850–d. 1928) (*-ae*, commemorating (dedication, eponym) (L.)) (*Hypsugo*).

caeruluscens L. *caerulescens* bluish (L. *caeruleus* dark blue) (J. Syn. Indian Grey Musk-rat or Musk Shrew or Musk-rat or Musk-shrew, see *Suncus murinus*)

coenosus/coenosa L. *coenosus* filthy, impure (J. Syn. Dull-coloured Rock-rat, see *Cremnomys cutchicus*/J. Syn. Assam Bush-rat, see *Golunda ellioti*).

caliginosus L. *caliginosus* dark, obscure, misty (L. *caligo*, *caliginis* mist, fog) (*Submyotodon*).

Callosciurus Mod. L. *callosciurus* beautiful squirrel (L. *callo* (Gr. *kalos*) beautiful; L. *sciurus* squirrel) (*C. erythraeus* and *C. pygerythrus*).

Calomyscus Mod. L. *calomyscus* beautiful mouse (Gr. *kalos* beautiful, Gr. *myscus* mouse) (*C. baluchi*, C. *hotsoni* and *C. elburzensis*).

camortae Belonging to Camorta Islands, Nicobar Archipelago, India (*-ae*, commemorating (dedication); geographic (location, toponym) (L.)) (Subsp. *Pipistrellus javanicus* Camorta Pipistrelle).

cana L. *cāna* hoary (*Vulpes*).

cancrivora Mod. L. *cancrivora* crab-eating (L. *cancer, canceris* crab; L. *-vorus* -eating (*vorare* to devour) J. Syn. Brown Crab Mongoose, see *Urva urva*).

caniceps Mod. L. *caniceps* grey-headed (L. *canus* grey; L. *-ceps* -headed (L. *caput* head)) (Subsp. *Petaurista elegans* Grey-headed Flying Squirrel).

canigula Mod. L. *canigula* hoary-necked, grey-necked (L. *canus* grey; L. *gula* throat) (Subsp. *Mustela sibirica* Hoary-necked Weasel).

Canis L. *canis* Dog (*C. aureus* and *C. lupus*).

caniscus Mod. L. *caniscus* light grey (L. *canus* grey; *-iscus*, diminutive (comparison) (L./Gr.)) (Subsp. *Paradoxurus jerdoni* Coorg Palm Civet).

cannomys Mod. L. *cannomys* bamboo rat (L. *canna* (Gr. *kannē*) reed, bamboo; Gr. *mӱs* mouse) (*C. badius*).

cantori Eponym after Theodore Edward Cantor (b. 1809–d. 1860), a Danish physician and natruralist (*-i*, commemorating (dedication, eponym) (L.)) (J. Syn. Cantor's Long-winged Bat, see *Taphozous longimanus*).

capensis L. adj. meaning from Cape of Good Hope, South Africa (*-ensis*, geographic, occurrence in (location, toponym) (L.)) (*Mellivora* and *Delphinus*).

caporiaccoi Eponym after Ludovico di Caporiacco (b. 1901–d. 1951), an Italian arachnologist (*-i*, commemorating (dedication, eponym) (L.)) (Subsp. *Mustela kathiah* Kashmir Yellow-bellied Weasel).

Capra L. *capra* she-goat (*C. aegagrus*, *C. falconeri*, and *C. sibirica*).

Capricornis L. *capricornis* with goatlike horns (L. *capra* she-goat, *cornu* horn) (*C. thar*).

caprimulga L. *caprimulgus* nightjar (L. *capra* nanny goat, L. *mulgere* to milk) (J. Syn. Ekllerman's Jerboa, see *Allactaga williamsi*).

Caprolagus Mod. L. *caprolagus* (Gr. *kapros* wild boar; L. *lagos* a hare) (*C. hispidus*).

Caracal/caracal Mod. L. *caracal* from Turkish *karakulak* (*kara* black, *kulak* ear), a native name of the species; in allusion to black-backed tufted ears (*C. caracal*).

carnatica/carnaticus L. adj. *carnatica*, *carnaticus* from Karnataka, India (*-ica*, *-icus*, belonging to, pertaining to (location, toponym) (L.)) (J. Syn. Carnatic False Vampire Bat, see *Lyroderma lyra*/J. Syn. Carnatic Mongoose, see *Urva edwardsii*).

carruthersi Eponym after Alexander Douglas Mitchell Carruthers (b. 1882–d. 1962), a British explorer and naturalist attached with the British Museum (Natural History), who collected along the borders of Afghanistan (*-i*, commemorating (dedication, eponym) (L.)) (J. Syn. Carruther's Juniper Vole, see *Neodon juldaschi*).

cashmiriensis L. adj. *cashmiriensis* meaning from the Cashmir (=Kashmir), India (*-ensis*, geographic, occurrence in (location, toponym) (L.)) (J. Syn. Pir Panjal Markhor or Kashmir Markhor, see *Capra falconeri*).

Cassistrellus Mod. L. *cassistrellus* helmeted pipistrelle (L. *cassis* wearer of a helmet; L. *pipistrello*, *vispitrello*, dim. of L. *vespertilio*, a bat) (*C. dimissus*).

castaneus L. *castaneus* chestnut-brown (L. *castanea* chestnut) (Subsp. *Mus musculus* Southwestern Asian House Mouse).

catodon Mod. L. *catodon* down-tooth (Gr. *katō* down; Gr. *ódoús*, *ódôn* tooth) (J. Syn. Giant Sperm Whale, see *Physeter macrocephalus*).

caudata/caudatus L. *caudata*, *caudatus* tailed, -tailed; having a (long/short) tail (L. *cauda* tail) (*Marmota/Episoriculua*).

caucasicus L. adj. *caucasicus* belonging to Caucasus mountains (*-icus* belonging to, pertaining to (L./Gr.)) (Subsp. *Hypsugo savii* Caucasian Pipistrelle).

caudiator Mod. L. *caudiator* -tailed (L. *cauda* tail) (*-ator* suffix used to form agent nouns) (J. Syn. Chestnut White-bellied Rat or Chestnut Rat, see *Niviventer fulvescens*).

cautus L. *cautus* shy, wary (L. *cavere* to beware) (J. Syn. Wells' Vole, see *Alticola roylei*).

cavifrons Mod. L. *cavifrons* hollow-foreheaded (L. *cavus*, *cavi* hollow, hole (*cavus* excavated); L. *-frons* forehead, front) (J. Syn. Gauri Bull, see *Bos gaurus*).

cavirostris Mod. L. *cavirostris* hollow-billed (L. *cavus*, *cavi* hollow, hole (*cavus* excavated); L. *-rostris* -billed (*rostrum* beak)) (*Ziphius*).

centralis L. *centralis* central, in the middle (L. *centrum* middle) (Subsp. *Ratufa indica* Central Indian Giant Squirrel/J. Syn. Central Indian Antelope, see *Antilope cervicapra*).

centrasiaticus L. adj. *centrasiaticus* belonging to Central Asia; type locality is located in northwest Tibet (*-icus* belonging to, pertaining to (L. /Gr.)) (Subsp. *Eptesicus gobiensis* Central Asian Serotine).

cervicapra Mod. L. *cervicapra* she-goat–like deer (L. *cervus* deer; L. *capra* she-goat) (*Antilope*).

cervicolor Mod. L. *cervicolor* fawn-coloured (L. *cervus* deer; L. *color* colour) (*Mus*).

Cervus L. *cervus* stag, deer (*C. elaphus*).

ceylonense/ceylonensis/ceylanicus/ceylonica/ceylonicus/ceylonus L. *ceylonense/ceylonensis/ceylanicus/ceylonica/ceylonicus/ceylonus* L. adj. belonging to Sri Lanka or Sri Lankan (*-ense*, *-ensis*, *-ica*, *-icus*, *-us*, geographic, occurrence in (location, toponym) (L.)) (Subsp. *Megaderma spasma* Sri Lankan False Vampire Bat/Subsp. *Cynopterus brachyotis* Sri Lankan Short-nosed Fruit Bat/*Pipistrellus ceylonicus* Kelaart's Pipistrelle/J. Syn. Ceylon Spotted Deer, see *Axis axis*; Ceylonese Elephant, see *Elephas maximus*; Ceylon Brown Mongoose, see *Urva fuscus*; Ceylon Gerbil, Ceylon Antelope-Rat, Sand-Rat, Kangaroo-Rat, see *Tatera indica*; Ceylon Mouse, see *Rattus rattus*; Ceylon Otter, see *Lutra lutra*).

Chaerephon Gr. *Chairephōn* a proper name, a loyal friend and follower of Socrates (*C. plicatus*).

chanco From Tib. *chanko*, a local name for the wolf (Subsp. *Canis lupus* Tibetan Wolf).

charltonii Eponym after Andrew Charlton (dates not known), of Liskard, Cheshire who presented the type specimen to British Museum (Natural History)

(*-ii*, commemorating (dedication, eponym) (L.)) (Subsp. *Pardofelis marmorata* Nepalese Marbled Cat).

chaus Apparently from common name *chaus* in Hindi (*Felis*).

chhattisgarhi L. adj. *chhattisgarhi* belonging to Chhattisgarh (*-i* commemorating (L./Gr.)) (Subsp. *Funambulus pennantii* Chhattisgarh Palm Squirrel).

chickara Mod. L. *chickara* from Hindi *chikara*, name for the species (J. Syn. Chikara, see *Tetracerus quadricornis*).

Chimarrogale Mod. L. *chimarrogale* montane water-shrew (Gr. *chimorros* mountain-torrent; Gr. *gale* weasel) (*C. himalayica*).

chinensis L. adj. *chinensis* meaning from China (*-ensis*, geographic, occurrence in (location, toponym) (L.)) (*Sousa*).

Chiropodomys Mod. L. *chiropodomys* (Gr. *kheir, kheiros* hand; Gr. *podos* foot; Gr. *mӯs* mouse) (*C. gliroides*).

chitralensis L. adj. *chitralensis* meaning from Chitral, Khyber Pakhtunkhwa, Pakistan (*-ensis*, geographic, occurrence in (location, toponym) (L.)) (J. Syn. Chitral Mole Rat, see *Nesokia indica*; Chitral Markhor, see *Capra falconeri*).

christyi Eponym after Dr. Alexander Turnbull Christie (dates not known), a medical officer of the Madras Establishment, who collected in India (*-i*, commemorating (dedication, eponym) (L.)) (Subsp. *Gazella bennettii* Gujarat Chinkara).

chrysogaster Mod. L. *chrysogaster* golden-bellied (Gr. *khrusos* gold; Gr. *gastēr* belly) (*Moschus*/J. Syn. Golden-bellied Marten, see *Martes flavigula*).

chrysothrix Mod. L. *chrysothrix* golden-haired (Gr. *khrusos* gold; Gr. *thríx* hair) (J. Syn. Golden-haired Pipistrelle, see *Pipistrellus ceylonicus*).

chrysurus Mod. L. *chrysurus* golden-tailed (Gr. *khrusos* gold; Gr. *-ouros* -tailed (Gr. *oura* tail) J. Syn. Fulvous-tailed Dog, see *Vulpes bengalensis*).

cinderella Mod. L. *cinderella*, after the German fairy tale character who slept in fireplace amongst cinders (from L. *cinis, cineris* ashes) (J. Syn. Cutch Spiny Mouse, see *Mus saxicola*).

cineraceus/cinerascens L. *cineraceus* ash-grey (*cinis, cineris* ashes)/Mod. L. *cinerescens* ashen (>L. *cinis, cineris* ashes, >*cinerescere* to turn to ashes) (*Hipposideros*/Subsp. *Cricetulus migratorius* Wagner's Little Grey Hamster).

cinerea/cinereus L. *cinereus* ash-grey, ash-coloured (L. *cinis, cineris* ashes) (*-eus* having the quality of) (*Aonyx*/*Eupetaurus*).

circumdatus L. *circumdatus* surrounded (L. *circumdare* to surround) (*Arielulus*).

civettina Diminutive of Fr. *civette* civet cat (*-ina*, belonging to (L.)) (*Viverra*).

Coelops Mod. L. *coelops* hollowed (Gr. *coilos* hollow; Gr. *óps* aspect); probably in allusion to large funnel-shaped ears (*C. frithii*).

coeruleoalba Mod. L. *coeruleoalba* dark blue and white (L. *caeruleus* dark blue; L. *albus* white); in allusion to the prominent dark blue and white stripes on its body (*Stenella*).

coffaeus Mod. L. *coffaeus* having the quality of coffee, perhaps due to its affinity to coffee plantations (*-eus*, made of, having the quality of (L.)) (J. Syn. Coffee Bush-rat, see *Golunda ellioti*).

cognatus L. adj. *cognatus* related, similar (L. *co* together; L. *gnātus* born) (*Rhinolophus*).

collaris L. *collaris* of the neck (L. *collum* neck) (*Arctonyx* and *Hemiechinus*).

collinus L. *collinus* found on a hill, hill- (L> *collis* hill) (J. Syn. Western Desert Gerbil, see *Meriones hurrianae*).

comorinus Mod. L. *comorinus* meaning from Cape Comorin (=Kanya Kumari) (*-inus*, belonging to, pertaining to; one who (L.)) (J. Syn. South Indian Common Palm Squirrel, see *Funambulus palmarum*).

concolor L. *concolor* uniform, similar in colour, plain (*Sicista*/Subsp. *Aonyx cinerea* Rafinesque's Small-clawed Otter/J. Syn. Little Burmese Rat, see *Rattus exulans*).

consul L. *cōnsul* (from Old L. *consol*) annually elected highest ranking official of the Roman empire; in allusion to its larger size in comparison to the nominate form (Subsp. *Arctonyx collaris* Large Hog-Badger).

cookii Eponym after J. Pemberton Cook (b. 1865–d. 1924), worked for the Burma Teak Company and collected from Moulmein for the Mammal Survey of British India (*-ii*, commemorating (dedication, eponym) (L.)) (*Mus*).

comberi Eponym after E. Comber (dates not known), a British officer associated with BNHS and collected the type at Nasik (*-i*, commemorating (dedication, eponym) (L.)) (J. Syn. Comber's Field Rat, see *Millardia meltada*).

coraginis No explanation provided (J. Syn. Coorg Bush-rat, see *Golunda ellioti*).

coromandra Mod. L. *coromandra* from the Coromandel (Cholamandalam) Coast, India (*Pipistrellus*).

corsac Mod. L. *corsac* from Russ. *korsak* (>Kirghiz *karsak*), a native name for the species (*Vulpes*).

craspedotis Mod. L. *craspedotis* large-eared (Gr. *craspedote* veil (*cràspedo* hem, fringe); Gr. *otis*, *ōtos* ear); in allusion to its large ears (Subsp. *Lepus tibetanus* Balochistan Desert Hare or Large-eared Hare or Afghan Hare or Kopet-Dagh Desert Hare).

crassicaudata Mod. L. *crassicaudata* heavy-tailed (L. *crassus* thick, heavy, dense; L. *caudatus* tailed, having a (long/short) tail (L. *cauda* tail)) (*Manis*).

crassidens Mod. L. *crassidens* heavy-toothed (L. *crassus* thick, heavy, dense; L. *dens* tooth) (*Pseudorca*).

crassipes Mod. L. *crassipes* heavy-footed (L. *crassus* thick, heavy, dense; L. *pes* foot, from Gr. *pous* foot) (J. Syn. Large-footed Mouse, see *Rattus rattus*).

crassus L. *crassus* thick, heavy, dense (*Meriones*/Subsp. *Saccolaimus saccolaimus* Indian Pouch-bearing Tomb Bat).

Cremnomys Mod. L. *cremnomys* rock rat (Gr. *cremnos* a steep bank; Gr. *mȳs* mouse) (*C. cutchicus* and *C. elvira*).

Cricetulus Mod. L. *cricetulus* dim. of *cricetus* hamster (Ital. *criceto*) (*-ulus*, diminutive (comparison), somewhat (adj.) (L.)) (*C. alticola* and *C. migratorius*/J. Syn. Short-tailed Mountain Vole, see *Alticola stoliczkanus*).

cristatus L. *cristatus* crested (L. *crista* crest) (Subsp. *Sus scrofa* Indian Wild Pig).

Crocidura Gr. *crocis* or *crocos* the flock or nap on the woolen cloth, a piece of wool, *oúrá* tail; in allusion to the tail which is covered with short hairs interspersed by long hairs (*C. andamanensis*, *C. attenuata*, *C. fuliginosa*,

C. gathornei, *C. gmelini*, *C. hikmiya*, *C. hispida*, *C. horsfieldii*, *C. jenkinsi*, *C. leucodon*, *C. miya*, *C. nicobarica*, *C. pergrisea*, *C. pullata*, *C. rapax*, *C. vorax,* and *C. zarudnyi*).

crossi Eponym after Edward Cross (B. year not known – d. 1854), an animal dealer and zoo proprietor (*-i*, commemorating (dedication, eponym) (L.)) (J. Syn. Cross's Palm Civet or Cross's Paradoxurus or Cross' Paguma, see *Paradoxurus hermaphroditus*).

crumpi Eponym after C. A. Crump (dates not known), who collected in Deccan for the Mammal Survey of British India (*-i*, commemorating (dedication, eponym) (L.)) (*Diomys*/J. Syn. Crump's Squirrel, see *Callosciurus erythraeus*).

cryptura Mod. L. *cryptura* hidden-tailed (Gr. *kruptos* hidden; Gr. *oura* tail) (J. Syn. Assam Short-tailed Mole, see *Euroscaptor micrura*).

csorbai Eponym after Dr. Gábor Csorba (b. 1961), curator of mammals at Hungarian Museum of Natural History, who collected bats in the Himalayas (*-i*, commemorating (dedication, eponym) (L.)) (*Myotis*).

Cuon Gr. *cyon* Dog (*C. alpinus*).

cunicularis Mod. L. *cunicularis* pertaining to rabbit (L. *cuniculus* mine, or an underground passage) (Mod. L. *-aris* a suffix meaning pertaining to or in allusion of) (J. Syn. Little Rabbit-mouse, see *Mus cervicolor*).

cupreus L. *cupreus* coppery (*cuprum* copper; *-eus* having the quality of (L.)) (*Moschus*).

curzoniae Eponym after 'Mrs. Curzon' (details not known) (*-ae*, commemorating (dedication), feminine (L.)) (*Ochotona*).

cutchicus/cutchensis L. adj. *cutchicus/cutchensis* meaning from Kutch, Gujarat, India (*-icus*, *-ensis*, belonging to, pertaining to (location, toponym) (L.)) (*Cremnomys*/J. Syn. Kutch Hare, see *Lepus nigricollis*).

cuvieri Eponym after M. F. Cuvier (b. 1773–d. 1838), a French zoologist and younger brother of Georges Cuvier, author of a monumental work on *Gerbillus* (*-i*, commemorating (dedication, eponym) (L.)) (Subsp. *Tatera indica* South Indian Gerbil).

cycloceros Mod. L. *cycloceros* spiral-horned (Gr. *kuklos* circle; Gr. *kéras* horn) (Subsp. *Ovis orientalis* Afghan Urial).

cyclotis L. *cyclotis* round-eared (Gr. *kuklos* circle; L. *otus*, *ōtos* ear) (*Murina*).

Cynopterus Mod. L. *cynopterus* 'winged dog' (Gr. *cynos* dog; L. *-pterus* -winged (>Gr. *pteron* wing)); in allusion to its doglike head (*C. brachyotis* and *C. sphinx*).

Dd

Dacnomys Mod. L. *Dacnomys* biting mouse (Gr. *dakno* to bite; Gr. *mӱs* mouse); owing to its unsually large maxillary toothrow (*D. millardi*).

dakhunensis/dukhunensis L. adj. *dakhunensis/dukhunensis* meaning from Dakhun/Dukhun (=Deccan – Indian peninsular plateau), India (*-ensis*, geographic, occurrence in (location, toponym) (L.)) (Subsp. *Cuon alpinus* Deccan Wild Dog or Indian Dhole or Nilgiri Dhole/J. Syn. Deccan

Elephant, see *Elephas maximus*; Sykes' Leaf-nosed Bat, see *Hipposideros speoris*).

dandolena Mod. L. *dandolena*, after Sinhalese name *dandolena* for the species (Subsp. *Ratufa macroura* Grizzled Giant Squirrel).

darjelingensis/darjilingensis L. adj. *darjelingensis/darjilingensis* meaning from Darjeeling, West Bengal, India (*-ensis*, geographic, occurrence in (location, toponym) (L.)) (*Barbastella*/S. Syn. Darjeeling Spiny Mouse, see *Mus cookie*/J. Syn. Darjeeling Bat, see *Myotis siligorensis*).

davidi Eponym after Reuben David (b. 1912–d. 1989), a founder and superintendent of the Ahmedabad Zoo, Gujarat, where the specimens of this taxon were first sighted by Colin Groves (*-i*, commemorating (dedication, eponym) (L.)) (Subsp. *Sus scrofa* Pakistan Wild Pig).

davidii Eponym after Father Armand 'Pere' David (b. 1826–d. 1900), missionary, priest, and naturalist; collected in China (*-i*, commemorating (dedication, eponym) (L.)) (*Myotis*).

dawsoni Eponym after Mr. Dawson (dates not known), taxidermist at the Trevandrum Museum who processed the Travancore specimens of cetaceans (*-i*, commemorating (dedication, eponym) (L.)) (J. Syn. Dawson's Dolphin, see *Tursiops truncatus*).

dayanus Eponym after Dr. Day (dates not known), the collector of the types; could be Dr. Francis Day (see below) (*Lepus nigricollis* Sindh Hare).

dayi Eponym after Dr. Francis Day (b. 1829–d. 1889), a zoologist, known for his work on fishes, Inspector General of Fisheries for India and Burma (*-i*, commemorating (dedication, eponym) (L.)) (*Suncus*).

deccanensis L. adj. *deccanensis* meaning from Deccan (Indian peninsular plateau), India (*-ensis*, geographic, occurrence in (location, toponym) (L.)) (J. Syn. Deccan Field-rat, see *Bandicota bengalensis*).

delesserti Eponym after Adolphe Delessert (d. 1843), a French naturalist and collector in India (between 1834 and 1839) (*-i*, commemorating (dedication, eponym) (L.)) (J. Syn. Delessert's Striped Squirrel, see *Funambulus sublineatus*).

Delphinus Gr. *delphinus* dolphin (*D. capensis*).

densirostris Mod. L. *densirostris* thick-billed (L. *densus* thick; L. *-rostris* -billed (L. *rostrum* beak)) (*Mesoplodon*).

deserti L. *deserti* desert, waste, solitude (Subsp. *Viverricula indica* Rajasthan Small Civet).

desertorum L. *desertorum* of the deserts (L. *desertum* desert) (Subsp. *Myotis emarginatus* Desert Myotis).

diadema L. *diadema* diadem (>Gr. *diadēma* diadem, royal head-dress) (*Hipposideros*).

Dicerorhinus Mod. L. *dicerorhinus* two horns on the nose (Gr. *di* two, *kéras* horn; Gr. *rhis*, *rhinos* nose); in allusion to the transversely paired nasal horns (*D. sumatrensis*).

dichrous Mod. L. *dichrous* bicolored (Gr. *dikhrous* two-coloured (*di-* two- (in comp.) (*dis* twice); Gr. *khroos* complexion) (J. Syn. Bicolored Marmot, see *Marmota caudata*).

diluta L. *dilutus* weak, diluted (L. *diluere* to dilute) (Subsp. *Macaca radiata* Pale-bellied Bonnet Macaque; also known as Agasthyamalai Bonnet Macaque, Southern Bonnet Macaque, Light-bellied Bonnet Macaque).

dimidiatus L. *dimidiatus* divided (L. *dimidius* halved; L. *di-* apart; L. *medius* middle) (*Acomys*).

dimissus L. *dimissus* abandon; originally thought to be an individual of *E. pachyotis*, later found to be larger and distinct basing on a combination of characters (*Eptesicus*).

Diomys Mod. L. *Diomys* God's mouse (Gr. *dios* god; Gr. *mÿs* mouse); as Oldfield Thomas described this new genus and new species based on an extracted skull, supposedly collected on Mt. Paresnath, Bihar (*D. crumpi*).

domesticus L. *domesticus* belonging to the house, domestic (L> *domus* house) (Subsp. *Mus musculus* Western House Mouse).

dormeri Eponym after Major General James Charlemagne Dormer (b. 1834–d. 1893), a British officer serving in the Madras Presidency (*-i*, commemorating (dedication, eponym) (L.)) *Scotozous dormeri* Dobson, 1875

draco L. *draco* dragon; no explanation provided, probably due to its origin in south China (*Apodemus*).

Dremomys Mod. L. *dremomys* fast-running squirrel (L. *dromos* running (from L. *drameïn* to run), Gr. *mÿs* mouse) (*D. lokriah*, *D. pernyi* and *D. rufigenis*).

dryas Gr. Myth. *Dryas*, tree-nymph, dryad (Subsp. *Myotis horsfieldii* Andaman Myotis).

Dryomys Mod. L. *dryomys* dormouse (Gr. *druos* (genitive singular of *drïs*) oak or tree; Gr. *mÿs* mouse) (*D. nitedula* and *D. niethammeri*).

dubius L. *dubius* doubtful, dubious (L. *duo* two; L. *habere* to have) (J. Syn. Nepal House Mouse, see *Mus musculus*).

dugon Possibly also from the same origin as of the generic name *Dugong* (see below) (*Dugong*).

Dugong Mod. L. *dugong* from the Malayan name *duyong* for the species (*D. dugon*).

dunni Eponym after Henry Nelson Dunn (b. 1864–d. 1952), a British army surgeon, naturalist and hunter who served in India (*-i*, commemorating (dedication, eponym) (L.)) (J. Syn. Dunn's Gerbil, see *Tatera indica*; Dunn's Field Rat, see *Millardia meltada*; Northern Field Mouse, see *Mus terricolor*).

durga Mod. L. *durga* for Durga, a major Hindu goddess (Subsp. *Trachypithecus pileatus* Orange-bellied Capped Langur, also Orange-bellied Capped Leaf Monkey, Orange-bellied Langur).

durgadasi Eponym after Durga Das (dates not known), father in-law of H. Khajuria describer of the species (*-i*, commemorating (dedication, eponym) (L.)) (*Hipposideros*).

dussumieri Eponym after Jean-Jacques Dussumier (b. 1792–d. 1883), a French collector, traveler, and trader (*-i*, commemorating (dedication, eponym) (L.)) (*Semnopithecus*/J. Syn. Dussumier's Palm Squirrel, see *Funambulus tristriatus*).

duvaucelii Eponym after Alfred Duvaucel (b. 1793–d. 1824), a French naturalist and explorer, collected widely in South and Souteast Asia for Paris Museum of Natural History (*-ii*, commemorating (dedication, eponym) (L.)). (*Rucervus* – Species description was based on drawings of the antlers sent by Alfred Duvaucel/J. Syn. Duvaucell's Goral, see *Naemorhedus goral*).

Ee

edeni Eponym after Ashley Eden (b. 1831–d. 1887), a British diplomat stationed in Burma (*-i*, commemorating (dedication, eponym) (L.)) (*Balaenoptera*).

edwardsi Eponym after Alphonse Milne-Edwards (b. 1835–d. 1900), a French mammalologist, ornithologist and carcinologist (*-i*, commemorating (dedication, eponym) (L.)) (*Leopoldamys*).

edwardsi/edwardsii Eponym after Henri Milne-Edwards (b. 1800–d. 1885), a French zoologist (*-i*, *-ii*, commemorating (dedication, eponym) (L.)) (J. Syn. Large Fox-bat or Milne-Edwards' Fruit Bat, see *Pteropus giganteus/Urva*).

eha Eponym after E. H. Aitken (b. 1851–d. 1909), Indian civil servant, co-founder of the Bombay Natural History Society, and writer – wrote under the pseudonym Eha (*Niviventer*).

eileenae Eponym after Ms. Eileen Wynell-Mayow (b. 1927–d. 2012), daughter of W. W. A. Phillips, a naturalist in Sri Lanka, author of Manual of the Mammals of Sri Lanka (*-ae*, commemorating (dedication, eponym) (L.)) (Subsp. *Murina cyclotis* Sri Lankan Round-eared Tube-nosed Bat).

elaphus Med. L. *elaphus* (Gr. *élaphos*) deer (*Cervus*).

elater Mod. L. *elater* (Gr. *elatēr*) one who elates, jumps (*Allactaga*).

elburzensis L. adj. *elburzensis* meaning from the Elburz (=Alborz) mountains, Iran (*-ensis*, geographic, occurrence in (location, toponym) (L.)) (*Calomyscus*).

eldii Eponym after Colonel Lionel Percy Denham Eld (b. 1808–d. 1863), an officer of Bengal Infantry, stationed in Assam (*-ii*, commemorating (dedication, eponym) (L.)) (*Rucervus*).

electra L. *electra* (Gr. *ēlektra*) shining one (*Peponocephala*).

elegans L. adj. *elegans* elegant, handsome or fine (*Petaurista* and *Nectogale*).

Elephas Gr. *elephas* Elephant (Gr. *ele* arch; Gr. *phas* huge; literally Gr. ivory) (*E. maximus*).

elphinstonei Eponym after Hon. Mountstuart Elphinstone (b. 1779–d. 1859), a Scottish diplomat in the Bengal civil service, who served in various positions in Bombay Presidency, India (J. Syn. Elphinstone's Red Squirrel, see *Ratufa indica*).

elissa Gr. Myth. *Elissa*, sister of Tyrian king Pygmalion, and founder of Carthage, from Virgil's Latin epic poem, *Aeneid* (J. Syn. Nagarhole Langur, see *Semnopithecus dussumieri*).

ellioti/elliotanus Eponym after Sir Walter Elliot (b. 1803–d. 1887), Indian Civil Servant and amateur naturalist (*-i*, commemorating (dedication, eponym) (L.); *-anus* belonging to, pertaining to (L./Gr.)) (*Anathana* and *Golunda*/J. Syn. Elliot's Mongoose, see *Urva edwardsii*; Elliot's Field-Rat or Bengal Bandicoot, see *Bandicota indica*; Elliot's Otter, see *Lutrogale perspicillata*).

Ellobius Mod. L. *ellobius* round-eared (Gr. *éllóvion* earring), from the rudimentary, somewhat circular, external ears (*E. talpinus* and *E. fuscocapillus*).

elvira Mod. L. *elvira* a female given name, originally means 'the white' (*Cremnomys*).

emarginatus L. *emarginatus* emarginated (L. *emarginare* to provide with a margin, *-atus* provided with (L.)) (*Myotis*/J. Syn. Large-eared Yellow Bat, see *Scotomanes ornatus*).

entellus Mod. L. *entellus* >Gr. *éntéllo* to command; species is held in veneration in India (*Semnopithecus*).

Eoglaucomys Mod. L. *eoglaucomys* Eastern blue-green mouse (L. *eos* East or pertaining to East, Gr. *glauco* blue green colour, Gr. *mȳs* mouse) (*E. fimbriatus*).

Eonycteris Mod. L. *eonycteris* dawn bat (Gr. *iós* dawn, Gr. *nikterís* bat) (*E. spelaea*).

Eothenomys Mod. L. *eothenomys* early mouse (Gr. *iós* dawn, Gr. *then* from, Gr. *mȳs* mouse) (*E. melanogaster*).

Episoriculus Mod. L. *episoriculus* somewhat like *soriculus* (Gr. *epi* somewhat, *soriculus* Dim. of L. *sorex* (>Gr. *ȳrax* shrew)) (*-ulus*, diminutive (comparison); somewhat (adj.) (L.)) (*E. baileyi E. caudatus*, *E. leucops*, *E. macrurus*, *E. sacratus,* and *E. soluensis*).

Eptesicus Mod. L. *eptesicus* 'house-flyer' (Gr. *éptin* from *pétomai* to fly, Gr. *oíkos* house) (*E. gobiensis*, *E. ognevi*, *E. pachyomus*, *E. pachyotis*, *E. serotinus* and *E. tatei*).

Eequus L. *equus* horse (*E. hemionus* and *E. kiang*).

eremtaga L. noun *eremtaga* derived from the Aka-Kora dialect of the Great Andamanese language, meaning 'forest-dweller' (Subsp. *Tylonycteris malayana* Malayan Greater Bamboo Bat).

erminea L. *erminea* ermine (*Mustela*).

erythraeus L. *erythraeus* red-coloured (Gr. *eruthros* red; *-eus* having the quality of) (*Callosciurus*).

erythrogaster Mod. L. *erythrogaster* red-bellied (Gr. *eruthros* red; Gr. *gastēr* belly) (Subsp. *Callosciurus erythraeus* Red-bellied Squirrel).

erythrotis/erythrotus Mod L. *erythrotis/erythrotus* red-eared (Gr. *eruthros* red, Gr. *-ōtis* -eared, Gr. *ōtos* ear) (Subsp. *Micromys minutus* Cherrapunji Red-eared Mouse/J. Syn. Red-eared Jungle Cat, see *Felis chaus*).

erythrourus Mod L. *erythrourus* red-tailed (Gr. *eruthros* red; Gr. *-ouros* -tailed (Gr. *oúrá* tail)) (J. Syn. Red-tailed Jerboa-Rat, see *Meriones libycus*).

etruscus Mod. L. *etruscus* Etruscan, from Etruscus, Italy (*Suncus*).

Eubalaena Mod. L. *eubalaena* true whale (Gr. *eȳ* well, typical; L. *balaena* whale) (*E. australis*).

Eupetaurus Mod. L. *eupetaurus* true flying squirrel (L. *eu* (>Gr. *eȳ*) true or well; L. *petaurus* (>Gr. *petauron*) a springboard or a perch) (*E. cinereus*).

Euroscaptor Mod. L. *euroscaptor* deep-digger (Gr. *eurus* broad or deep; Gr. *scaptor* to dig). (*E. micrura*).

everesti Commemorating Mt. Everest, type locality of the taxon, Mt. Everest is named after CoL. Sir George Everest (b. 1790–d. 1866), a Welsh surveyor and geographer and served as Surveyor General of India (*-i*, commemorating (dedication, eponym) (L.)) (J. Syn. Everest Vole, see *Phaiomys leucurus*).

excelsus L. *excelsus* lofty; owing to its high altitudinal habitat (*Sorex*).

exulans L. *exulans* wandering (L. *exsulare* to be exiled) (*Rattus*).

Ff

falconeri Eponym after Dr. Hugh Falconer (b. 1808–d. 1865), a Scottish geologist, botanist, palaeontologist and palaeoanthropologist, and worked in the Himalayas and Burma (*-i*, commemorating (dedication, eponym) (L.)) (*Capra*).

Falsistrellus Mod. L. *falsistrellus* false pipistrelle (Ital. *falsi* fake; L. *pipistrello, vispitrello*, dim. of L. *vespertilio*, a bat) (*Falsistrellus*).

famulus L. *famulus* servile (*Mus*/Subsp. *Rhinolophus cognatus* Andersen's Horseshoe Bat).

fascicularis Mod. L. *fascicularis* small-banded (L. *fasciculus* a band from L. *fascia*; Mod. L. *-aris* a suffix meaning pertaining to or in allusion of) (*Macaca*).

faunulus From L. *faunus* (Rom. Myth. A horned deity of the forest, plains, and fields) (*-ulus,* diminutive (comparison); somewhat (adj.) (L.)); in allusion to its habit of living in a dense forest and having pointed ears (*Pteropus*).

favonicus No explanation was given (J. Syn. Striped Squirrel, Western Ceylon Palm Squirrel, Submontane Ceylon Palm Squirrel, see *Funambulus palmarum*).

feae Eponym Leonardo Fea (b. 1852–d. 1903), Italian naturalist, collected in Burma in 1885 (*-e*, commemorating (dedication, eponym) (L.)) (*Murina*).

Felis L. *felis* cat (probably from root *fe*, to produce, bear young) (*F. chaus*, *F. margarita* and *F. silvestris*).

fellowesgordoni Eponym after Ms. Marjory Tutein-Nolthenius *née* Fellowes-Gordon (dates not known), wife of A. C. Tutein-Nolthenius, estate owner and collector in Sri Lanka (*-i*, commemorating (dedication, eponym) (L.)) (*Suncus*).

Feresa Mod. L. *feresa* from Fr. *feres*, a name for this dolphin (*F. attenuata*).

ferghanae After Ferghana, Kazakhstan (*-ae*, geographic (location, toponym)) (Subsp. *Mustela erminea* Ferghana Ermine).

fergusoni Eponym after Harold S. Ferguson (b. 1851–d. 1921), Scottish Zoologist who served as the Director of Travancore Museum, India (*-i*, commemorating (dedication, eponym) (L.)) (J. Syn. Ferguson's Dolphin, see *Sousa chinensis*).

fernandoni Eponym after E. C. Fernando (b. 1900–d. 1966), a collector and taxidermist at the Colombo Museum, Sri Lanka (*-i*, commemorating (dedication, eponym) (L.)) (*Mus*).

Feroculus/feroculus L. *feroculus* (dim. of *ferox* fierce), somewhat fierce or spirited (*-ulus*, diminutive (comparison), somewhat (adj.) (L.)) (*Feroculus*).

ferrilata Mod. L. *ferrilata* iron-grey sided (L. *ferrum* iron; L. *latus* flank); in allusion to its colour on the lateral sides (*Vulpes*).

ferruginea/ferrugineus L. *ferruginea* rusty-coloured, ferruginous (L. *ferrugo, ferruginis* iron rust) (Subsp. *Otocolobus manul* Rufous Pallas's Cat; Subsp. *Urva edwardsii* Desert Grey Mongoose/J. Syn. Ceylon Rufescent Shrew, see *Suncus montanus*).

ferrumequinum L. *ferrumequinum* horseshoe (L. *ferrum* iron; *eqinum*, of horses) (*Rhinolophus*).

fertilis L. *fertilis* productive, fertile (L. *ferō* carry, bear) (*Hyperacrius*).

fimbriatus Med. L. *fimbriatus* bordered (L. *fimbriae* border or fringe; *-atus*, provided with (L.)) (*Eoglaucomys*).

flavidens Mod. L. *flavidens* yellow-toothed (L. *flavus* yellow; L. *dens* tooth) (Subsp. *Urva fuscus* Ceylon Wetzone Brown Mongoose).

flavidus L. *flavidus* yellowish (L. *flavus* yellow, golden) (J. Syn. Yellowish Spiny Mouse, see *Acomys dimidiatus*).

flavigula Mod. L. *flavigula* yellow-throated (L. *flavus* yellow; *gula* throat) (*Martes*).

flavus L. *flavus* yellow (J. Syn. Yellow Birch Mouse, see *Sicista concolor*).

fluminalis Mod. L. *fluminalis* pertaining to river (L. *flumen*, *fluminis* river) (*-alis*, pertaining to, having the nature of (L.)) (J. Syn. Fresh-water Round-headed dolphin, see *Orcaella brevirostris*).

foina Mod. L. *foina* (Fr. *fouine*, marten from L. *fagina* pertaining to a beech tree) marten (*Martes*).

formosus L. *formosus* beautiful (L> *forma* beauty) (*Myotis*).

forresti Eponym after George Forrest (b. 1873–d. 1932), a Scottish botanist, who collected extensively in Yunnan beginning 1904 (*-i*, commemorating (dedication, eponym) (L.)) (*Ochotona*).

frithii Eponym after R. W. G. Frith (dates not known), Indigo Planter in northeast India (*-i*, commemorating (dedication, eponym) (L.)) (*Coelops*).

fulgens L. *fulgens* glittering (L. *fulgere* to shine) (*Ailurus*).

fuliginosa/fuliginosus Mod. L. *fuliginosus* sooty (>L. *fuligo*, *fuliginis* soot) (*Crocidura/Miniopterus*).

fulvescens L. *fulvescens* somewhat tawny (L. *fulvus* tawny; *-escens*, somewhat) (*Niviventer*).

fulvidiventris Mod. L. *fulvidiventris* tawny-bellied (L. *fulvidus* more tawny; L. *venter*, *ventris* belly) (J. Syn. Ceylon Field Mouse, see *Mus booduga*).

fulvida/fulvidus Mod. L. *fulvida* or *fulvidus* (L. *fulvus* fulvous, tawny) (*Tylonycteris*/J. Syn. Fulvous Long-winged Bat, see *Taphozous longimanus*).

fulvinus Mod. L. *fulvinus* (L. *fulvus* fulvous, tawny) (*-inus*, belonging to, pertaining to; one who (L.)) (J. Syn. Western Himalayan Flying Squirrel, see *Petaurista petaurista*).

fulvus L. *fulvus* tawny, yellow (*Hipposideros* and *Spermophilus*/Subsp. *Bubalus arnee* Assam Wild Buffalo and *Cricetulus migratorius* Sandy Hamster/J. Syn. Red Rhesus Monkey, see *Macaca mulatta*).

Funambulus Mod. L. *funambulus* 'rope-walker' (L. *funis* rope or cord; L. v. *ambulo* I walk) (*F. layardi*, *F. palmarum*, *F. pennantii*, *F. sublineatus* and *F. tristriatus*).

fusca L. *fuscus* brown, dusky (Subsp. *Panthera pardus* Indian Leopard/J. Syn. Dusky Fruit Bat, see *Rousettus leschenaultii*; Brown Crab Mongoose, see *Urva urva*).

fuscifrons Mod. L. *fuscifrons* brown-foreheaded (L. *fuscus* dusky, brown; L. *frons* forehead, front) (Subsp. *Gazella bennettii* Baluchistan Chinkara).

fuscocapillus Mod. L. *fuscocapillus* dusky-headed (L. adj. *fusco* dark or black; L. *-capillus* headed (L. *capillus* hair of the head) (*Ellobius* and *Petinomys*).

fuscus L. *fuscus* brown, dusky (*Urva* and *Moschus*/Subsp. *Urva fuscus* South Indian Mongoose, Nilgiri Brown Mongoose/J. Syn. Brown Gibbon, see *Hoolock hoolock*).

fusiformis L. adj. *fusiformis* spindle-shaped (L. *fusus* spindle; L. *-formis* -shaped (>*forma* shape) (J. Syn. Spindle-shaped Dolphin, see *Peponocephala electra*).

Gg

galeritus L. *galeritus* hooded (from *galera* helmet, *galerum* bonnet, cap) (*Hipposideros*).

gangetica/gangeticus L. adj. *gangetica/gangeticus* belonging to the Ganges river, India (*-ica*, *-icus* belonging to, pertaining to (L./Gr.)) (*Platanista*/J. Syn. Gangetic Short-nosed Fruit Bat, see *Cynopterus sphinx*).

gangutrianus L. adj. *gangutrianus* belonging to Gangotri (*-anus* belonging to, pertaining to (L./Gr.)) (Subsp. *Funambulus pennantii*, Gangotri Palm Squirrel).

garoensis/garonum L. adj. *garoensis/garonum* meaning from Garo Hills, Meghalaya, India (*-ensis*, *-unum*, belonging to, pertaining to (location, toponym) (L.)) (Subsp. *Dremomys lokriah* Garo Hill's Squirrel/J. Syn. Garo Hills Giant Rat, see *Leopoldamys edwardsi*; Garo Hills Horseshoe Bat, see *Rhinolophus subbadius*).

gathornei Eponym after Gathorne Gathorne-Hardy (b. 1933), V Earl of Cranbrook, a British Biologist and author of many books on Southeast Asian Mammals (*-i*, commemorating (dedication, eponym) (L.)) (*Crocidura*).

gaurus Mod. L. *gaurus* from Hin. *gaur* (Sans. *gaura-mriga* name of the buffalo) the name of this species (*Bos*).

Gazella Mod. L. *gazelle* >Fr. *gazelle* gazelle (>Arab. *ghazal* wild goat, gazelle) (*G. bennettii* and *G. subgutturosa*).

gedrosianus Mod. L. *gedrosianus* belonging to Gedrosia (hellenized name of an area that corresponds to today's Balochistan) (*-anus*, belonging to, pertaining to (L.) geographic (location, toponym)) (Subsp. *Ursus thibetanus* Balochistan Black Bear).

geei Eponym after Edward Prichard Gee (ca. 1910–1966), a British tea planter, who first observed this species (*-i*, commemorating (dedication, eponym) (L.)) (*Trachypithecus*).

geminorum Mod. L. *geminorum* (L. *geminus* twin, double) (Subsp. *Pteropus hypomelanus* South Twin Island Flying Fox).

gentilis L. *gentilis* of or belonging to the same clan (L. *gens* clan, race) (Subsp. *Hipposideros pomona* Andersen's Leaf-nosed Bat).

Gerbillus/gerbillinus L. *gerbillus* gerbil (dim. of *gerbua* or *jerboa*, >Arab. *yarbū* the flesh of the back and loins, an oblique descending muscle); in allusion to its strong muscles or its hind legs (*-inus* pertaining to (L.)) (*G. aquilus*, *G. gleadowi* and *G. nanus*/J. Syn. Blyth's House Mouse, see *Mus musculus*).

gigantea/giganteus L. *giganteus* gigantic (L. *gigas* giant) (Subsp. *Ratufa bicolor*- McClelland's Giant Squirrel/J. Syn. Gigantic Bandicoot Rat, see *Bandicota indica*)/(*Pteropus*).

gilgitensis/gilgitianus Mod. L. *gilgitensis/gilgitianus* belonging to Gilgit (*-ensis*, *-anus*, belonging to, pertaining to geographic (location, toponym) (L.)) (J. Syn. Gilgit Markhor or Hazara Markhor, see *Capra falconeri*; Gilgit Rat, see *Rattus pyctoris*).

girensis L. adj. *girensis* meaning from the Gir Forest, Gujarat, India (*-ensis*, geographic, occurrence in (location, toponym) (L.)) (J. Syn. Gir House Rat, see *Rattus rattus*).

gleadowi Eponym after F. Gleadow (dates not known), a naturalist and painter, worked in north India (*-i*, commemorating (dedication, eponym) (L.)) (*Gerbillus* and *Millardia*).

gliroides Mod. L. *gliroides* glires-like (L. *glires* squirrel); *-oides* (>Gr. *eidos*) apparent shape (*Chiropodomys*).

Globicephala Mod. L. *globicephala* round-headed (L. *globus* ball; Gr. *kephalus* head); in allusion to the globular shape of the head (*G. macrorhynchus*).

gmelini Eponym after Johann Friedrich Gmelin (b. 1748–d. 1804), a German naturalist, botanist, entomologist, herpetologist, and malacologist (*-i*, commemorating (dedication, eponym) (L.)) (*Crocidura*).

gobiensis L. adj. *gobiensis* meaning from Gobi desert, Mongolia (*-ensis*, geographic, occurrence in (location, toponym) (L.)) (*Eptesicus*).

Golunda Mod. L. *golunda* from Can. *gulandi*, the native name for Indian Bush-Rat (*G. ellioti*).

goojaratensis L. adj. *goojaratensis* meaning from Gujarat, India (*-ensis*, geographic, occurrence in (location, toponym) (L.)) (J. Syn. Gujerat Lion, Indian Lion, Maneless Lion, see *Panthera leo*).

goral Mod. L. *goral* from Nep. *ghural*, the name of the species (*Naemorhedus*).

gour Mod. L. *gour* from Hin. *gaur* (Sans. *gaura-mriga*, name of the buffalo) the name of Indian Bison (J. Syn. Wild Bull, see *Bos gaurus*).

gracilis L. *gracilis* slender, elegant, slim; in allusion to its small forearm (Subsp. *Rhinolophus pusillus* Andersen's Least Horseshoe Bat/J. Syn. Ceylon Mole Rat or Lesser Bandicoot or Paddy-field rat, see *Bandicota bengalensis*).

gracilicauda Mod. L. *gracilicauda* slender-tailed (L. *gracilis* slender, elegant, slim; L. *cauda* tail) (J. Syn. Sikkim Mountain Shrew, see *Episoriculus caudatus*).

grahami Eponym after J. A. Graham (dates not known), who collected in North Coorg, India (*-i*, commemorating (dedication, eponym) (L.)) (J. Syn. Coorg Hill Spiny Mouse, see *Mus platythrix*).

Grampus Mod. L. *grampus*, possibly a corruption of the Fr. *grand poisson*, 'great fish' (*G. griseus*).

grandis L. *grandis* large (*Hipposideros*/*Loris lydekkerianus* Highland Slender Loris).

grayi/grayii Eponym after John Edward Gray (b. 1800–d. 1875), a British zoologist, Keeper of Zoology in the British Museum (Natural History) (*-i*, *-ii*, commemorating (dedication, eponym) (L.)) (Subsp. *Paguma larvata* Gray's Masked Palm Civet/J. Syn. Gray's Hedgehog, see *Hemiechinus grayii*).

griffithi/grifithii Eponym after William Griffith (b. 1810–d. 1845), an English doctor, naturalist and botanist, who made the first scientific collection of mammals in Afghanistan (*-i*, commemorating (dedication, eponym) (L.)) (Subsp. *Vulpes vulpes* Griffith's Red Fox/J. Syn. Hazara Nesokia, see *Nesokia indica*; Griffith's Shrew or Large Black Shrew, see *Suncus murinus*).

grisea/griseus Med. L. *griseus* grey (>Old French *gris* grey) (*Harpiola/Grampus, Naemorhedus*/J. Syn. Gray Field Mouse, see *Apodemus rusiges*; Grey Pika, see *Ochotona macrotis*; Madras Mongoose, see *Urva edwardsii*).

gruberi Eponym after U. F. Gruber (dates not known), a German zoologist (*-i*, commemorating (dedication, eponym) (L.)) (J. Syn. Gruber's Shrew, see *Episoriculus baileyi*).

gorkhali Mod. L. *gorkhali*, toponym meaning from Gorkha, Nepal (J. Syn. Gorkha Flying Squirrel, see *Petaurista elegans*).

guhai Eponym doubtedly after B. S. Guha (dates not known) (*-i*, commemorating (dedication, eponym) (L.)) (J. Syn. Guha's Rat, see *Rattus nitidus*).

gujerati Mod. L. *gujerati* meaning from the State of Gujarat, India (J. Syn. Gujarat Bush-rat, see *Golunda ellioti*).

guilleni Eponym after Antonio Guillen-Servent (b. 1964), a Spanish zoologist, now at Institute of Ecology, Veracruz, Mexico, who worked on bats in southeast Asia (*-i*, commemorating (dedication, eponym) (L.)).

gurkha Mod. L. *gurkha*, toponym meaning from Gorkha, Nepal (*Apodemus*/Subsp. *Mus saxicola* Kumaon Spiny Mouse, for this taxon no explanation as to why this *nomen* has been given; perhaps owing to its collector C.A. Crump, who collected extensively in Nepal and Himalayas in Kumaon region of India).

gwatkinsii Eponym after R. Gwatkins (dates not known), who collected in the Himalayas, but not in the Nilgiris (*-ii*, commemorating (dedication, eponym) (L.)), wrong credit, the correct collector's name is Walter Elliot (*Martes*).

gyi Eponym after Khin Maung Gyi (dates not known), retired professor of Zoology at University of Mandalay, Myanmar (*-i*, commemorating (dedication, eponym) (L.)) (Subsp. *Hipposideros lankadiva* Stanley's Leaf-nosed Bat).

Hh

habibi Eponym after Kushal Habibi (dates not known), a wildlife biologist from Afghanistan (*-i*, commemorating (dedication, eponym) (L.) (J. Syn. Habibi's Hare, see *Lepus tolai*).

Hadromys Mod. L. *hadromys* stout-mouse (Gr. *hadrōs* thick, stout; Gr. *mȳs* mouse) (*H. humei*).

hagenbecki Eponym after John Hagenbeck (b. 1866–d. 1940), an animal dealer and zoo owner, based in Sri Lanka (*-i*, commemorating (dedication, eponym) (L.)) (J. Syn. Hagenbeck's Antelope, see *Antilope cervicapra*).

hanglu Mod. L. *hanglu* after Kashmiri name *hanglu* for the species (Subsp. *Cervus elaphus* Kashmir Stag).

hannyngtoni Eponym after F. Hannyngton (b. 1874–d. 1919), a British civil servant and an amateur entomologist, who encouraged Indian Mammal Survey

(*-i*, commemorating (dedication, eponym) (L.)) (J. Syn. Coorg Lowland Spiny Mouse, see *Mus platythrix*).

hardwickii/hardwickei Eponym after Major-General Thomas Hardwicke (b. 1756–d. 1835), an English soldier and naturalist, travelled extensively throughout the Indian subcontinent (*-i*, commemorating (dedication, eponym) (L.)) (*Rhinopoma* and *Kerivoula*/J. Syn. Hardwicke's Gerbil, see *Tatera indica*; Hardwicke's Field Rat or Short-tailed Mole Rat, see *Nesokia indica*; Nepal Marten, see *Martes flavigula*).

harpia Gr. Myth. *harpya* harpy a Mythological winged monster, ravenous and filthy, with the head of a woman and the wings of a bird of prey (*Harpiocephalus*).

Harpiocephalus Mod. L. *harpiocephalus* bearing head of a harpy (Gr. Myth. *harpya* harpy a mythological winged monster, ravenous and filthy, with the head of a woman and the wings of a bird of prey; Gr. *kephalus* head) (*H. harpia*).

Harpiola Mod. L. *harpiola* somewhat like *Harpiocephalus* (see above) (*-ola*, diminutive (comparison); somewhat (adj.) (L.)) (*H. grisea*).

hasseltii Eponym after Johan Coenraad van Hasselt (b. 1797–d. 1823), a Dutch physician, zoologist, and botanist, worked and collected in Java (*-i*, commemorating (dedication, eponym) (L.)) (*Myotis*).

heathii Eponym after Josiah Marshall Heath (d. 1851), an English metallurgist, businessman, and ornithologist, worked at Porto Novo in Madras Presidency (*-ii*, commemorating (dedication, eponym) (L.)) (*Scotophilus*).

hector Gr. Myth. *Hector*, prince of Troy – a major character in Homer's *Iliad* (*Semnopithecus*).

Helarctos Mod. L. *helarctos* sun bear (Gr. *hel* the sun's heat, Gr. *arktos* a bear) (*H. malayanus*).

Hemiechinus Mod. L. *hemiechinus* semi-hedgehog (Gr. *hēmi-* half-; *ékhïnos* hedgehog) (*H. auritus* and *H. collaris*).

hemionus L. *hemionus* half-ass (L. *hemi-* half- (>Gr. *hēmi-* half-); *onus* (Gr. *ónos*) ass) (*Equus*).

Hemitragus Mod. L. *hemitragus* half-goat (Gr. *hēmi-* half-; Gr. *trágos* goat) (*H. jemlahicus*).

hemprichii Eponym after Wilhelm Friedrich Hemprich (b. 1796–d. 1825), a German naturalist and explorer, collected in the Middle-East (*-ii*, commemorating (dedication, eponym) (L.)) (*Otonycteris*).

heptneri Eponym after Dr. V.G. Heptner (dates not known), curator of Zoological Collections at Moscow State University (*-i*, commemorating (dedication, eponym) (L.)) (Subsp. *Capra falconeri* Bukharan Markhor).

hermaphroditus Gr. Myth. *hermaphroditus* (Gr. *hermaphróditos*, from *Hermês* and *Aphrodítē*), the son of Hermes and Aphrodite who merged bodies with a naiad (*Paradoxurus*).

Hesperoptenus Mod. L. *hesperoptenus* a crepuscular winged creature, a bat (Gr. *ésperos* evening, west; Gr. *ptènós* feathered or winged) (*H. tickelli*).

hikmiya Mod. L. hik*miya* shrew from Sinhala *hik-miya* musk-rat (*Crocidura*).

himalayana/himalayanus/himalayica Mod. L. *himalayana/himalayanus/himalay ica* from Himalayas or belonging to the Himalaya (Sans. *Himalaya* abode of snow (*him* snow, *alaya* abode), (*-ana*, *-anus*, *-ica,* belonging to, pertaining to (L.) geographic (location, toponym)) (*Marmota* and *Ochotona*/ Subsp. *Rhinolophus affinis* Himalayan Horseshoe Bat/*Chimarrogale*/J. Syn. Warwick's Cat, see *Prionailurus viverrinus*).

hippelaphus Mod. L. *hippelaphus*, generic combination of *hippos* (Gr. *hippos* horse) and *elaphus* (Gr. *élaphos* deer) (J. Syn. Nilagau, see *Boselaphus tragocamelus*).

Hipposideros/hipposideros Gr. *ïppo-sídèros* horseshoe (Gr. *ïppos* horse; Gr. *sídèros* iron); in allusion to the anterior part of the complicated noseleaf (*H. armiger*, *H. ater*, *H. cineraceus*, *H. diadema*, *H. durgadasi*, *H. fulvus*, *H. galeritus*, *H. grandis*, *H. hypophyllus*, *H. lankadiva*, *H. nicobarulae*, *H. pomona*, and *H. speoris*/*Rhinolophus*).

hirsutus L. *hirsutus* hairy, bristled (L. *hirtus* hairy) (J. Syn. Nepal Palm Civet, see *Paradoxurus hermaphroditus*).

hispida/hispidus L. *hispida* (>L. *hispidus*) rough, hairy (*Crocidura*/*Caprolagus*).

hodgsoni Eponym after Brian Houghton Hodgson (b. 1801–d. 1894), a British Civil Servant, naturalist and ethnologist stationed in Nepal (*-i*, commemorating (dedication, eponym) (L.)) (*Pantholops*/Subsp. *Hystrix brachyura* Nepal Porcupine; Subsp. *Mustela sibirica* Hodgson's Weasel; Subsp. *Ovis ammon* Tibetan Argali; Subsp. *Petaurista magnificus* Hodgson's Giant Flying Squirrel/J. Syn. Hodgson's Pika, see *Ochotona roylei*; Nepal Pygmy Shrew or Darjeeling Pygmy Shrew or Himalayan Pygmy Shrew, see *Suncus etruscus*; Hodgson's Fox, see *Vulpes bengalensis*; Nepal Goral or Hodgson's Goral, see *Naemorhedus goral*).

holchu From Nicobarese *holchu* friend (J. Syn. Nicobar Brown Rat, see *Rattus andamanensis*).

homochrous Gr. *homokhrous* uniform, of one colour (Gr. *homos* common; Gr. *khroa* colour) (*Plecotus*).

homorous Mod. L. *homorous* equal-tailed (Gr. *homos* common, equal; Gr. *-ouros* -tailed (Gr. *oúrá* tail)) (J. Syn. Hill Mouse or Himalayan House Mouse, see *Mus musculus*).

Hoolock/hoolock Mod. L. >Burmese *Huluk*, the native name for gibbon (*H. hoolock* and *H. leuconedys*).

horeites Gr. *oreiōtes* mountaineer (J. Syn. Hodgson's Grey-bellied Rat, see *Rattus nitidus*).

horsfieldii Eponym after Dr. Thomas Horsfield (b. 1773–d. 1859), an American physician and naturalist, later curator of the East India Company Museum in London (*-ii*, commemorating (dedication, eponym) (L.)) (*Crocidura* and *Myotis*/Subsp. *Megaderma spasma* Horsfield's False Vampire Bat and *Prionailurus bengalensis* Horsfield's Leopard Cat).

hosei Eponym after Dr. Charles Hose (b. 1863–d. 1929), a British naturalist stationed at Sarawak, Borneo who collected the skull in 1895 (*-i*, commemorating (dedication, eponym) (L.)) (*Lagenodelphis*).

hotaula Mod. L. *hotaula* from Sinh. *Hota ul*, a common name for dolphins and whales (*Mesoplodon*).

hotsoni Eponym after John Ernest Buttery Hotson (b. 1877–d. 1944), an Indian Civil Servant, subsequently the Governor of the Bombay Presidency, naturalist, collected extensively in Balochistan and Persia (*-i*, commemorating (dedication, eponym) (L.)) (*Allactaga* and *Calomyscus*).

howelli Eponym after E. B. Howell (dates not known), a British army officer and naturalist posted in Burma, who collected the type specimen from Ma Chang Kai, S. W. of Tengyueh (*-i*, commemorating (dedication, eponym) (L.)) (Subsp. *Dremomys pernyi* Howell's Long-nosed Squirrel).

humei Eponym after Allan Octavian Hume (b. 1829–d. 1912), a civil servant, political reformer, ornithologist and horticulturalist in British India, one of the founders of Indian National Congress and co-founders of the Bombay Natural History Society (*-i*, commemorating (dedication, eponym) (L.)) (*Hadromys*/J. Syn. Hume's Forest Goat, see *Capricornis thar*).

hurrianae Mod. L. *hurrianae* from Hurriana district (presently Haryana State), India (*-ae*, geographic (location, toponym) (L.)) (*Meriones*).

huttoni Eponym after Captain Thomas Hutton (b. 1806–d. 1875), an English officer in the Bengal Native Infantry, served in Afghan War (*-i*, commemorating (dedication, eponym) (L.)) (*Murina*/J. Syn. Hutton's Mole Rat, see *Nesokia indica*).

Hyaena/hyaena Gr. *ÿaina* hyena (*H. hyaena*).

hydrophillus Mod. L. *hydrophillus* water-loving (Gr. *hudro-* water-; Gr. *philos* -loving) (J. Syn. Small Nepal Water Rat, see *Bandicota indica*).

hylocrius Mod. L. *hylocrius* forest-dweller (>Gr. *ÿlè* forest, wood; Gr. *ákrios* inhabitant) (*Nilgiritragus*).

Hylopetes Mod. L. *hylopetes* forest flying squirrel (L. *hylo* (Gr. *ÿlè*) forest, wood; Gr. *petēs* winged, flyer) (*H. alboniger*).

Hyperacrius Mod. L. *hyperacrius* high-altitude vole (Gr. *oíÿperákrios* inhabitants of the heights); due to its high altitude range (*H. fertilis* and *H. wynnei*).

hypoleucos Mod. L. *hypoleucos* somewhat white, from Gr. *hupoleukos* whitish (>Gr. *hupo* beneath; Gr. *leukos* white) (*Semnopithecus*).

hypomelanus Mod. L. *hypomelanus* black underneath (Gr. *hupo* beneath; Gr. *melanos* black) (*Pteropus*).

hypomelas Mod. L. *hypomelas* black-ventered (Gr. *hupo* beneath; Gr. *melas* black) (*Paraechinus*).

hypophyllus Mod. L. *hypophyllus* leafletted (Gr. *hupo* beneath; Gr. *phÿllon* leaf); in allusion to it bearing a thin supplementary leaflet (*Hipposideros*).

hypsibius Mod. L. *hypsibius* (Gr. *ÿpsi* on high; Gr. *bios* living); in allusion to the high elevation range of the animal (Subsp. *Lepus oiostolus* Ladakh Woolly Hare).

Hypsugo Mod. L. *hypsugo* high-flying bat (Gr. *ÿpsi* on high, aloft; ending *-ugo*; formed in analogy of L. *vesperugo* a bat) (*H. cadornae* and *H. savii*).

Hystrix L. *hystrix* (Gr. *ýstriz*) porcupine (apparently from Gr. *ýs* hog and Gr. *trikh* hair) (*H. brachyura* and *H. indica*).

Ii

Ia L. *Ia* a young woman of classical times, like many women of those times, a bat is essentially flighty (*I. io*).

ilaeus Specific epithet meaning unclear; presumed L. origin *ileum* groin, flank, lower part of the body (*Microtus*).

imbrensis L. adj. *imbrensis* meaning from 'the hills' (L. *imber*, *imbris* rain, cloud); in allusion to the type locality being in the hills experiencing heavy rainfall (*-ensis*, geographic, occurrence in (location, toponym) (L.)) (Subsp. *Scotomanes ornatus* Jantia Hills Harlequin Bat).

imphalensis L. adj. *imphalensis* meaning from Imphal, Manipur, India (*-ensis*, geographic, occurrence in (location, toponym) (L.)) (J. Syn. Imphal Field Mouse, see *Mus cervicolor*).

inauritus L. *inauritus* without ears (L. *in-* not; L. *auritus* -eared) (Subsp. *Mellivora capensis* Nepal Ratel).

indica/indicus L. adj. *indica/indicus* belonging to India (*-ica*, *-icus* belonging to, pertaining to (L./Gr.)) (*Bandicota*, *Hystrix*, *Moschiola*, *Nesokia*, *Ratufa*, *Tatera* and *Viverricula*/Subsp. *Bandicota indica* Indian Bandicoot; Subsp. *Balaenoptera musculus* Pygmy Blue Whale; Subsp. *Canis aureus* Indian Jackal; Subsp. *Elephas maximus* Indian Elephant; Subsp. *Mellivora capensis* Indian Ratel and Subsp. *Pipistrellus ceylonicus* Indian Pipistrelle/J. Syn. Indian long-eared Hedgehog or North-Indian Hedgehog, see *Hemiechinus collaris*; Afghan Small Five-toed Jerboa, see *Allactaga elater*; Eastern Indian Mole Rat, see *Nesokia indica*; Indian Fox, see *Vulpes bengalensis*; Indian Lion, see *Panthera leo*; Northern Indian Otter, see *Lutra lutra*; Indian Striped Hyaena, see *Hyaena hyaena*; Indian Rhinoceros, see *Rhinoceros unicornis*; Ocean Cuvier's Beaked Whale, see *Ziphius cavirostris*).

indigitatus Mod. L. *indigitatus* without digits (L. *in-* without; L. *digitatus* having fingers or toes (L. *digitus* finger, toe)) (J. Syn. Himalayan Small-clawed Otter, see *Aonyx cinerea*).

Indopacetus Mod. L. *indopacetus* Indo-Pacific whale, generic *nomen* combination of *Indopa* from Indo-Pacific and L. *cetus* whale (*I. pacificus*).

indorouxii Mod. L. *indorouxii* Indian-Roux's Horseshoe Bat; specific *nomen* is a combination to refer to populations of Rufous Horsehsoe Bat from peninsular India (*Rhinolophus*).

indosinicus Mod. L. *indosinicus* Indo-Chinese (J. Syn. Indo-chinese Arboreal White-bellied Rat or Indo-Chinese Arboreal Niviventer, see *Niviventer langbianis*).

indus L. *indus* Indian (Subsp. *Hipposideros lankadiva* Indian Leaf-nosed Bat/J. Syn. Indian Dwarf Gerbil, see *Gerbillus nanus*).

inermis Mod. L. *inermis* unarmed, without weapons (L. *in* prefix for no; L. *arma* arm) (Subsp. *Rhinoceros sondaicus* Lesser Indian Rhinoceros, extinct from South Asia).

infralineatus Mod. L. *infralineatus* lined-below (L. *infra* below; L. *lineatus* lined, >L. *linea* line) (J. Syn. Striped-bellied Rat, see *Rattus rattus*).

infuscata L. *infuscatus* dusky (L. *infuscare* to make dark >*infuscus* dusky, blackish) (J. Syn. Obscure Fruit Bat, see *Rousettus leschenaultii*).

inornatus L. *inornatus* plain, unadorned (Subsp. *Urva vitticollis* Pocock's Striped-necked Mongoose; *Melursus ursinus* Sri Lankan Sloth Bear/J. Syn. Himalayan Flying Squirrel, see *Petaurista petaurista*).

insularis L. *insularis* of an island (L. *insula* island) (Subsp. *Chaerephon plicatus* Sri Lankan Wrinkle-lipped Free-tailed Bat/J. Syn. Paddy-field rat or Northern Ceylon Mole Rat, see *Bandicota bengalensis*).

intermedia/intermedius L. *intermedia/intermedius* intermediate (*cf.* Med. L. *intermediatus* intermediate) (Subsp. *Callosciurus erythraeus* Intermediate Squirrel; *Martes foina* Himalayan Beech Marten/J. Syn. Intermediate Hedgehog, see *Paraechinus micropus*).

io L. *Io* a nymph who was seduced by Zeus (*Ia*).

iodes From Gr. *iṓdēs* 'violet-coloured' or 'rusty-red'; in allusion to its pelage (Subsp. *Tetracerus quadricornis* Himalayan Four-horned Antelope).

isabellinus Mod. L. *isabellinus* fawn, greyish-yellow (>Fr. *Isabelle* or Spanish *Isabella*) (Subsp. *Lynx lynx* Tibetan Lynx and *Ursus arctos* (Linnaeus, 1758) Himalayan Brown Bear).

iulus Gr. Myth. *Iulus*, also known as Julus or Ascanius, son of Aeneas, character from Virgil's Latin epic poem, *Aeneid* (J. Syn. Jog Langur, see *Semnopithecus dussumieri*).

Jj

jacksoniae Eponym after Mrs. Jackson of Tura (dates not known) who rendered help to H. W. Wells, collector of types (*-ae*, commemorating (dedication, eponym) (L.)) (J. Syn. Mrs. Jackson's Hill Mouse, see *Mus pahari*).

jacquemonti Eponym after Victor Vincelas Jacquemont (b. 1801–d. 1832), a French botanist and geologist (*-i*, commemorating (dedication, eponym) (L.)) (J. Syn. Jacquemont's Jungle Cat, see *Felis chaus*).

Jaculus L. *jaculus* that which is thrown, a dart; in allusion to animals dart-like leaps (*J. blanfordi*).

jamrachi Eponym after W. Jamrach (dates not known), of Kalimpong, Darjeeling, India from whom the type was purchased (*-i*, commemorating (dedication, eponym) (L.)) (J. Syn. Mainland Serow or Jamrach's Serow, see *Capricornis thar*).

javanicus L. adj. *javanicus* Javan (*-icus* belonging to, pertaining to (L./Gr.) (*Pipistrellus*).

jemlahicus L. adj. *jemlahicus* meaning from Jemla Hills, Nepal (*-icus*, belonging to, pertaining to (location, toponym) (L.) (*Hemitragus*).

jenkinsi Eponym after Dr. Paulina D. Jenkins (b. 1966), a British zoologist, systematist and senior curator of mammals at the Natural History Museum, London (*-i*, commemorating (dedication, eponym) (L.)) (*Crocidura*).

jerdoni Eponym after Thomas Caverhill Jerdon (b. 1811–d. 1872), a British physician, zoologist, and botanist, worked extensively in the Madras Presidency (*-i*, commemorating (dedication, eponym) (L.)) (*Paradoxurus*/J. Syn. Himalayan

Spiny Field-Mouse, see *Niviventer fulvescens*; Jerdon's Hedgehog, see *Paraechinus hypomelas*; Lesser Leopard Cat, see *Prionailurus bengalensis*; Straight-horned Markhor or Trans-Indus Markhor, see *Capra falconeri*).

johnii Eponym after Rev. Dr. Christoph Samuel John (b. 1747–d. 1813), a medical missionary at Tranquebar (now Tharangambadi) in Tamil Nadu, India (*-ii*, commemorating (dedication, eponym) (L.)) (*Semnopithecus*).

joongshaiensis L. adj. *joongshaiensis* meaning from Jungshahi, Thatta, Sindh Province, Pakistan (*-ensis*, geographic, occurrence in (location, toponym) (L.)) (J. Syn. Joongshai Hare, see *Lepus nigricollis*).

jubatus New L. *jubatus* crested, having a mane or crest (L. *iuba* mane, crest) (*-ātus* provided with or having (L.)) (*Acinonyx*).

juldaschi Origin not clear, could be an eponym, Juldasch is an Uzbek name (*-i*, commemorating (dedication, eponym) (L.)) (*Neodon*).

Kk

kachhensis L. adj. *kachhensis* meaning from Kutch, Gujarat (*-ensis*, geographic, occurrence in (location, toponym) (L.)) (Subsp. *Taphozous nudiventris* Dobson's Naked-rumped Tomb Bat).

kandianus Mod. L. *kandianus* belonging to Kandy, Central Province, Sri Lanka (*-anus*, belonging to, pertaining to (L.) geographic (location, toponym)) (J. Syn. Common Ceylon House Rat or White-bellied Tree Rat or Bunglow Rat, see *Rattus rattus*; Kandyan Shrew, see *Suncus murinus*).

kashgaricus L. adj. *kashgaricus* belonging to Kashgar, Xinjiang, China (*-icus* belonging to, pertaining to (L. /Gr.)) (Subsp. *Eptesicus gobiensis* Kashgar Serotine).

kathiah Mod. L. *kathiah* from native name *Káthiah nyúl* of the Yellow-bellied Weasel in Nepal and India in its range (*Mustela*).

kathleenae Eponym after Ms. Kathleen V. Ryley (dates not known), a British woman and unofficial worker at British Museum (Natural History) who assisted Oldfield Thomas (*-ae*, commemorating (dedication, eponym) (L.)) (J. Syn. Ceylon Dusky-striped Squirrel, see *Funambulus sublineatus*).

kathygre Mod. L. *kathygre* living in wet places (Gr. *kathygre*, *kathygros* >Gr. *kata* down from, on account of; Gr. *hygros* wetness); with reference to its wet zone habitat in Sri Lanka (*Moschiola*).

kekrimus Mod. L. *kekrimus* Kekrima Mouse (combination of type locality name Kekrima, Nagaland, India and Gr. *mȳs* mouse) (J. Syn. Naga Hills Manipur Rat, see *Berylmys manipulus*).

kelaarti Eponym after Lieutenant Colonel Edward Frederick Kelaart (b. 1819–d. 1860), a Ceylonese-born physician and naturalist in Sri Lanka, author of *Prodromus fauna Zeylanica*, the first description of Ceylonese fauna (*-i*, commemorating (dedication, eponym) (L.)) (Subsp. *Felis chaus* Kelaart's Jungle Cat/J. Syn. Striped Squirrel or Eastern Ceylon Palm Squirrel or Lowland Ceylon Palm Squirrel or 'Tree Rat'. see *Funambulus palmarum*; Ceylon Highland Rat or Common Highland Rat or Kelaart's Rat, see *Rattus rattus*; Ceylon Flying Fox, see *Pteropus giganteus*).

Kerivoula Mod. L. *kerivoula* from the Sinhala *kehelvoulha* name for this bat (*K. hardwickii*, *K. lenis* and *K. picta*).

khasiana Mod. L. *khasiana* from Khasi Hills, Meghalaya, India (*-ana*, belonging to, pertaining to (L.) geographic (location, toponym) (*Hipposideros*).

khumbuensis L. adj. *khumbuensis* meaning from Khumbu (Mt. Everest region), Nepal (*-ensis*, geographic, occurrence in (location, toponym) (L.)) (J. Syn. Khumbu Spiny Mouse, see *Mus saxicola*; Khumbu Rat, see *Rattus pyctoris*).

khur Mod. L. *khur* after Hin. name *ghor-khur* for the species (Subsp. *Equus hemionus* Indian Wild Ass).

kiang Mod. L. *kiang* from Tib. *kyang*, a name for the species (*Equus*).

kingiana Eponym after Dr. King (dates not known), from Sikkim, India, who collected the type (*-ana*, commemorating (dedication, eponym) (L.)) (J. Syn. King's Shrew, see *Crocidura attenuata*).

kinneari Eponym after Sir Norman Boyd Kinnear (b. 1882–d. 1957) British ornithologist at the British Museum (Natural History), and curator of the Bombay Natural History Museum 1907 (*-i*, commemorating (dedication, eponym) (L.)) (Subsp. *Rhinopoma microphyllum* Kinnear's Mouse-tailed Bat).

Kogia Possibly latinized form of *codger*, a barborous word, but it might be a tribute to Cogia Effendi, a Turk, who observed whales in the Mediterranean (*K. breviceps* and *K. sima*).

kok Mod. L. *kok*, possibly latinized form of *kok*, possibly >TeL. *kokku*, rat (J. Syn. Kok Bandicoot, see *Bandicota bengalensis*).

kokree Mod. L. *kokree* from Marathi name *kokree*, name of the Bengal Fox (J. Syn. Dukhun Fox, see *Vulpes bengalensis*).

kola Mod. L. *kola* from Marathi name *kola*, name of the Jackal (J. Syn. Western India Jackal, see *Canis aureus*).

kondana Mod. L. *kondana* from the Kondana fort (also known as Sinhgad fort), Pune, Maharashtra, India (*-ana*, belonging to, pertaining to (L.) geographic (location, toponym)) (*Millardia*).

kotiya Mod. L. *kotiya* >Sinh. *kotiya* (meaning tiger) the name of the leopard in Sri Lanka (Subsp. *Panthera pardus* Sri Lankan Leopard).

kuhlii Eponym after Heinrich Kuhl (b. 1797–d. 1821), a German naturalist and zoologist, collected and described species from South and Southeast Asia (*-ii*, commemorating (dedication, eponym) (L.)) (*Scotophilus* and *Pipistrellus*).

kurrachiensis L. adj. *kurrachiensis* meaning from Karachi, Pakistan (*-ensis*, geographic, occurrence in (location, toponym) (L.)) (J. Syn. Karachi Finless Porpoise, see *Neophocaena phocaenoides*).

kutab Mod. L. *kutab* possibly from Urdu *kutab*, local name for the pole-cat (Subsp. *Lutra lutra* Kashmir Otter).

kutas Mod. L. *kutas* possibly from the local name for this animal (Subsp. *Felis chaus* Bengal Jungle Cat).

kutchicus L. adj. *kutchicus* meaning from Kutch, Gujarat, India (*-icus*, belonging to, pertaining to (location, toponym) (L.)) (J. Syn. Kutch Hedgehog, see *Paraechinus micropus*).

kura Mod. L. *kura* from Sinhalese *kura* dwarf, pygmy (J. Syn. Ceylon White-toothed Shrew, see *Suncus etruscus*).

Ll

labiata L. adj. *labiata* meaning two-lipped (L. *labia* lip) (Subsp. *Nyctalus noctula* Thick-lipped Noctule).

ladacensis L. adj. *ladacensis* from Ladakh, India (-*ensis*, geographic, occurrence in (location, toponym) (L.)) (*Ochotona*).

Lagenodelphis Mod. L. *lagenodelphis* bottle-nosed dolphin (Gr. *lágènos* flagon, bottle; Gr. *delphis* dolphin); this *nomen* is a combination of two generic names, *Lagenorhynchus* and *Delphinus*, with which it shares some characters (*L. hosei*).

lahulius Mod. L. *lahulius* pertaining to Lahul, Himachal Pradesh (-*ius* commemorating or dedication to (L./Gr.)) (J. Syn. Lahul Vole, see *Alticola argentatus*).

langbianis Mod. L. *langbianis* meaning from Lang Bian Peak, on Lâm Viên Plateau, Vietnam (-*is*, geographic, occurrence in (location, toponym) (L.)) (*Niviventer*).

laniger L. *laniger* woolly (*Cuon* and *Myotis*/Subsp. *Ursus thibetanus* Himalayan Black Bear).

lanka Mod. L. *lanka* meaning belonging to or from Sri Lanka (Subsp. *Urva edwardsii* Sri Lankan Grey Mongoose or Silvery Mongoose/J. Syn. Large Ceylon Flying Squirrel, Large Grey Flying Squirrel, Ceylon Grey Flying-Squirrel, see *Petaurista philippensis*; Ceylon Jackal, see *Canis aureus*).

lankadiva Mod. L. *lankadiva* meaning belonging to Sri Lanka (*Hipposideros*).

laosiensis L. adj. *laosiensis* meaning from Laos (-*ensis*, geographic, occurrence in (location, toponym) (L.)) (Subsp. *Bos gaurus* Southeast Asian Gaur).

larvata L. *larvata* masked (L. *larva* mask) (*Paguma*).

lasiotis Mod. L. *lasiotis* hairy-eared (Gr. *lásios* hairy; L. *otus, ōtos* ear); in allusion to its hairy ears (Subsp. *Dicerorhinus sumatrensis* Northern Sumatran Rhinoceros).

lasiurus/lasyurus Mod. L. *lasiurus/lasyurus* hairy-tailed (Gr. *lásios* hairy; Gr. *oúrá* tail); in allusion to its hairy tail (*Platacanthomys*)/the entire interfemoral membrane is hairy on the dorsal side (Subsp. *Harpiocephalus harpia* Hodgson's Hairy-winged Bat).

Latidens Mod. L. *latidens* concealed teeth (L. *latens* hidden, secret, unknown (>L. *latere* to be concealed); L. *dens* tooth) (*L. salimalii*).

latronum Mod. L. *latronum* barking (L. *latro* bark (>Gr. *lātris*)) (*Apodemus*).

layardi Eponym after E. L. Layard (b. 1824–d. 1900), Civil Servant in Ceylon (-*i*, commemorating (dedication, eponym) (L.)) (*Funambulus*/J. Syn. Ceylon Small Flying-Squirrel, Small Ceylon Flying-Squirrel, Small Flying-squirrel, see *Petinomys fuscocapillus*).

leathemi Eponym after Major G. H. Leathem (dates not known), a British army officer and collector (-*i*, commemorating (dedication, eponym) (L.)) (J. Syn. Leathem's Birch Mouse, see *Sicista concolor*).

leisleri Eponym after Johann Philipp Achilles Leisler (b. 1772–d. 1813), a German physician and naturalist (-*ii*, commemorating (dedication, eponym) (L.)) (*Nyctalus*).

lenis L. *lenis* soft, smooth (*Kerivoula*).

lentiginosa L. *lentiginosa* freckled, dotted (L. *lentigo* freckle) (J. Syn. Freckled Dolphin or Speckled Dolphin or Spotted Dolphin, see *Sousa chinensis*).

leo L. *leo* lion (>Gr. *léòn*) (*Panthera*).

leonina L. *leoninus* belonging to lion (>Gr. *léònina* tawny-coloured, Gr. *léòn* lion) (*Macaca*).

Leopoldamys Mod. L. *Leopoldamys* Leopold's mouse (Leopold and Gr. *mȳs* mouse) (*L. edwardsi* and *L. sabanus*).

lepcha Mod. L. *lepcha* meaning from Sikkim (Lepcha also used to identify the people of Eastern Nepal, Sikkim and Darjeeling in India and their language) (J. Syn. Lepcha Rat, see *Niviventer niviventer*).

lepidus L. *lepidus* charming, elegant (*Rhinolophus*/Subsp. *Pipistrellus kuhlii*/J. Syn. Small Spiny Mouse, see *Mus booduga*).

leptodactylus Mod. L. *leptodactylus* slender-fingered (Gr. *leptos* fine or slender; L. *dactylus* (Gr. *daktulōs*) finger or toe) (*Spermophilopsis*).

leptophyllus Mod. L. *leptophyllus* slender-leafletted (Gr. *leptos* fine or slender; Gr. *phýllon* leaf); in allusion to its slender supplementary leaflets (Subsp. *Hipposideros grandis*).

Lepus L. *lepus* rabbit, hare (genitive *Leporis* a hare) (*L. nigricollis*, *L. oiostolus*, *L. tibetanus* and *L. tolai*).

leschenaultii Eponym after Jean-Baptiste Louis Claude Théodore Leschenault de La Tour (b. 1773–d. 1826), a French botanist and ornithologist, collected plants and animals in Madras, Bengal, China, India and Nepaland Ceylon (*-ii*, commemorating (dedication, eponym) (L.)) (*Rousettus*).

leucodon Gr. *leukodon* white-toothed (Gr. *leukos* white; Gr. *ódoús*, *ódôn* tooth) (*Crocidura*).

leucogaster Mod. L. *leucogaster* white-bellied (Gr. *leukogastēr* white-belly, from Gr. *leukos* white; Gr. *gastēr* belly) (*Murina* and *Moschus*).

leucogenys Mod. L. *leucogenys* white-cheeked (Gr. *leukogenus* white-cheek, from Gr. *leukos* white; L. *genys* (>Gr. *genus*) cheek or jaw). (*Macaca*/ J. Syn. White-cheeked Shrew, see *Suncus stoliczkanus*).

leuconedys Mod. L. *leuconedys* pale-ventered (Gr. *leukos* white; Gr. *nedys* stomach) (*Hoolock*).

leucops Mod. L. *leucops* white-faced (Gr. *leukos* white; Gr. *ōps* face, eye) (*Episoriculus*).

leucotis Mod. L. *leucotis* white-eared (Gr. *leukos* white; Gr. *-ōtis* -eared (>Gr. *ous*, *ōtos* ear)) (J. Syn. White-eared Bat, see *Pipistrellus kuhlii*).

leucophaea Mod. L. *leucophaea* whitish (Gr. *leukos* white, *pháos* light) (*Otonycteris*)

leucura/leucurus Mod. L. *leucura*/L. *leucurus* white-tailed (>Gr. *leukouros* white-tailed) (*Parascaptor*/*Phaiomys*/J. Syn. Western Asiatic Crested Porcupine, see *Hystrix indica*).

lhasaensis L. adj. *lhasaensis* from Lhasa, Tibet (*-ensis*, geographic, occurrence in (location, toponym) (L.)) (Subsp. *Ochotona nubrica* Lhasa Pika).

libyicus L. adj. *libycus* Libyan, belonging to Libya (*-icus* belonging to, pertaining to (L./Gr.)) (*Meriones*).

limitaris L. *limitaris* limiting (L. *limitare* to enclose) (J. Syn. Pakistan Bush-rat, see *Golunda ellioti*).

listeri Eponym after R.S. Lister (dates not known), from Darjeeling who has assisted N.A. Baptista, Bombay Natural History Society's animal collector during the Mammal Survey of British India (*-i*, commemorating (dedication, eponym) (L.)) (J. Syn. Lister's Giant Rat, see *Leopoldamys edwardsi*).

listoni Eponym after Major W.G. Liston, Indian Medical Service (b. 1872–d. 1950), from Konkan, who collected the types (*-i*, commemorating (dedication, eponym) (L.)) (J. Syn. Liston's Field Rat, see *Millardia meltada*).

littledalei Eponym after Major Harold A.P. Littledale (b. 1853–d. 1930), an English collector, worked in Nepal (*-i*, commemorating (dedication, eponym) (L.)) (Littledale's Marmot, see *Marmota caudata*).

lobatus Mod. L. *lobatus* lobed (>Mod. L. *lobus* lobe; >Gr. *lobos* lobe) (J. Syn. Lobe-eared Bat, see *Pipistrellus kuhlii*).

lokriah Mod. L. *lokriah* >Nep. *locriah*, the local name of this squirrel (*Dremomys*).

lokroides Mod. L. *lokroides* resembling *lokriah* (*lokriah* >Nep. *locriah*, the local name of Orange-bellied Himalayan Squirrel; *-oides* (>Gr. *eidos*) apparent shape) (Subsp. *Callosciurus pygerythrus* Hoary-bellied Himalayan Squirrel).

longicauda/longicaudata Mod. L. *longicauda/longicaudata* long-tailed (L. *longus* long; L. *cauda* tail) (J. Syn. Long-tailed Crestless Porcupine, see *Hystrix brachyuran*/J. Syn. Long-tailed Leopard, see *Panthera pardus*).

longimanus Mod. L. *longimanus* long-handed (L. *longus* long; L. *manus* hand); in allusion to its long wings (*Taphozous*).

longipes L. *longipes* long-footed (L. *longus* long; L. *pes* foot, >Gr. *pous* foot) (*Myotis*).

longirostris Mod. L. *longirostris* long-billed (L. *longus* long; L. *-rostris* -billed (L. *rostrum* bill)) (*Stenella*).

longstaffi Eponym after Dr. T.G. Longstaff (b. 1875–d. 1964), a British doctor, explorer and mountaineer, who collected the type (*-i*, commemorating (dedication, eponym) (L.)) (J. Syn. Longstaff's Weasel, see *Mustela altaica*).

lordi Eponym after Rev. W. Lord (b. 1875–d. 1964), a British doctor, explorer and mountaineer, who collected the type (*-i*, commemorating (dedication, eponym) (L.)) (J. Syn. Konkan Mole Rat, see *Bandicota bengalensis*).

Loris Mod. L. *loris* from French *loris* for the species (from Dutch *loeris* meaning simpleton) (*L. lydekkerianus* and *L. tardigradus*).

luctus L. *luctus* mourning (>L. *lugere* to mourn) (*Rhinolophus*).

lupus L. *lupus* Wolf (*Canis*).

lutescens Mod. L. *lutescens* somewhat yellowish (>L. *luteus* saffron-yellow) (*cf.* L. *lutescens* muddy-coloured, >*lutum*, *luti* mud) (Subsp. *Funambulus pennantii* Gujarat Palm Squirrel).

luteus L. *luteus* saffron-yellow (J. Syn. Bengal Yellow Bat, see *Scotophilus heathii*).

Lutra/lutra L. *lutra* otter (*L. lutra*).

Lutrogale Mod. L. *lutrogale* otter-weasel, generic name combination (L. *lutra* otter; Gr. *gale* weasel) (*L. perspicillata*).

lydekkerianus Eponym after Richard Lydekker (b. 1849–d. 1915), English Zoologist and Palaentologist (*-anus*, belonging to, pertaining to (L.)) (*Loris*).

Lynx/lynx Gr. *lýgé* lynx, probably from its bright eyes (>Gr. root *lyk* eye or *lýkhnos* lamp) (*L. lynx*).

lyra L. *lyra* lyre (*Lyroderma*).

Lyroderma Mod. L. *lyroderma* lyre-skinned (Gr. *lyro* lyre; Gr. *derma* skin); in allusion to the prominent lyre-shaped noseleaf (*L. lyra*).

Mm

Macaca Mod. L. *macaca* macaque (>Fr. *macaque* or Portugese *macaco* meaning 'a monkey') (*M. arctoides, M. assamensis, M. fascicularis, M. leonina, M. leucogenys, M. mulatta, M. munzala, M. radiata, M. silenus* and *M. sinica*).

maccarthiae Eponym after Mrs. MacCarthy (dates not known), a naturalist and collector, wife of Sir Charles J. MacCarthy, a British Governor-General of Sri Lanka (*-ae*, commemorating (dedication, eponym) (L.)) (Subsp. *Urva fuscus* Northern Ceylon Brown Mongoose).

macclellandi Eponym after Dr. John McClelland (b. 1805–d. 1875), geologist and amateur naturalist, worked in Assam (*-i*, commemorating (dedication, eponym) (L.)) (*Tamiops*).

mackenziei Eponym after J.M.D. Mackenzie (dates not known), an officer with Burmese Forest Service, collected from Burma for the Mammal Survey of British India (*-i*, commemorating (dedication, eponym) (L.)) (*Berylmys*).

macmillani Eponym after S.A. Macmillan (d. 1915), a British army officer and naturalist, who collected accompanied CoL. G.C. Shortridge on his Chindwin trip (*-i*, commemorating (dedication, eponym) (L.)) (Subsp. *Dremomys lokriah* Macmillan's Squirrel).

macrocephalus Gr. *makrokephalos* (Gr. *makros* great, big; Gr. *kephalus* head) great-headed (*Physeter*).

macrodens Mod. L. *macrodens* (Gr. *makrodens*) large-toothed (Gr. *makros* long; L. *dens* tooth) (Subsp. *Miniopterus magnater* Large-toothed Long-fingered Bat).

Macroglossus Mod. L. *macroglossus* (>Gr. *makroglōssus*) long-tongued (Gr. *makros* long; Gr. *glōssa* tongue) (*M. sobrinus*).

macropus Gr. *makropous* long-footed (Gr. *makros* long; Gr. *pous* foot) (J. Syn. Large-footed Bandicoot Rat, see *Bandicota indica*; Large-footed Kashmir Myotis, see *Myotis longipes*).

macrorhynchus Mod. L. *macrorhynchus* (>Gr. *makrorhunkhos*) (Gr. *makros* long; Gr. *rhunkhos* bill) long-billed (*Globicephala*).

macrosceloides Probably after Gr. adj. *macroscelous* long-legged (*-oides* >Gr. *eidos*, apparent shape) (Subsp. *Neofelis nebulosa* Tibetan Clouded Leopard).

macrotis Mod. L. *macrotis* big-eared (>Gr. *makrōtēs*) long-eared (Gr. *makros* long; Gr. *ous, ōtos* ear)) (*Rhinolophus/Ochotona*).

macroura/macrourus/macrurus/macrura Mod. L. *macroura/macrourus/macrurus* long-tailed (Gr. *makros* long, Gr. *-ouros* -tailed (Gr. *oúrá* tail)) (*Ratufa/Atherurus/Episoriculus*/J. Syn. Himalayan Long-tailed Mole, see *Euroscaptor micrura*).

maculiventer Mod. L. *maculiventer* spotted-bellied (L. *macula* spot; L. *ventris* belly) (J. Syn. Spot-bellied Dolphin, see *Sousa chinensis*).

madrassius Mod. L. *madrassius* pertaining to Madras Presidency in British India (*-ius* commemorating or dedication to (L./Gr.)) (Subsp. *Harpiocephalus harpia* South Indian Hairy-winged Bat).

Madromys Mod. L. *Madromys* Madras mouse (Madras and Gr. *mȳs* mouse); owing to the fact that the type was collected from Cuddapah (=Kadapa), Madras Presidency (*M. blanfordi*).

magnater Mod. L. *magnater* large (L. *magnus* great); in allusion to its large body and cranial size in comparison to *M. schreibersii* (*Miniopterus*).

magnificus L. adj. *magnificus* magnificent (*Petaurista*).

mahadeva Hindu Myth. Mahadeva, a Great Spirit or great god, synonym of Siva, one of the Hindu holy trinity (J. Syn. Pachmarhi Hare, see *Lepus nigricollis*).

maimanah Toponym after Maimana, Afghanistan (Subsp. *Felis chaus* Maimanah Jungle Cat).

majus L. *maius* great (*Megaderma spasma* Great False Vampire Bat).

malabarica/malabaricus L. adj. *malabarica/malabaricus* meaning from Malabar – a region in southern India lying between the Western Ghats and the Arabian Sea corresponding to the Malabar District of the Madras Presidency in British India (presently includes areas under five northern districts of Kerala) (*-ica*, *-icus*, belonging to, pertaining to (location, toponym) (L.)) (Subsp. *Loris lydekkerianus* Malabar Grey Slender Loris; *Bandicota indica* Malabar Bandicoot-rat and *Muntiacus vaginalis* South Indian Muntjak/J. Syn. Malabar Porcupine, see *Hystrix indica*).

malayanus/malayana L. *malayanus/malayana* belonging to Malay or Malaysia (*-anus*, *-ana*, belonging to, pertaining to geographic (location, toponym) (L.)) (*Helarctos*/Subsp. *Tylonycteris malayana* Malayan Greater Bamboo Bat/J. Syn. Malay Dolphin, see *Stenella attenuata*).

malpasi Eponym after A. H. Malpas (dates not known), a marine biologist in service to Government of Ceylon (*-i*, commemorating (dedication, eponym) (L.)) (J. Syn. Malpas's Bat or Small Brown Bat, see *Kerivoula hardwickii*).

manei Mod. L. *manei* house (>Tam. *manei* courtyard or house) (J. Syn. Ceylon House Mouse, see *Mus musculus*).

manipulus L. adj. *manipulus* belonging to Manipur, India (*Berylmys*).

Manis L. *manis* ghost; in allusion to the animal's nocturnal habit (*M. crassicaudata* and *M. pentadactyla*).

manul Mod. L. *manul* >Mongol. *manuul*, the native name of the species (*Otocolobus*).

margarita Eponym after Jean Auguste Margueritte (b. 1823–d. 1870), a French General, the leader of the expedition during which Victor Loche (b. 1806–d. 1863) found this animal (*Felis*).

masoni Eponym after Francis Mason (b. 1799–d. 1874), an American missionary and naturalist, who worked in British Burma (=Myanmar) (*-i*, commemorating (dedication, eponym) (L.)) (Subsp. *Hipposideros diadema* Mason's Diadem Roundleaf Bat).

marginatus L. *marginatus* bordered, edged (L. *marginare* to emarginated; >L. *margo* border, edge) (J. Syn. Margined Short-nosed Fruit Bat or Small Fox-bat, see *Cynopterus sphinx*).

marica Rom. Myth. *marica* a river- or water-nymph (J. Syn. Bengal Tree Mouse *Vandeleuria oleracea*).

maris L. *maris* sea (Subsp. *Pteropus hypomelanus* Maldive Island Flying Fox).

marmorata L. *marmoratus* marbled (L. *marmor*, *marmoris* marble) (*Pardofelis*).

Marmota Mod. L. *marmota* marmot (Ital. *marmotta* >*murmont* of local dialect which is >*muris* (genitive singular of *mȳs* mouse) and *montis* (genitive singular of *mons*)) (*M. caudata* and *M. himalayana*).

Martes L. *martes* marten (*M. flavigula*, *M. foina* and *M. gwatkinsii*).

matugamaensis L. adj. meaning from the Matugama, Kalutara, Western Province, Sri Lanka (*-ensis*, geographic, occurrence in (location, toponym) (L.)) (*Funambulus palmarum* Matugama Palm Squirrel).

maxima/maximus L. *maxima*/Gr. *maximus* greatest or largest (super. >*magnus* great) (Subsp. *Ratufa indica* Malabar Giant Squirrel/J. Syn. Greater Bandicoot Rat, see *Bandicota indica*) (*Elephas*).

mayori Eponym after Major E.W. Mayor (dates not known), collected in Sri Lanka for the Mammal Survey of British India (*-i*, commemorating (dedication, eponym) (L.)) (*Mus*/Subsp. *Viverricula indica* Ceylon Small Civet).

mcmahoni Eponym after Lt.-Col. Sir Arthur Henry McMahon (b. 1862–d. 1949), a British army officer and diplomat, who served as Chief Commissioner of Balochistan (*-i*, commemorating (dedication, eponym) (L.)) (J. Syn. McMahon's Rhesus Macaque, see *Macaca mulatta*).

mechukaensis L. adj. *mechukaensis* meaning from the Mechuka Valley, West Siang District, Arunachal Pradesh, India (*-ensis*, geographic, occurrence in (location, toponym) (L.)) (*Petaurista*).

medius L. *medius* middle (J. Syn. Kathiawar Rock-rat, see *Cremnomys cutchicus*/J. Syn. Indian Ternate Bat or Indian Pteropus or Warbagool, see *Pteropus giganteus*).

megaceros Mod. L. *megaceros* big-horned (Gr. *megas* great; Gr. *kéras* horn); in allusion to the big horns (Subsp. *Capra falconeri* Sulaiman Markhor).

Megaderma Mod. L. *megaderma* large-skinned (Gr. *megas* great; Gr. *derma* skin); in allusion to the large wings and interfemoral membrane (*M. spasma*).

Megaerops Mod. L. *megaerops* like Megaera (Gr. Myth. *Megaera* one of the Three 'Furies'; *óps* aspect) (*M. niphanae*).

megalotis Mod. L. *megalotis* great-eared (Gr. *megalos* great; Gr. *-ōtis* -eared (Gr. *ous*, *ōtos* ear) (Subsp. *Hemiechinus auritus* Great-eared Afghan Hedgehog).

Megaptera Mod. L. *megaptera* great winged (Gr. *megas* great; Gr. *ptera* >Gr. *pteron* wing); in allusion to the strong dorsal fin (*M. novaeangliae*).

meinertzhageni Eponym after Colonel Richard Meinertzhagen (b. 1878–d. 1967), a British soldier, intelligence officer and ornithologist (*-i*, commemorating (dedication, eponym) (L.)) (J. Syn. Meinertzhagen's Bat, see *Myotis nipalensis*).

melanochra Mod. L. *melanochra* >Gr. *melanokhrōs* black-skinned (Gr. *melas* black; Gr. *khrōs* colour) (Subsp. *Ratufa macroura* Western Ceylon Giant Squirrel).

melanodon Mod. L. *melanodon* black-tooth (Gr. *melas* black; Gr. *ódoús*, *ódôn* tooth) (J. Syn. Black-toothed Pygmy Shrew, see *Suncus etruscus*).

melanogaster Gr. *melanogastēr* black-belly (Gr. *melanos* black; Gr. *gastēr* belly) (*Eothenomys*).

melanopogon Mod. L. *melanopogon* black-bearded (Gr. *melas* black; Gr. *pōgōn* beard); in allusion to the characteristic black beard present in males (*Taphozous*).

melanotus Mod. L. *melanotus* black-eared (Gr. *melas*, *melanos* black; Gr. *otus* ear) (*Pteropus*).

Meles/meles L. *meles* badger (*M. meles*).

Mellivora L. *mellivora* honey-eating (L. *mel* honey; L. *voro* to devour), from the animal's favourite food (*M.capensis*).

Melogale Mod. L. *melogale* badger-weasel, generic name combination (L. *meles* badger; Gr. *gale* weasel) (*M. moschata* and *M. personata*).

meltada Mod. L. *meltada* metad from the Wadur *mettade*, a name by which this species in known (*Millardia*).

Melursus L. *melursus* honey-bear (L. *mel* honey; L. *ursus* a bear), from its fondness for honey (*M. ursinus*).

meminna Mod. L. *meminna* after Sinhalese name *meeminna* for the species (*Moschiola*).

mentalis Mod. L. *mentalis* pertaining to the chin (>Fr. *mental* of the chin >L. *mentum* chin) (J. Syn. Black-chinned Hedgehog, see *Paraechinus micropus*).

meridianus L. *meridianus* of noon, southern (L. *meridies* south); in allusion to its activity during the middle of the day (*Meriones*).

Meriones Mod. L. *meriones* jird (Gr. *mèrós* thigh); in allusion to the development of its hind legs (*M. crassus*, *M. hurrianae*, *M. libycus*, *M. meridianus*, *M. persicus* and *M. zarudnyi*).

Mesoplodon Mod. L. *mesoplodon* armed with a tooth in the middle (Gr. *mésos* middle; Gr. *ópla* arms; Gr. *ódoús*, *ódôn* tooth); in allusion to a prominent tooth in the lower jaw (*M. densirostris* and *M. hotaula*).

meyeroehmi Eponym after Dr. D. Meyer-Oehme (dates not known), a German mammologist who collected in Middle East (*-i*, commemorating (dedication, eponym) (L.)) (Subsp. *Rhinolophus blasii* Meyer-Oehm's Horseshoe Bat).

michaelis Eponym after Michael FitzGibbon (dates not known), an English animal importer, in whose consignment from Balochistan, the types of this species were discovered (*Salpingotulus*).

Micromys Mod. L. *micromys* small mouse (Gr. *mikros* small; Gr. *mŷs* mouse) (*M. minutus*).

micronyx Mod. L. *micronyx* small-clawed (Gr. *mikros* small; Gr. *ónyx* nail, claw) (J. Syn. Small-clawed Pygmy Shrew or Kumaon Pygmy Shrew, see *Suncus etruscus*).

microphyllum Mod. L. *microphyllum* small-leafed (Gr. *mikros* small; Gr. *phýllon* leaf) (*Rhinopoma*).

micropus Mod. L. *micropus* small-footed (Gr. *mikros* small; Gr. *pous* foot) (*Paraechinus*/J. Syn. Least Leaf-nosed Bat, see *Hipposideros cineraceus*).

Microtus Mod. L. *microtus* small-eared (Gr. *mikros* small; Gr. *otus* ear) (*M. ilaeus*).

micrura Mod. L. *micrura* small-tailed (Gr. *mikros* small; Gr. *-ouros* -tailed) (*Euroscaptor*).

midas Mod. L. *midas* Gr. Myth. Midas, king of Phrygia, favoured by Bacchus with the gift that everything he touched turned to gold (Subsp. *Rhinolophus hipposideros* Persian Lesser Horseshoe Bat).

migratorius Med. L. *migratorius* migrating (L. *migrator, migratoris* migrant, wanderer (L. *migrare* to migrate)) (*-ius*, having the nature of (L. /Gr.)) (*Cricetulus*).

Millardia/millardi Eponym after Dr. Walter Samuel Millard (b. 1864–d. 1952), a British naturalist, Honorary Secretary of the Bombay Natural History Society, and along with Oldfield Thomas commenced the Mammal Survey of British India (*-i*, commemorating (dedication, eponym) (L.)) (*M. gleadowi, M. kondana* and *M. meltada/Dacnomys*/J. Syn. Kashmir Leopard, see *Panthera pardus*).

millsi Eponym after J. Phillip Mills (b. 1890–1960), British Imperial Civil Service officer serving in Nagaland (*-i*, commemorating (dedication, eponym) (L.)) (Subsp. *Arctogalidia trivirgata* Indian Small-toothed Palm Civet; *Melogale moschata* Nagaland Ferret-Badger/J. Syn. Mills' Crestless Porcupine, see *Hystrix brachyura*).

mimus L. *mimus* mimic (Subsp. *Pipistrellus tenuis* Wroughton's Least Pipistrelle).

Miniopterus Gr. *miniopterus* small-winged (Gr. *miniós* small; Gr. *pterón* wing); from the very short phalanx of the third or longest finger (*M. fuliginosus, M. pusillus,* and *M. magnater*).

minor L. *minor* smaller (Subsp. *Platanista gangetica* Indus River Dolphin and *Soriculus nigrescens* Dobson's Large-clawed Shrew).

minutus L. *minutus* small, little (*Micromys* and *Sorex*).

mishmiensis L. adj. *mishmiensis* from Mishmi Hills, Arunachal Pradesh, India (*-ensis*, geographic, occurrence in (location, toponym) (L.)) (*Petaurista*/Subsp. *Hoolock hoolock* Mishmi Hills Hoolock Gibbon).

mitchelli Eponym after Richard Merle Mitchell (dates not known), an American zoologist who collected mammals in Nepal (*-i*, commemorating (dedication, eponym) (L.)). (J. Syn. Mitchell's Pika, see *Ochotona roylei*).

mitratus Mod. L. *mitratus* mitred, having a hood (Gr. *mítra* hood; *-atus*, provided with (L.)); origin not known, probably due to long hair on head forming a hood (*Rhinolophus*).

mixtus L. *mixtus* mixed or mingled (L. *miscere* to mix) (J. Syn. Mysore Leaf-nosed Bat, see *Hipposideros lankadiva*).

miya Mod. L. *miya* shrew from Sinhala *miya* musk-rat (*Crocidura*).

modesta L. *modestus* plain, modest, unassuming (L. *modus* measure, standard) (J. Syn. Kumaon Tree Mouse, see *Vandeleuria oleracea*).

molagan Mod. L. *molagan* from Tamil, a name for this species (J. Syn. Molagan Finless Porpoise, see *Neophocaena phocaenoides*).

moormenensis L. adj. *moormenensis* meaning from Moormi, Maharashtra (*-ensis*, geographic, occurrence in (location, toponym) (L.)) (J. Syn. Moormi Cat, see *Pardofelis temminckii*).

montana/montanus L. *montana/montanus* of the mountains, mountain- (L. *mons*, *montis* mountain) (Subsp. *Vulpes vulpes* Montane Red Fox/*Nyctalus*, *Paradoxurus*, *Rattus* and *Suncus*/Subsp. *Urva edwardsii* Montane Grey Mongoose).

monticola L. *monticola* mountain-dweller, mountaineer (L. *mons*, *montis* mountain; L. *-cola* inhabitant >L. *colere* to dwell) (Subsp. *Lutra lutra* Hill Otter; Subsp. *Rhinolophus lepidus* Montane Horseshoe Bat; Subsp. *Semnopithecus vetulus* Montane Purple-faced Langur (also Highland Purple-faced Leaf-monkey, Bear Monkey, Highland Wanderoo/J. Syn. Ghose's white-bellied Rat, see *Niviventer niviventer*).

montivagus L. *montivagus* mountain roaming (L. *mons*, *montis* mountain; *L. vagari* to wander) (*Myotis*).

montosa L. *montosus* mountainous (*Alticola*).

mordax L. adj. *mordax* biting, snappish (L. *mordeo* bite, nibble or gnaw) (Sub. J. Syn. Myanmar Hairy-winged Bat or Greater Hairy-winged Bat, see *Harpiocephalus harpia*).

moschata Mod. L. *moschatus* musky (Gr. *móskhos* musk >Pers. *mušk*) (*Melogale*).

Moschiola Mod. L. *moschiola* moschus-like (Gr. *móskhos* musk >Pers. *mušk*); in allusion to the musk glands of the male (*-ola* dim. meaning in comparison, or somewhat like (adj.) (L.)) (*M. indica*, *M. kathygre*, and *M. meminna*).

Moschus Med. L. *moschus* (Gr. *móskhos* musk >Pers. *mušk*) musk; in allusion to the musk glands of the male (*M. chrysogaster*, *M. cupreus*, *M. fuscus*, and *M. leucogaster*).

mulatta L. *mulus* a mule, from where '*mulatto*' is derived that pertains to the offspring of a black person and a white person (*Macaca*).

Muntiacus L. *muntiacus* derivative meaning pertaining to *muntjak*, a native name of this animal in Sunda language (*-acus*, pertaining to (L. /Gr.)) (*M. putaoensis* and *M. vaginalis*).

munzala Mod. L. *munzala* from *mun zala* ('monkey of the deep forest') in the dialect spoken by Dirang Monpa (*Macaca*).

muricola Mod. L. *muricola* wall-dweller (L. *murus* wall; L. *-cola* inhabitant >*colere* to dwell) (*Myotis*).

Murina L. *murina* mouse-like (L. *muris* mouse); in allusion to the shape of its ear and head (*M. aurata*, *M. cyclotis*, *M. huttoni*, *M. leucogaster*, and *M. tubinaris*).

murinus L. *murinus* of mice, mouse- (>L. *mus*, *muris* mouse); in allusion to its mouse-like appearance (*Suncus* and *Vespertilio*/J. Syn. Mouse-coloured Horseshoe Bat or Little Horseshoe Bat, see *Hipposideros fulvus*; Mouse-like Bat, see *Myotis blythi*).

murinoides Mod. L. *murinoides* mouse-like (L. *murinus* of mice, mouse- (>L. *mus*, *muris* mouse); *-oides* (>Gr. *eidos*) apparent shape); in allusion to its mouse-like appearance (J. Syn. Small Mouse-like Bat, see *Myotis blythi*).

murraiana Eponym after James A. Murray (dates not known), zoologist and curator Karachi Museum, Pakistan (*-ana*, commemorating (dedication, eponym) (L.)) (Subsp. *Asellia tridens* Murray's Trident Bat).

Mus Gr. *mÿs* mouse (*M. booduga*, *M. cervicolor*, *M. cookii*, *M. famulus*, *M. fernandoni*, *M. mayori*, *M. musculus*, *M. pahari*, *M. phillipsi*, *M. platythrix*, *M. saxicola*, and *M. terricolor*).

muscatellum Mod. L. *muscatellum* meaning from Muscat (*Rhinopoma*).

musculus Mod. L. *musculus* small mouse (Gr. *mÿs* mouse, *-ulus*, diminutive (comparison); somewhat (adj.) (L.)) (*Balaenoptera* and *Mus*).

<ustela L. *mustela* weasel (*M. altaica*, *M. erminea*, *M. kathiah*, *M. nivalis*, *M. sibirica*, and *M. strigidorsa*).

mustersi Eponym after James Lawrence Chaworth-Musters (b. 1901–d. 1948), a geographer, collector, volunteer, and Assistant Keeper at British Museum (Natural History) (J. Syn. Kabul Mouse-like Hamster, Muster's Mouse-like Hamster, see *Calomyscus baluchi*).

mutus L. *mutus* silent, dumb; in allusion to its silent nature in comparison to the domestic yak (*B. grunniens*, L. *grunniens* grunting) (*Bos*).

myoides Mod. L. *myoides* mouse-like (Gr. *mÿs* mouse; Gr. *-oides* (>Gr. *eidos*), apparent shape) (J. Syn. Mouse-like Shrew, see *Crocidura horsfieldii*).

myothrix Mod. L. *myothrix* mouse-haired (Gr. *mÿs* mouse; Gr. *thríx* hair) (J. Syn. Himalayan Bush-rat, see *Golunda ellioti*).

Myotis Mod. L. *myotis* mouse-eared bat (Gr. *mÿs* mouse; Gr. *otus*, *ōtos* ear); in allusion to its mouse-like ears (*M. annectans*, *M. blythi*, *M. csorbai*, *M. davidii*, *M. emarginatus*, *M. formosus*, *M. hasseltii*, *M. horsfieldii*, *M. laniger*, *M. longipes*, *M. montivagus*, *M. muricola*, *M. nipalensis*, *M. peytoni*, *M. sicarius*, and *M. siligorensis*).

Nn

Naemorhedus Mod. L. *naemorhedus* young goat of the woods (L. *nemus*, *nemoris* wood; L. *hedus* a young goat); in allusion to its montane woody habitat (*N. baileyi*, *N. griseus*, and *N. goral*).

nagarum Mod. L. *nagarum* after Naga Hills in Nagaland, India on the borders of India and Myanmar (J. Syn. Naga Squirrel, see *Callosciurus erythraeus*/J. Syn. Naga Spiny Mouse, see *Mus cookii*).

nair From Can. *nair*, a local name for the otter (*nair* water dog) (Subsp. *Lutra lutra* Indian Otter).

nallamalaensis L. adj. *nallamalaensis* meaning from Nallamala Hills, Andhra Pradesh (*-ensis*, geographic, occurrence in (location, toponym) (L.)) (Subsp. *Hipposideros ater* Nallamala Dusky Leaf-nosed Bat).

nanus L. *nanus* dwarf (Gr. *nanos* dwarf) (*Gerbillus*).

narbadae Mod. L. *narbadae*, after Narbada (=Narmada) river in northern peninsular India, (*-ae*, geographic (location, toponym) (L.)) (J. Syn. Narmada Rat, see *Rattus rattus*).

naria From Can. *nariya*, a local name for the jackal (Subsp. *Canis aureus* Wroughton's Jackal).

nasutus L. *nasutus* large-nosed (L. *nasus* nose) (*Rhyneptesicus*).

nayaur Mod. L. *nayaur* from Nep. *nahoor* or *naur*, a name for this species (*Pseudois*).

nebulosa L. *nebulosus* misty, cloudy (L. *nebula* cloud, mist) (*Neofelis*).

Nectogale Mod. L. *nectogale* swimming-weasel (Gr. *nektōs* swimming; Gr. *gale* weasel); in allusion to its broad-webbed hind foot, an excellent swimmer (*N. elegans*).

neglecta L. *neglectus* ignored, overlooked, neglected, disregarded (L. *neglegere* to neglect) (Subsp. *Paguma larvata* Naga Hills Masked Palm Civet).

nemoralis L. *nemoralis* of woods or groves (L. *nemus* grove) (J. Syn. Ceylon Large Tree Rat, see *Rattus rattus*).

nemorivaga/nemorivagus L. *nemorivaga/nemorivagus* forest-roaming, wood-roaming (L. *nemus*, *nemoris* wood; L. *vagari* to wander) (Subsp. *Bandicota indica* Forest Large Bandicoot-rat/J. Syn. Nepal Wood Shrew, see *Suncus murinus*).

Neodon Mod. L. *neodon* new-tooth (Gr. *neos* new; Gr. *ódoús*, *ódôn* tooth) (*N. juldaschi* and *N. sikimensis*).

Neofelis Mod. L. *neofelis* new cat (Gr. *neos* new; L. *felis* cat; probably from root *fe*, to produce, bear young) (*N. nebulosa*).

Neophocaena Gr. *neophocaena* new porpoise (Gr. *neos* new; Gr. *phôkaina* porpoise) (*N. phocaenoides*).

nepalensis L. adj. *nepalensis* from Nepal (*-ensis*, geographic, occurrence in (location, toponym) (L.)) (*Ochotona roylei* Nepalese Pika/J. Syn. Nepal Lesser Striped Shrew, see *Sorex bedfordiae*; Nepal Otter, see *Lutra lutra*).

Nesokia Mod. L. *nesokia*, >*nesoki*, a local name for the species (*N. indica*).

nestor Gr. Myth. *Nestor*, grey-haired wise old king of Pylos at the siege of Troy – a character in Homer's *Iliad* (Subsp. *Semnopithecus vetulus* Southern Lowland Wet Zone Purple-faced Langur – also Western Purple-faced Leaf-monkey, Wanderoo).

nicobarensis/nicobarica/nicobaricus L. adj. *nicobarensis/nicobarica/nicobaricus* Nicobarese or belonging to Nicobar Islands, India (*-ensis*, *-ica*, *-icus*, belonging to, pertaining to (L./Gr.)) (Subsp. *Hipposideros diadema* Nicobar Diadem Leaf-nosed Bat/*Crocidura* and *Tupaia*/J. Syn. Nicobar Black-eared Flying Fox, see *Pteropus melanotus*).

nicobarulae L. adj. *nicobarulae* belonging to Nicobar Islands, India (*-ae*, geographic (location, toponym) (L.)) (*Hipposideros*).

nictitatans Mod. L. *nictitatans* bearing nictitating membrane, after well-developed nictitating membrane (L. *membrane nictitans*) (Subsp. *Paradoxurus hermaphroditus* Taylor's Palm Civet).

niethammeri Eponym Jochen Neithammer, mammalogist, an expert on dormice (*-i* commemorating (dedication, eponym) (L.)) (*Dryomys*).

niger L. *niger* black, dark-coloured, shining black (*Suncus*/J. Syn. Black Wolf, see *Canis lupus*).

nigra L. *niger* black, dark-coloured, shining black (*cf. ater* dull black) (J. Syn. Indian Toddy Cat, see *Paradoxurus hermaphroditus*).

nigrescens/nigriscens L. *nigrescens/nigriscens* blackish (L. *nigrescere* to become black >L. *niger* black) (*Soriculus*/J. Syn. Sooty Jungle Cat, see *Felis chaus*).

nigricollis L. *nigricollis* black-collared (L. *nigri* (from *niger*) black; Mod. L. *-collis* -throated (>L. *collum* neck)) (*Lepus*).

nigrifrons Mod. L. *nigrifrons* black-foreheaded (L. *niger* black; L. *frons* forehead) (J. Syn. Black-fronted Paradoxure, see *Paradoxurus hermaphroditus*).

nigripecta Mod. L. *nigripecta* black-breasted (L. *nigri* (>L. *niger*) black; L. *pectus* breast) (Subsp. *Otocolobus manul* Himalayan Pallas's Cat).

nilagirica L. adj. *nilagirica* belonging to Nilgiri (Sans. *nil*, *neel* blue; Sans. *giri* mountain) Hills, Tamil Nadu, India (*-ica* belonging to, pertaining to (L. /Gr.)) (*Vandeleuria*/J. Syn. Nilagiri Pygmy Shrew, see *Suncus etruscus*).

Nilgiritragus Mod. L. *nilgiritragus* Nilgiri-goat (Nilgiri (Sans. *nil*, *neel* blue; Sans. *giri* mountain) Hills; Gr. *trágos* goat) (*N. hylocrius*).

nipalensis L. adj. *nipalensis* meaning from Nepal (*ni* holy, *pal* land) (*-ensis*, geographic, occurrence in (location, toponym) (L.)) (*Myotis*/Subsp. *Melogale personata* Nepal Ferret-Badger/J. Syn. Nepal Rhesus Monkey, see *Macaca mulatta*; Nepal Masked Palm Civet, see *Paguma larvata*).

niphanae Eponym after Chanthawanich Ratanaworabhan Niphan (dates not known), a Thai Diptera taxonomist (*-ae*, commemorating (dedication), feminine (L.)) (*Megaerops*).

nirnai From Can. *nirnai*, a local name for the otter (*nir* water, *nai* dog) (Subsp. *Aonyx cinerea* South Indian Small-clawed Otter).

nitedula L. *nitedula* dormouse (*Dryomys*).

nitidus L. *nitidus* shining, glittering (L. *nitere* to shine) (*Rattus*).

nitidofulva Mod. L. *nitidofulva* bright-yellow (L. *nitidus* shining, glittering; L. *fulvus* yellow) (J. Syn. Bright-yellow Pygmy Shrew, see *Suncus etruscus*).

nivalis L. *nivalis* snowy, snow-white (L. *nix*, *nivis* snow) (*Mustela*).

nivicolus Mod. L. *nivicolus* snow-dweller (L. *nix*, *nivis* snow; L. *-cola* dweller (*colere* to inhabit)) (J. Syn. Alpine Bat, see *Scotomanes ornatus*).

Niviventer*/*niviventer Mod. L. *niviventer* white-bellied (*L. nivis* snow; *ventris* belly); in allusion to the whitish belly (*N. brahma*, *N. eha*, *N. fulvescens*, *N. langbianis,* and *N. niviventer*).

nobilis L. adj. *nobilis* noble (*Petaurista*/Subsp. *Petaurista nobilis*).

noctula New L. *noctula* a bat >Fr. *noctule* common name for a bat (>L. *noct-* or *nox* night) (*Nyctalus*).

nolthenii Eponym after A.C. Tutein-Nolthenius (b. 1898–d. 1953), estate owner and collector in Sri Lanka (*-i*, commemorating (dedication, eponym) (L.)) (*Vandeleuria*).

nordicus L. adj. *nordicus* northern (Fr. *nordique*, *nord* north) (*-icus* belonging to, pertaining to (L. /Gr.)) (Subsp. *Loris lydekkerianus* Northern Ceylonese Slender Loris).

norvegicus L. adj. *norvegicus* meaning from Norway (*-icus*, belonging to, pertaining to (location, toponym) (L.)). This appears to be a case of simple misunderstanding of Latinization of vernacular name of the rat, which was common in Great Britain and Ireland (*Rattus*).

novaeangliae Mod. L. *novaeangliae* New Englander (L. *novae* new; L. *angliae* England) (*-ae*, geographic (location, toponym) (L.)) (*Megaptera*).

nubrica L. adj. *nubrica* from Nubra Valley, Jammu and Kashmir, India (*-ica* belonging to, pertaining to (L./Gr.)) (*Ochotona*).

nudiventris Med. L. *nudiventris* bare-bellied (L. *nudus* bare; L. *venter, ventris* belly) (*Paraechinus* and *Taphozous*).

nudipes L. *nudipes* bare-footed (L. *nudus* bare; L. *pes* foot >Gr. *pous* foot) (J. Syn. Naked-footed Pygmy Shrew, see *Suncus etruscus*).

numarius L. *numarius* of or belonging to money (Subsp. *Funambulus tristriatus* Prater's Striped Squirrel).

newara Mod. L. *newara* from Nuwara Eliya (historically Newara Eliya), Central Province, Sri Lanka (a toponym) (Subsp. *Golunda ellioti* Ceylon Highland Bush-Rat/J. Syn. Newara Eliya Bush-rat, see *Golunda ellioti*).

Nyctalus Mod. L. *nyctalus* (>Gr. *niktalós* drowsy), in allusion to its crepuscular habits (*N. leisleri*, *N. montanus*, and *N. noctula*).

nycticeboides L. adj. *Nycticeboides* nycticebus-like (from Mod. L. *nycticebus* night-monkey (L. genitive *nuktos* night >Mod. L. *nox* or Gr. *nux*; Gr. *kēbos* long-tailed monkey, here simply 'a monkey', (*-oides* >Gr. *eidos*) apparent shape)) (Subsp. *Loris tardigradus* Horton Plains Slender Loris).

Nycticebus Mod. L. *nycticebus* night-monkey (L. genitive *nuktos* night >Mod. L. *nox* or Gr. *nux*; Gr. *kēbos* long-tailed monkey, here simply 'a monkey') (*N. bengalensis*).

nyula Mod. L. *nyula* >Hin. *nyul*, the name of the mongoose in Gangetic plains in India (Subsp. *Urva edwardsii* North Indian Grey Mongoose).

Oo

obscurus L. *obscurus* dark, dusky (Subsp. *Funambulus sublineatus* Sri Lankan Dusky-striped Squirrel).

Ochotona Mod. L. *ochotona* >Mongolian *ochodona*, a native name for pika (*O. curzoniae*, *O. forresti*, *O. himalayana*, *O. ladacensis*, *O. macrotis*, *O. nubrica*, *O. roylei*, *O. rufescens*, and *O. thibetana*).

ochraceus Mod. L. *ochraceus* ochraceous (>L. *ochra* ochre >Gr. *ōkhra* yellow-ochre) (J. Syn. Nepalese Red Panda, see *Ailurus fulgens*).

ogilbii Eponym, possibly after William Ogilby (b. 1808–d. 1873), an Irish barrister and naturalist (*-ii*, commemorating (dedication, eponym) (L.)) (J. Syn. Sikkim Marbled Cat, see *Pardofelis marmorata*).

ognevi Eponym after Prof. Sergei Ivanovich Ognev (b. 1886–d. 1951), a Russian zoologist who worked on taxonomy and distribution of mammals in Soviet Union and Central Asia (*-i*, commemorating (dedication, eponym) (L.)) (*Eptesicus*).

ohiensis L. adj. *ohiensis* from Ohiya, Sri Lanka (*-ensis*, geographic, occurrence in (location, toponym) (L.)) (*Srilankamys*).

oiostolus Mod. L. *oiostolus* woolly (Gr. *óiis* sheep fleece; L. *stola* (Gr. *stolē*) robe, vestment) (*Lepus*).

oleracea Etymology not known (*Vandeleuria*).

olympius Mod. L. *olympius* (L. *olympus*/Gr. *olympos*), the highest mountain in Greece, probably in allusion to the altitude of the type locality (J. Syn. Striped Squirrel, Highland Ceylon Palm Squirrel, see *Funambulus palmarum*).

oniops Mod. L. *oniops* resembling Oni of Jap. myth (Gr. *-opsis* -appearance) (J. Syn. Oniops, see *Macaca mulatta*).

opimus L. adj. *opimus* fat (*Rhombomys*).

opisthomelas Mod. L. *opisthomelas* black-backed (Gr. *opisthe* behind, rear; Gr. *melas* black) (Subsp. *Macaca sinica* Highland Toque Macaque).

orca L. *orca* a kind of whale (*Orcinus*).

Orcaella Mod. L. *orcaella* dim. of L. *orca* a kind of whale (*-ella*, diminutive (comparison); somewhat (adj.) (L.)) (*O. brevirostris*).

Orcinus Mod. L. *orcinus* (>L. *orca*) a kind of whale (*O. orca*).

orientalis L. *orientalis* eastern (*oriens*, *orientis* east) (*-alis*, pertaining to (L.)) (*Ovis*/J. Syn. Oriental Civet-Cat, see *Viverra zibetha*).

ornata/ornatus L. *ornatus* ornate, adorned (*ornare* to adorn) (Subsp. *Felis silvestris* Asian Wild Cat/*Scotomanes*).

oryzus L. adj. *oryzus* pertaining to paddy or rice (L. *oryza* rice >G. *óruza* rice) (J. Syn. Ceylon Hog-Deer or Paddyfield Deer, see *Axis porcinus*).

Otocolobus Mod. L. *otocolobus* mutilated-ears (Gr. *otus* ear; Gr. *kolobos* mutilated); in allusion to its short ears (*O. manul*).

Otomops Mod. L. *otomops* dog-eared (Gr. *otus* ear, L. *mops* (>Ger. *mops* roly-poly, usually a dog)); in allusion to the doglike appearance of bat's face (*O. wroughtoni*).

Otonycteris Mod. L. *otonycteris* eared-bat (Gr. *otus* ear; Gr. *nikterís* bat); in allusion to its long ears (*O. hemprichii* and *O. leucophaea*).

Ovis L. *ovis* sheep (*O. ammon* and *O. orientalis*).

Pp

paccerois Mod. L. *paccerois* full-horn (Gr. adj. *pakh* all, equal; Gr. *kéras* horn); in allusion to all the four horns being equal in size and shape (J. Syn. Full-horned Antelope, see *Tetracerus quadricornis*).

pachyomus L. *pachyomus* (Gr. *pakhus* large, thick; Gr. *mȳs* mouse); no explanation was given, maybe in allusion to its thick and obtuse muzzle or stout mouse-like appearance (*Eptesicus*).

pachyotis Mod. L. *pachyotis* thick-eared (Gr. *pakhus* thick, dense; Gr. *-ōtis* -eared (*ous*, *ōtos* ear)) (*Eptesicus*).

pacificus L. adj. *pacificus* meaning from the Pacific Ocean (*-icus*, belonging to, pertaining to (location, toponym) (L.)) (*Indopacetus*).

Paguma Origin unknown, probably a coined word (*Paguma*).

pahari Mod. L. *pahari* from Hin. *pahari* someone from the hills; owing to the fact that the types were collected in the Himalayas (*Mus*/J. Syn. Sikkim Brown-toothed Shrew, see *Soriculus nigrescens*).

pallasi Eponym after Peter Simon Pallas (b. 1741–d. 1811), a German zoologist, and botanist who worked in Russia (*-i*, commemorating (dedication, eponym) (L.)) (Subsp. *Paradoxurus hermaphroditus* Pallas's Palm Civet).

pallida/pallidus L. *pallida*, *pallidus* pallid, pale (*Scotoecus*/Subsp. *Hipposideros fulvus* Pallid Leaf-nosed Bat/J. Syn. Pale Treeshrew, see *Anathana ellioti*; Pale Grey Mongoose, see *Urva edwardsii*).

pallidior L. *pallidior* paler (comp. from *pallidus* pale) (J. Syn. Pallid Field Rat, see *Millardia meltada*).

pallipes Mod. L. *pallipes* pale-footed (L. *pallidus* pale; L. *pes* foot >Gr. *pous* foot) (*Apodemus*/Subsp. *Canis lupus* Indian Wolf, *Urva auropunctatus* Pale-footed Small Mongoose and *Lepus oiostolus* White-footed Woolly Hare/S. Syn. Pale-footed Langur, see *Semnopithecus priam*).

palmarum L. *palmarum* (>L. *palmarius*) pertaining to palm trees (*Funambulus* and *Rattus*).

palnica L. adj. *palnica* belonging to the Palni Hills, Tamil Nadu, India (*-ica* belonging to, pertaining to (L. /Gr.)) (J. Syn. Palni Hills Spiny Mouse, see *Mus cookii*).

palustris L. *palustris* marshy (L. *palus*, *paludis* swamp, marsh) (*Urva*).

Panthera L. *panthera* (Gr. *pánthēr*) panther (*P. leo*, *P. pardus*, *P. tigris*, and *P. uncia*).

Pantholops Mod. L. *panthalops* all-antelope (Gr. *pan* all; Gr. *ánthólops* antelope) (*P. hodgsoni*).

Paradoxurus Mod. L. *paradoxurus* strange tail (Gr. *paradoxus* strange, marvelous; Gr. *oúrá* tail) (*P. aureus*, *P. hermaphroditus*, *P. jerdoni*, *P. montanus*, and *P. stenocephalus*).

Paraechinus Mod. L. *paraechinus* similar to the hedgehog (Gr. *para* near to; Gr. *ékhïnos* hedgehog) (*P. hypomelas*, *P. micropus*, and *P. nudiventris*).

Parascaptor Mod. L. *parascaptor* shallow-digger (Gr. *para* near to; Gr. *scaptor* to dig) (*P. leucura*).

pardicolor Mod. L. *pardicolor* leopard-coloured (L. *pardi*, *pardus* leopard; L. *color* colour) (*Prionodon*).

Pardofelis Mod. L. *pardofelis* leopard-cat (L. *pardus* (Gr. *párdos*) leopard; L. *felis* cat), generic combination (*Pardus* and *Felis*) (*P. marmorata* and *P. temminckii*).

pardus L. *pardus* (Gr. *párdos*) leopard (*Panthera*).

parvidens Mod. L. *parvidens* small-toothed (L. *parvus* small; L. *dens* tooth) (J. Syn. Small-toothed Vole, *Alticola argentatus*).

pashtonus Toponym after Pashto, Khyber-Pakhtunkhwa, Pakistan (*-anus*, geographic (location, toponym); belonging to, pertaining to (L.)) (Subsp. *Eptesicus serotinus* Gaisler's Serotine).

paterculus L. *paterculus* paternal (L. *paterculus* dim. from Mod. L. *pater* father) (*Pipistrellus*).

paupera L. *paupera* poor, scanty (J. Syn. Ambala Bush-rat, see *Golunda ellioti*).

pearsoni Eponym after Dr. Joseph Pearson (b. 1881–d. 1971), a marine biologist and Director of the Colombo Museum, Sri Lanka (*-i*, commemorating (dedication, eponym) (L.)) (*Solisorex*).

pearsonii Eponym after Dr. John Thomas Pearson (dates not known), fellow medical student of J. E. Gray, subsequently collected in India (*-i*, commemorating (dedication, eponym) (L.)) (*Belomys* and *Rhinolophus*).

peguensis L. adj. *peguensis* meaning from Pegu (now Bago), Burma (*-ensis*, geographic, occurrence in (location, toponym) (L.)) (*Pipistrellus javanicus* Pegu Pipistrelle/J. Syn. Pegu Tree Mouse, see *Chiropodomys gliroides*).

pelops L. *pelops* (Gr. *pélops*) dark face, eye (Subsp. *Macaca assamensis* Western Assamese Macaque).

penicillatus Mod. L. *penicillatus* with brushlike tufts (>L. *penicillus* brush, dim. from *peniculus* brush (>L. *penis* tail)) (J. Syn. Khurree, see *Funambulus palmarum*).

pennantii Eponym after T. Pennant (b. 1726–d. 1798), British naturalist (*-ii*, commemorating (dedication, eponym) (L.)) (*Funambulus*/Kuttaas or Pennant's Palm Civet or Pennant's Paradoxurus, see *Paradoxurus hermaphroditus*).

pentadactyla Mod. L. *pentadactyla* five-fingered (L. *penta-* five- (in comp.); L. *dactylus* (>Gr. *daktulos*) finger or toe) (*Manis*).

pentax Mod. L. *pentax* inclining towards five (L. *penta* five; L. *-ax* inclining towards) (J. Syn. Hazara Field Mouse, see *Apodemus pallipes*).

Peponocephala Mod. L. *peponocephala* melon-headed whale (Mod. L. *pepon* >Spanish *pepón* melon; Gr. *kephalus* head) (*P. electra*).

peregusna Mod. L. *peregusna* >Ukranian *pereguznya*, a name for the species (*Vormela*).

perforatus L. *perforatus* pierced through (L. *per-* through; L. *forare* pierce); in allusion to its tail, that emerges out of the interfemoral membrane on the dorsal side in mid point (*Taphozous*).

pergrisea Mod. L. *pergrisea* pale grey (L. *per-* very- (in comp.) (*per* through); Med. L. *griseus* grey (>Old French *gris* grey)) (*Crocidura*).

perniger L. *perniger* very black (Subsp. *Rhinolophus luctus* Himalayan Woolly Horseshoe Bat/J. Syn. Tibetan Leopard or Ni Boer Leopard or Nepal Leopard, see *Panthera pardus*; Black Dolphin, see *Tursiops aduncus*).

pernyi Eponym after Paul-Hubert Perny (b. 1818–d. 1907), a French missionary (*-i*, commemorating (dedication, eponym) (L.)) (*Dremomys*).

perrotettii Eponym after Georges Samuel Perrotett (b. 1793–d. 1870), a Swiss-born French botanist, who collected from many parts of the world and in India from 1834 to 1839 and 1843 to 1870 (*-ii*, commemorating (dedication, eponym) (L.)) (J. Syn. Nilgiri Pygmy Shrew, see *Suncus etruscus*).

persica/persicus L. adj. *persica/persicus* Persian, belonging to Persia (*-ica*, *-icus* belonging to, pertaining to (L./Gr.)) (Subsp. *Panthera leo* Asiatic Lion/*Meriones* and *Triaenops*).

personata L. *personatus* masked (L. *persona* mask) (*Melogale*).

perspicillata Mod. L. *perspicillatus* spectacled (>L. *perspicillum* lens, spectacle >L. *perspicere* to see through) (*Lutrogale*).

peshwa No explanation was given; *peshwa* (>Arab. *pēshwā*, foremost, leader) is the titular equivalent of a modern Prime Minister, possibly in allusion to its type locality Poona (= Pune), Maharashtra, the capital of Mahrata Kingdom (Subsp. *Myotis horsfieldii* Wroughton's Myotis).

Petaurista/petuarista Mod. L. *petaurista* flying squirrel (L. *petaurus* (>Gr. *petauron*) a springboard or a perch; *-ista* L. suffix denoting ability); alternatively from Gr. *petauristís* 'rope-dancer' (*P. elegans*, *P. magnificus*, *P. nobilis*, *P. petaurista*, and *P. philippensis*).

petersi Eponym after Wilhem Karl Hartwig Peters (b. 1815–d. 1883), a German zoologist and explorer, who collected in Africa and Southeast Asia (*-i*, commemorating (dedication, eponym) (L.)) (Peter's Horseshoe Bat, see *Rhinolophus rouxii*).

Petinomys Mod. L. *petinomys* flying squirrel (L. *petino*, >Gr. *petauron* a spring-board or a perch; Gr. *mŷs* mouse) (*P. fuscocapillus*).

petulans L. *petulans* petulant, impudent (J. Syn. Whitehead's Vole, see *Phaiomys leucurus*).

peytoni Eponym after General Walter Peyton (dates not known), an English soldier, naturalist and Conservator of Forests stationed at Kanara (*-i*, commemorating (dedication, eponym) (L.)) (*Myotis*).

Phaiomys Mod. L. *Phaiomys* grey-mouse (Gr. *phaiós* grey; Gr. *mŷs* mouse) (*P. leucurus*).

phayrei Eponym after Arthur Purves Phayre (b. 1812–d. 1885), a British Indian army officer and First Commissioner of British Burma and naturalist (*-i*, commemorating (dedication, eponym) (L.)) (*Trachypithecus*).

philbricki Eponym after A. N. Philbrick (dates not known), a manager of the Tea Estate at Mousakande, Sri Lanka, where W.W.A. Phillips worked (*-i*, commemorating (dedication, eponym) (L.)) (Subsp. *Semnopithecus vetulus* Dry Zone Purple-faced Langur – also Northern Purple-faced Leaf-monkey, Black Wanderoo).

Philetor Gr. *philêtòr* lover (*P. brachypterus*).

philippensis L. adj. *philippensis* meaning from the Philippines (*-ensis*, geographic, occurrence in (location, toponym) (L.)). Erroneous, type locality later reallocated as near Madras, India (in original description no exact location was given but the paper in which it was dealt was about mammals of South Mahratta Country) (*Petaurista*).

phillipsi Eponym after Dr. William Watt Addison Phillips (b. 1892–d. 1981), a naturalist in Sri Lanka, author of *Manual of the Mammals of Sri Lanka* (*-i*, commemorating (dedication, eponym) (L.)) (Subsp. *Prionailurus rubiginosus* Sri Lankan Rusty-spotted Cat/J. Syn. Central Ceylon Brown Mongoose, see *Urva fuscus*). Eponym after R. M. Phillips (dates not known), District Superintendent of Police, Dharwar, India, assisted C.A. Crump in collecting for the Mammal Survey of British India (*-i*, commemorating (dedication, eponym) (L.)) (*Mus*).

phocaenoides Mod. L. *phocaenoides* porpoise-like (>Gr. *phôkaina* porpoise) (*-oides* (>Gr. *eidos*), apparent shape) (*Neophocaena*).

physalus Mod. L. *physalus* (Gr. *phýsalos*) whale (*Balaenoptera*).

Physeter Gr. *physètêr* blowpipe, a whale; from the single spiracle or blowpipe (*P. macrocephalus*).

picta/pictus L. *pictus* painted (L. *pingere* to paint) (*Kerivoula*/J. Syn. Nilagau, see *Boselaphus tragocamelus*/Subsp. *Dryomys nitedula* Painted Forest Dormouse/J. Syn. Painted Hedgehog, see *Paraechinus micropus*).

picticaudata Mod. L. *picticaudata* painted-tailed (L. *pictus* painted (L. *pingere* to paint); Mod. L. *caudatus* -tailed (L. *cauda* tail)) (*Procapra*).

pileatus L. *pileatus* -capped (L. *pileus* felt-cap) (*-atus*, provided with (L.)) (*Trachypithecus*/Subsp. *Trachypithecus pileatus* Blond-bellied Capped Langur, also Yellow-bellied Capped Langur, Blond-bellied Langur).

Pipistrellus/pipistrellus Mod. L. *pipistrellus* pipistrelle bat (>Ital. *pipistrello, vispitrello* dim. of L. *vespertilio* a bat; *-ellus*, diminutive (comparison), somewhat (adj.))L.)) (*P. abramus, P. ceylonicus, P. coromandra, P. javanicus, P. kuhlii, P. paterculus, P. pipistrellus* and *P. tenuis*).

planiceps Mod. L. *planiceps* flat-headed (L. *planus* plain, level ground; L. *-ceps* -headed (L. *caput* head)) (*Sorex*).

Platacanthomys Mod. L. *platacanthomys* flat-spine bearing mouse (Gr. *platōs* broad, flat; Gr. *akantha* spine, thorn; Gr. *mȳs* mouse); in allusion to the flattened spines mingled with fur (*P. lasiurus*).

Platanista Mod. L. *platanista* given name for Gangetic Dolphin (>Gr. *platanistês*, a fish of the Ganges, probably this species) (*P. gangetica*).

platythrix Mod. L. *platythrix* broad-spined (Gr. *platōs* broad; Gr. *thríx* hair); in allusion to broad spine-like hairs on its body (*Mus*).

Plecotus Gr. *plecotus* twisted-ears (Gr. *plékò* to twine, to twist; Gr. *otus* ear) (*P. homochrous, P. strelkovi*, and *P. wardi*).

plicatus L. *plicatus* folded, doubled (L. *plicare* to fold) (*Chaerephon*).

plumbea L. *plumbeus* leaden, plumbeous, lead-coloured (*plumbum* lead) (J. Syn. Dark Humpbacked Dolphin or Plumbeous Dolphin or Indian Humpback Dolphin or Plumbeous Humpback Dolphin, see *Sousa chinensis*).

plurimammis L. *pliurimammis* many-breasted (L. *plus, pluris* many; L. *mammis* (pl. of *mamma*) breast) (J. Syn. Nepal Rat, see *Bandicota bengalensis*).

pococki Eponym after Reginald Innes Pocock (b. 1863–d. 1947), a British zoologist who worked in the British Museum (Natural History) (*-i*, commemorating (dedication, eponym) (L.)) (J. Syn. Pocock's Bicoloured Spiny Rat or Pocock's Bicoloured Coelomys, see *Mus cookii*).

polii Eponym after Marco Polo (b. 1254–d. 1324), a Venetian merchant traveler who first mentioned this sheep (*-i*, commemorating (dedication, eponym) (L.)) (Subsp. *Ovis ammon* Pamir Argali).

polyodon Gr. *poluodous* with many teeth (Gr. *polus* many; Gr. *odous, odontos* tooth), owing to two extra molars in the upper jaw (Subsp. *Equus kiang* Hodgson's Kiang).

pomona Gr. Myth. *Pomona*, goddess of fruit trees, gardens, and orchards (*Hipposideros*).

ponticeriana L. adj. *ponticeriana* meaning from Pondicherry (Pondichéry), French India (= Puduchery) (*-ana*, belonging to, pertaining to (L.) geographic (location, toponym) (J. Syn. Pondicherry Mongoose, see *Urva edwardsii*).

porcinus Mod. L. *porcinus* hoglike (L. *porca* a sow) (*Axis*).

Porcula Mod. L. *porcula* little piglike (L. *porca* a sow) (*P. salvania*).

povensis L. adj. *povensis* from *Powah* (=Caravanserai) original description states that the type was collected from the *Powah* or Caravansery of Jaher Sing (*-ensis*, geographic, occurrence in (location, toponym) (L.)) (J. Syn. Powah Mouse, see *Vandeleuria oleracea*).

prateri Eponym after Stanley Henry Prater (b. 1890–d. 1960), a British naturalist in India, who was the curator of mammal collections at Bombay Natural History Society and Prince of Wales Museum of Western India, Bombay (*-i*, commemorating (dedication, eponym) (L.)) (Subsp. *Felis chaus* Indian Jungle Cat).

prehensilis L. adj. *prehensilis* grasping (L. *prehendo* grasp; *-ilis*, pertaining to (adj.) (L.)) (J. Syn. Prehensile Paradoxurus, see *Paradoxurus hermaphroditus*).

priam Gr. Myth. *Priam*, King of Troy – a major character in Homer's *Iliad* (*Semnopithecus*).

priamellus Mod. L. *priamellus* somewhat like *priam* (from Gr. Myth. *Priam*, King of Troy – a major character in Homer's *Iliad*, a specific name given to Coromandel Gray Langur) (*-ellus*, diminutive (comparison); somewhat (adj.) (L.)) (J. Syn. Shernelly Langur, see *Semnopithecus dussumieri*).

primaevus L. *primaevus* early in life, youthful (L. *primus* first; L. *aevum* an age) (Subsp. *Cuon alpinus* Himalayan Wild Dog).

primula L. *primula* first (dim. from *primus* first, foremost, super. >L. *prior* former) (J. Syn. Thomas' Myotis, see *Myotis annectans*).

Prionailurus Mod. L. *prionailurus* cat bearing sawtooth (Gr. *príòn* saw; Gr. *aílouros* cat (later a weasel)) (*P. bengalensis*, *P. rubiginosus*, and *P. viverrinus*).

Prionodon Mod. L. *prionodon* sawtoothed (Gr. *príòn* saw; Gr. *ódôn* tooth) (*P. pardicolor*).

problematicus L. *problematicus* problematic (L. *problema* riddle, enigma >Gr. *problēma* puzzle, enigma) (J. Syn. Problematic Assamese Macaque, see *Macaca assamensis*).

Procapra Mod. L. *procapra* before-goat (L. *pro* before; L. *capra* she-goat) (*P. picticaudata*).

proximus L. *proximus* very near, nearest, most like (super. >L. *propior* nearer) (Subsp. *Rhinolophus ferrumequinum* Kashmir Horseshoe Bat).

providens L. *providens* caring (J. Syn. Southern India Field Rat, see *Bandicota bengalensis*).

pruinosus L. *pruinosus* frosty, cold (L. *pruina* hoar-frost) (*Rhizomys*).

Pseudois Mod. L. *pseudois* false sheep (Gr. *pseudos* false; Gr. *ois* from L. *ovis* sheep); this species lacks facial gland and has a goatlike tail (*P. nayaur*).

Pseudorca Mod. L. *pseudorca* false whale (Gr. *pseudos* false; L. *orca* a kind of whale) (*P. crassidens*).

Pteropus Mod. L. *pteropus* wing-footed (Gr. *pterópous* wing-footed); in allusion to the wing membrane that attaches to the back of the second toe (*P. faunulus*, *P. giganteus*, *P. hypomelanus*, and *P. melanotus*).

pullata L. *pullatus* clad in black garments (*Crocidura*).

punctatissimus Mod. L. *punctatissimus* very spotted, heavily spotted (super. >L. *punctatus* spotted >L. *punctum* spot) (J. Syn. Gray's Squirrel, see *Callosciurus erythraeus*).

punjabiensis L. adj. *punjabiensis* meaning from Punjab (*-ensis* geographic, occurrence in (location, toponym) (L.)) (Subsp. *Ovis orientalis* Punjab Urial).

pusilla/pusillus L. *pusillus* tiny, very small (Subsp. *Vulpes vulpes* Little Red Fox/*Miniopterus* and *Rhinolophus*).

putaoensis L. adj. *putaoensis* meaning from Putao, Kachin State, Myanmar (*-ensis* geographic, occurrence in (location, toponym) (L.)) (*Muntiacus*).

pyctoris Mod. L. *pyctoris* flat-nose (Gr. *puktēs* boxer; Gr. *rhis* nose) (*Rattus*).

pygerythrus Mod. L. *pygerythrus* red-backed (Mod. L. *pygi* (>Gr. *pugé*) buttock or rump; L. *erythrus* (Gr. *eruthrós*) red) (*Callosciurus*).

pygmaeus L. *pygmaeus* dwarf, pygmy (>Gr. *pugmaios* dwarfish) (J. Syn. Small Spiny Mouse, see *Mus saxicola*).

pygmeoides L. adj. *pygmeoides* pygmy-like (from L. *pygmaeus* dwarf, pygmy (>Gr. *pugmaios* dwarfish), *-oides* (>Gr. *eidos*) apparent shape) (J. Syn. Anderson's Pygmy Shrew, see *Suncus etruscus*).

pyrivorus Mod. L. *pyrivorus* fruit-eating (L. *pyrus* fruit; L. *-vorus* eating) (J. Syn. Nepal Fruit Bat, see *Rousettus leschenaultii*).

Qq

quadricornis Mod. L. *quadricornis* four-horned (L. *quadri* four; L. *cornu* horn) (*Tetracerus*).

Rr

radiata L. *radiatus* furnished with rays, a disc (*-ata*, provided with (L.)) (*Macaca*/ Subsp. Northern Bonnet Macaque).

rajput Mod. L. *rajput* after Rājput (from Sans. *raj-putra* – son of a king), a member of clans living in the present-day Indian state of Rajasthan and also parts of Madhya Pradesh and Gujarat, and some parts of Sindh in Pakistan (J. Syn. Mt. Abu Rock-rat, see *Cremnomys cutchicus*; Rajput Hare, see *Lepus nigricollis*).

rajputanae Mod. L. *rajputanae* after Rājputāna, a historical region that included present day Indian state of Rajasthan and also parts of Madhya Pradesh and Gujarat (Subsp. *Antilope cervicapra* Rajputana Blackbuck).

rammanika No explanation was given (Kelaart's Horseshoe Bat, see *Rhinolophus rouxii*).

ramnadensis L. adj. *ramnadensis* meaning from Ramnad, Tamil Nadu (*-ensis*, geographic, occurrence in (location, toponym) (L.)) (J. Syn. Ramnad Spiny Mouse, see *Mus saxicola*).

ranjiniae Eponym after Ms. P.V. Ranjini (dates not known), collector of the species (*-ae*, commemorating (dedication), feminine (L.)) (*Rattus*).

ranjitsinhi Eponym after M. K. Ranjitsinh (dates not known), Indian Civil Servant, naturalist, and author of many books (*-i*, commemorating (dedication, eponym) (L.)) (Subsp. *Rucervus duvaucelii* Assam Swamp Deer).

rapax L. *rapax* rapacious (L. *rapere* to seize) (*Crocidura*).

rattus/rattus L. *rattus* rat (*R. andamanensis*, *R. burrus*, *R. exulans*, *R. montanus*, *R. nitidus*, *R. norvegicus*, *R. palmarum*, *R. pyctoris*, *R. ranjiniae*, *R. rattus*, *R. satarae*, *R. stoicus* and *R. tanezumi*).

rattoides Mod. L. *rattoides* ratlike (L. *rattus* rat; *-oides* (>Gr. *eidos*) apparent shape) (J. Syn. Black Nepal Rat, see *Rattus pyctoris*).

ratufa Mod. L. *ratufa* from *ratuphar*, the local name of this squirrel in northern Bihar, India (*R. bicolor*, *R. indica*, and *R. macroura*).

regulus L. *regulus* prince, kinglet (dim. from *rex*, *regis* king) (J. Syn. Large Himalayan Horseshoe Bat, see *Rhinolophus ferrumequinum*).

retusa L. *retusus* blunt (L. *retundere* to blunt) (J. Syn. Shrew-mouse, see *Crocidura horsfieldii*).

rhesus From Gr. Myth. *Rhêsos*, a legendary Prince of Thrace – a character from *Iliad* (J. Syn. Bengal Monkey, see *Macaca mulatta*).

Rhinoceros Mod. L. *rhinoceros* (>Gr. *rhinōkéras*) horned-nose (Gr. *rhis*, *rhinos* nose; Gr. *kéras* horn) (*R. sondaicus* and *R. unicornis*).

Rhinolophus Mod. L. *rhinolophus* crest-nosed (>Gr. *rhinolōphos* (Gr. *rhis*, *rhinos* nose; Gr. *lōphos* crest)); in allusion to the complicated nose leaf consisting of three different parts and sella (*R. affinis*, *R. beddomei*, *R. blasii*, *R. bocharicus*, *R. cognatus*, *R. ferrumequinum*, *R. hipposideros*, *R. lepidus*, *R. luctus*, *R. macrotis*, *R. mitratus*, *R. pearsonii*, *R. pusillus*, *R. rouxii*, *R. shortridgei*, *R. sinicus*, *R. subbadius*, *R. trifoliatus*, and *R. yunanensis*).

Rhinopoma Mod. L. *rhinopoma* lid-nosed (Gr. *rhis*, *rhinos* nose; Gr. *pōma* lid, cover); in allusion to the valvular nostrils, which open through a narrow transverse slit (*R. hardwickii*, *R. microphyllum*, and *R. muscatellum*).

Rhizomys Mod. L. *rhizomys* root rat (L. *rhizo* (>Gr. *rhiza*) root; Gr. *mӱs* mouse); after its preference to roots of bamboo (*R. pruinosus*).

Rhombomys Mod. L. *rhombomys* (Gr. *rómbos* rhombus; Gr. *mӱs* mouse); in allusion to the shape of the upper molars (*R. opimus*).

Rhyneptesicus Mod. L. *rhyneptesicus* combination of *rhyn* 'cape' and *eptesicus* 'house-flyer' (Gr. *éptin* from *pétomai* to fly; Gr. *oíkos* house). No explanation was given (*R. nasutus*).

robertsoni Eponym after Laurence Roberston (dates not known), a British civil servant posted at Junagadh, a naturalist who was Honorary Treasurer of the Bombay Natural History Society when the species was described (*-i*, commemorating (dedication, eponym) (L.)) (Subsp. *Funambulus palmarum* Robertson's Palm Squirrel).

rodoni Eponym after Major G.S. Rodon (dates not known), a British working in the Himalayas who collected the type (*-i*, commemorating (dedication, eponym) (L.)) (J. Syn. Chamba Serow, see *Capricornis thar*).

rogersi Eponym after C.G. Rogers (dates not known), who collected the type on South Andaman Island (*-i*, commemorating (dedication, eponym) (L.)) (J. Syn. Roger's Andaman Rat, see *Rattus stoicus*).

Rousettus Mod. L. *rousettus* reddish (>Fr. *rousette* from *rousset* reddish); in allusion to the characteristic colour (*R. aegyptiacus* and *R. leschenaultii*).

rouxii Eponym after Jean Louis Florent Polydore Roux (b. 1792–d. 1833), curator of the Museum of Marseille, collected in India and Ceylon (*-ii*, commemorating (dedication, eponym) (L.)) (*Rhinolophus*).

roylei Eponym after Dr. John Forbes Royle (b. 1799–d. 1858), an Indian-born British botanist, surgeon and collector (*-i*, commemorating (dedication, eponym) (L.)) (*Alticola* and *Ochotona*).

rubex Mod. L. *rubex* red (L. *ruber* red) (Subsp. *Murina leucogaster* Thomas' Greater Tube-nosed Bat).

rubida L. *rubidus* ruddy, red, dark-red (J. Syn. Rufous Tree Mouse, see *Vandeleuria oleracea*).

rubidior Mod. L. *rubidior* redder, ruddier (*cf.* L. *rubidus* red, ruddy) (Subsp. *Urva fuscus* Western Ceylon Brown Mongoose).

rubidus L. *rubidus* ruddy, red, dark-red (Subsp. *Rhinolophus rouxii* Sri Lankan Rufous Horseshoe Bat).

rubiginosus L. *rubiginosus* rusty, reddish (L. *rubigo, rubiginis* rust) (*Prionailurus*/J. Syn. Rusty Striped-necked Mongoose, see *Urva vitticollis*).

rubricosa Mod. L. *rubricosa* full of red (L. *ruber* rusty, reddish (*rubigo, rubiginis* rust)) (J. Syn. Anderson's Assam Shrew, see *Crocidura attenuata*).

Rucervus Mod. L. *rucervus* given name; generic combination of first two letters of *rusa*, Malay name for sambar and L. *cervus* stag, deer (*R. eldii* and *R. duvaucelii*).

rueppelli Eponym after Wilhelm Peter Edward Simon Rueppell (b. 1794–d. 1884), a German naturalist, worked in northeast Africa (*-i*, commemorating (dedication, eponym) (L.)) (*Vulpes*).

rufescens L. *rufescens* reddish (L. *rufescere* to become reddish >*rufus* red; *-escens*, somewhat) (*Ochotona*/J. Syn. Indian House Rat or Common Indian Rat or Rufescent Tree Rat or Rufous Rat, see *Rattus rattus*).

ruficaudatus Mod. L. *ruficaudatus* rufous-tailed (L. *rufus* rufous; L. *-caudatus* -tailed (L. *cauda* tail)) (Subsp. *Lepus nigricollis* Rufous-tailed Hare).

rufigenis Mod. L. *rufigenis* red-cheeked (L. *rufi* (from *rufus*) red; L. *genys* cheek or jaw) (*Dremomys*).

Rusa L. *rusa* from Malay name for sambar (*R. unicolor*).

rusiges No explanation provided; possibly after L. *russeus* reddish (L. *russus* red), owing to the coloration of the dorsal fur being somewhat reddish (*Apodemus*).

Ss

sabanus Mod. L. *sabanus* from Sabah peninsula, Malaysia (*-anus*, geographic, belonging to, pertaining to (L.) geographic (location, toponym)) (*Leopoldamys*).

Saccolaimus/saccolaimus Gr. *saccolaimus* gular sac (Gr. *sákkos* sac; Gr. *laimūs* throat, gullet); in allusion to the well-developed gular sacs (*S. saccolaimus*).

sacratus L. *sacratus* sacred (L. *sacer* sacred) (*Episoriculus*).

sadhu L. adj. *sadhu* (from Sans. *sādhu* an ascetic, holy man), perhaps owing to its drab grey pelage in allusion to the ash-smeared ascetics (Subsp. *Mus saxicola* Wroughton's Spiny Mouse).

sadiya Mod. L. *sadiya* from Sadiya, Assam (*Lepus nigricollis* Assam Hare).

salimalii Eponym after Dr. Salim Ali (b. 1896–d. 1987), an Indian ornithologist and conservationist (*-i*, commemorating (dedication, eponym) (L.)) (*Latidens*).

salinarum L. *salinarum* (L. *salina* salt pans), of salt pans (Subsp. *Gazella bennettii* Salt Works Chinkara).

Salpingotulus Mod. L. *salpingotulus* dim. of *salpingotus* trumpet-eared pygmy jerboa (Gr. *salpingos* trumpet; L. *otus* ear) (*-ulus*, diminutive (comparison), somewhat (adj.) (L.)) (*S. michaelis*).

salvania Origin not known, possibly either from L. *silvanus* of the woods (L. *silva* woodland) or L. *silvania* wood-nymph (L. *silva* wood, forest) (*Porcula*).

satarae Mod. L. *satarae* from Satara, Maharashtra, India (*-ae*, geographic (location, toponym) (L.)) (*Rattus*).

satuni Eponym after Konstantin Alekseevich Satunin (b. 1863–d. 1915), a Russian zoologist and naturalist who collected the type specimens (*-i*, commemorating (dedication, eponym) (L.)) (J. Syn. Satunin's Cat, see *Otocolobus manul*).

saturatior L. *saturatior* richly coloured (L. *satur*, *satura* rich, copious >L. *satis* enough) (J. Syn. Brown Capped Langur, see *Trachypithecus pileatus*; Dark-brown Shrew, see *Suncus murinus*).

satyrus L. *satyrus* satyr, horned sylvan deity (*Pteropus hypomelanus* Narcondam Island Flying Fox).

savii Eponym after Pablo Savi (b. 1798–d. 1871), a Professor of Zoology at University of Pisa, Italy (*-ii*, commemorating (dedication, eponym) (L.)) (*Hypsugo*).

saxicola Mod. L. *saxicola* stone-dweller (>L. *saxum* stone; L. *-cola* dweller (>L. *colere* to dwell)) (*Mus*).

saxicolor Mod. L. *saxicolor* stone-coloured (>L. *saxum* stone, *color* colour) (Subsp. *Panthera pardus* Persian Leopard).

scheffeli Eponym after Walter Scheffel (dates not known), a fellow colleague of Prof. Helmut Hemmer, University of Mainz, Germany (Subsp. *Felis margarita* Pakistan Sand Cat).

scherzeri Eponym after Karl Ritter von Scherzer (b. 1821–d. 1903), an Austrian explorer, diplomat, and naturalist (*-i*, commemorating (dedication, eponym) (L.)) (Subsp. *Cynopterus sphinx* Nicobar Short-nosed Fruit Bat).

schistaceus Mod. L. *schistaceus* slate-grey (>Mod. L. *schistus* slate) (*-aceus*, pertaining to, having the nature of (L.)) (*Semnopithecus*/J. Syn. Southern India Leaf-nosed Bat or Split Roundleaf Bat, see *Hipposideros lankadiva*; Grey Greater False Vampire Bat, see *Lyroderma lyra*).

schmidi Eponym after Dr. Fernand Schmid (b. 1924–d. 1998), an authority on Trichoptera of the world (*-i*, commemorating (dedication, eponym) (L.)) (*Anourosorex*).

schmitzi Eponym after Ernst Johann Schmitz (b. 1845–d. 1922), a German naturalist, ornithologist, and entomologist who collected the type specimens (*-i*, commemorating (dedication, eponym) (L.)) (Subsp. *Caracal caracal* Schmitz's Caracal).

scindiae Eponym after Madho Rao Scindia (b. 1876–d. 1925), the fifth Maharaja of Gwalior belonging to the Scindian dynasty of the Marathas, a benefactor who helped start the Mammal Survey of British India (*-ae*, commemorating (dedication, eponym) (L.)) (Subsp. *Paradoxurus hermaphroditus* Central Indian Palm Civet).

Scotoecus Mod. L. *scotoecus* 'dweller of darkness' (Gr. *skōtos* darkness; Gr. *ōiko* to dwell); in allusion to its crepuscular habit (*S. pallidus*).

Scotomanes Gr. *skōtománes* slave of darkness (Gr. *skōtos* darkness, Gr. *mánes* slave); in allusion to its crepuscular habit (*S. ornatus*).

Scotophilus Mod. L. *scotophilus* 'lover of darkness' (Gr. *skōtos* darkness; Gr. *philos* loving); in allusion to its crepuscular habit (*S. heathii* and *S. kuhlii*).

Scotozous Mod. L. *scotozous* 'one who lives in darkness' (Gr. *skōtos* darkness; *zòós* living); in allusion to its crepuscular habit (*S. dormeri*).

scrofa L. *scrofa* sow (*Sus*).

scyritus Mod. L. *scyritus* pale, probably from Gr. *skuros* stucco, gypsum (J. Syn. White Whiskered Gibbon, see *Hoolock hoolock*).

secatus L. adj. *secatus* meaning to cut (L. *secare* to cut off); in allusion to the fur not extending on to the dorsal side of the wing-membrane and the interfemoral membrane, unlike the nominate subspecies *Taphozous theobaldi* (Subsp. *Taphozous theobaldi* Central Indian Tomb Bat).

seianum Toponym after Seistan, a region in Balochistan area of Afghanistan, Iran, and Pakistan (*-anum*, geographic (location, toponym); belonging to, pertaining to (L.)) (Subsp. *Rhinopoma muscatellum* Seistan Mouse-tailed Bat).

seminudus Mod. L. *seminudus* body slightly covered with fur (L. *semi-* half- (in comp.) (*semis* half); L. *nudus* bare) (Subsp. *Rousettus leschenaultii* Sri Lankan Rousette or Large Fruit-Bat or Ceylon Fruit Bat or Dog-faced Bat).

Semnopithecus Mod. L. *semnopithecus* 'sacred-ape' (Gr. *semnos* solemn, sacred, august; *pithecus* ape) (*S. ajax*, *S. entellus*, *S. dussumieri*, *S. hector*, *S. hypoleucos*, *S. johnii*, *S. priam*, *S. schistaceus,* and *S. vetulus*).

senex L. *senex* old person (i.e. grey-haired, white-haired, querulous) (J. Syn. Grey Flying Squirrel, see *Petaurista elegans*).

serotinus L. *serotinus* that which comes late (or that which happens in the evening) (L. *sero*, *serus* late) (*Eptesicus*).

serpentarius Mod. L. *serpentarius* snake-like (L. *serpens*, *serpentis* serpent, snake; *-arius* having the natue of (L.)) (J. Syn. Dark Brown Shrew or Rufescent Shrew, see *Suncus murinus*).

servalina Mod. L. *servalina* servaline or serval-like (*-ina*, belonging to, pertaining to; one who (L.)) (J. Syn. Servaline Chaus, see *Felis silvestris*).

sherrini Eponym after W.R. Sherrin (dates not known), curator of the British Museum (Natural History) (*-i*, commemorating (dedication, eponym) (L.)) (J. Syn. Sherrin's Gerbil, see *Tatera indica*).

shigaricus L. adj. *shigaricus* belonging to Kashgar, Xinjiang, China (*-icus* belonging to, pertaining to (L./Gr.)) (J. Syn. Kashgar Rat, see *Rattus pyctoris*).

shortridgei Eponym after Guy Chester Shortridge (b. 1880–d. 1949), collected for the British Museum (Natural History), and later for the Mammal Survey of British India in southern India and Burma (*-i*, commemorating (dedication, eponym) (L.)) (*Rhinolophus*).

siangensis L. adj. *siangensis* from Siang basin, Siang District, Arunachal Pradesh, India (*-ensis*, geographic, occurrence in (location, toponym) (L.)) (*Petaurista*).

sibirica L. adj. *sibirica* belonging to Siberia, Asiatic Russia (*-ica* belonging to, pertaining to (L./Gr.)) (*Capra* and *Mustela*).

sicarius L. *sicarius* assassin, murderer (*Myotis*).

siccatus L. *sicca* dry place, dry land (Subsp. *Urva fuscus* Ceylon Dryzone Brown Mongoose).

Sicista Mod. L. *sicista* from Tartar name *sikistan* 'gregarious mouse' (*S. concolor*).

sifanicus L. adj. *sifanicus* meaning from Sifan, Gansu, China (*-icus*, belonging to, pertaining to (location, toponym) (L.)) (Subsp. *Moschus chrysogaster* Kansu Musk Deer).

signatus L. *signatus* distinct, well marked (*signare* to notice >*signum* sign) (Flame-striped Jungle Squirrel, see *Funambulus layardi*).

sikimaria L. adj. *sikimaria* belonging to Sikkim, India (*-aria* pertaining to (L./Gr.)) (Subsp. *Ochotona thibetana* Sikkim Pika).

sikimensis/sikhimensis/sikkimensis L. adj. *sikimensis/sikhimensis/sikkimensis* meaning from Sikkim, India (*-ensis*, geographic, occurrence in (location, toponym) (L.)) (*Neodon*/J. Syn. Sikkim Water Shrew or Finger-tailed Water Shrew, see *Nectogale elegans*; Sikkim Rat, see *Rattus andamanensis*).

silenus Gr. Myth. Silenus, companion of the roman god *Bacchus*, one among the *Sileni* (plural), for gods of the woods (*Macaca*).

siligorensis L. adj. *siligorensis* meaning from Siliguri, West Bengal, India (*-ensis*, geographic, occurrence in (location, toponym) (L.)) (*Myotis*).

silvestris Mod. L. *silvestris* of or pertaining to wood or forest, wild, living in forests (L. *silva* forest, wooded) (*Felis*).

sima L. *sima*, *simus* (Gr. *simos*) snubnosed (*Kogia*).

simcoxi Eponym after A.H.A. Simcox, Indian Civil Service (dates not known), Collector of Khandesh, where the type locality is located (Subsp. *Lepus nigricollis* Khandesh Hare).

sindhica/sindica/sindicus L. adj. *sindhica/sindica/sindicus* belonging to Sindh, Pakistan (*-ica*, *-icus* belonging to, pertaining to (L./Gr.)) (Subsp. *Tadarida aegyptiaca* Sindh Free-tailed Bat/*Lutrogale perspicillata* Sindh Smooth-coated Otter; Sind Leopard or Baluchistan Leopard or Shun Tak Leopard, see *Panthera pardus*/J. Syn. Sindh Mole Rat, see *Bandicota bengalensis*).

singuri Mod. L. *singuri* from Singur, West Bengal (*-i*, commemorating (L.)) (J. Syn. Singur Field Rat, see *Millardia meltada*).

sinhaleyus Mod. L. *sinhaleyus* meaning from *Sinhala* (= Ceylon = Sri Lanka) (J. Syn. Eastern Ceylon Elephant, see *Elephas maximus*).

singhala Mod. L. *singhala* meaning from *Sinhala* (= Ceylon = Sri Lanka) (Subsp. *Lepus nigricollis* Ceylon Hare).

singhei Eponym after Jigme Singye Wangchuck (b. 1955), the king of Bhutan (*-i*, commemorating (dedication, eponym) (L.)) (Subsp. *Petaurista nobilis* Bhutan Giant Flying Squirrel).

sinica/sinicus L. adj. *sinica* meaning Chinese or from China or belonging to China (*-ica*, *-icus* belonging to, pertaining to (L. /Gr.)) (a misnomer, in case of *Macaca*, as the species is endemic to Sri Lanka) (*Macaca/Rhinolophus*).

siva L. adj. *siva* (Sans. *siva*, the 'auspicious one', one of the Hindu trinity), nomen commemorating the type locality Sivasamudram, Tamil Nadu, India (J. Syn. Sivasamudram Rock-rat, see *Cremnomys cutchicus*/Sivasamudram Small Spiny Mouse, see *Mus phillipsi*).

smithii Eponym after Charles Hamilton Smith (b. 1776–d. 1859), an English naturalist and soldier (*-ii*, commemorating (dedication, eponym) (L.)) (*Urva*).

sobrinus L. *sobrinus* maternal cousin (*Macroglossus*/Subsp. *Rhinolophus beddomei* Sri Lankan Horseshoe Bat).

soccatus L. *soccatus* wearing slippers or socks (L. *socca* slipper) (J. Syn. Hairy-footed Shrew, see *Suncus murinus*).

Solisorex Mod. L. *solisorex* sun-shrew (L. *solus* sun; *sorex* (>Gr. *ÿrax*) shrew) (*S. pearsoni*).

soluensis L. adj. *soluensis* meaning from Solu, Solukhumbu district, Nepal (*-ensis*, geographic, occurrence in (location, toponym) (L.)) (*Episoriculus*).

sondaicus L. adj. *sondaicus* meaning from Sunda, Indonesia (*-icus*, belonging to, pertaining to (location, toponym) (L.)) (*Rhinoceros*).

Sorex L. *sorex* (>Gr. *ÿrax*) shrew (*S. bedfordiae*, *S. excelsus*, *S. minutus*, and *S. planiceps*).

Soriculus Dim. of L. *sorex* (>Gr. *ÿrax*) shrew (*-ulus*, diminutive (comparison); somewhat (adj) (L.)) (*S. nigrescens*).

Sousa Mod. L. *sousa* from Bengali name *súsúk* or *sishúk*, the name of the Gangetic Dolphin (*S. chinensis*).

spadicea L. *spadix*, *spadicis* chestnut-coloured, date-coloured, chestnut-brown (>Gr. *spadix*, *spadikos* palm-coloured) (J. Syn. Gujarat Tree Mouse, see *Vandeleuria oleracea*).

spasma L. *spasma* sudden burst of energy or activity (>Gr. *spasmós* convulsion) (*Megaderma*).

spatangus Mod. L. *spatangus* (>Gr. *spatangēs*) a kind of sea urchin; as to why this nomen was used is unclear (J. Syn. Himalayan Hedgehog, see *Hemiechinus collaris*).

speciosa L. *speciosus* splendid, beautiful (L. *species* beauty) (S. Syn. Brown Stump-tailed Macaque, see *Macaca arctoides*).

spectrum L. *spectrum* appearance, image, apparition (>L. *specio* look at, view) (J. Syn. Kashmir Vampire Bat, see *Lyroderma lyra*).

spelaea/spelaeus Gr. *spēlaion* cave; owing to its habit of dwelling/roosting in caves (*Eonycteris*/S. Syn. Dawn Bat, see *Eonycteris spelaea*).

speoris Mod. L. *speoris* (Gr. *spēlaion* cave, *rhis*, *rhinos* nose) meaning cave-like nose; in allusion to the deep depression in which nostrils are located.

Spermophilopsis Mod. L. *spermohilopsis* spermophilus-like, ground squirrel-like (L. *spermophilus* ground squirrel; Gr. *-opsis* -appearance) (*S. leptodactylus*).

Spermophilus Mod. L. *spermophilus* seed-lover (Gr. *sperma* seed; Gr. *philos* loving) (*S. fulvus*).

Sphaerias No explanation provided (*S. blanfordi*).

Sphinx Gr. Myth. *Sphinx*, a monster of varied appearance at Thebes who set riddles to travellers and killed those who could not answer correctly (*Cynopterus*).

spinulosus Mod. L. *spinulosus* very spined (L. *spinus* thorn; *-osus*, abundance, fullness (L.)) (J. Syn. Dusky Spiny Mouse, see *Mus saxicola*).

squamipes Mod. L. *squamipes* scaly-footed (L. *squameus* scaly, from *squama* scale; L. *pes* foot, >Gr. *pous* foot) (*Anourosorex*).

Srilankamys Mod. L. *Srilankamys* Sri Lanka mouse (Sri Lanka and Gr. *mȳs* mouse) (*S. ohiensis*).

Stenella Dim. of *steno*, genus *nomen* in honour of Dr. Nikolaus Steno (b. 1638–d. 1687), a celebrated Danish anatomist (*-ella*, diminutive (comparison), somewhat (adj.) (L.)) (*S. attenuata*, *S. coeruleoalba* and *S. longirostris*).

Steno Genus *nomen* in honour of Dr. Nikolaus Steno (b. 1638–d. 1687), a celebrated Danish anatomist (*S. bredanensis*).

stenocephalus Mod. L. *stenocephalus* narrow-headed (Gr. *stēnos* narrow, thin; Gr. *kephalus* head) (*Paradoxurus*).

stevensi Eponym after H. Stevens (dates not known), an English army officer, who collected the type from Beni-Chang, Abor-Miri Hills, northern frontier of Upper Assam (*-i*, commemorating (dedication, eponym) (L.)) (Subsp. *Callosciurus pygerythrus* Steven's Squirrel).

stirlingi Eponym after Captain H.O. Stirling (dates not known), a British army officer, stationed at Chitral (J. Syn. Stirling's Marmot, see *Marmota caudata*).

stoicus Mod. L. *stoicus* (>Gr. *stōikos*) stoic, enduring, uncomplaining; in allusion to its silent disposition when trapped (*Rattus*).

stoliczkana/stoliczkanus Eponym after Dr. Ferdinand Stoliczka (b. 1838–d. 1874), a British paleontologist, who participated in Yarkand Missions (Subsp. *Mustela nivalis* Stoliczka's Weasel/*Alticola* and *Suncus*).

stracheyi Eponym after Lt.-Gen. Sir Richard Strachey (b. 1817–d. 1908), a British army engineer, explorer, botanist, collector, and geologist (J. Syn. Ladhak Mountain Vole, Strachey's Mountain Vole, see *Alticola stoliczkanus*).

strelkovi Eponym after Petr Petrovich Strelkov (b. 1931–d. 2012), a Russian biologist known for his work on *Plecotus* taxonomy (*-i*, commemorating (dedication, eponym) (L.)) (*Plecotus*).

strigidorsa Mod. L. *strigidorsa* back-striped (L. *strigis* furrow, groove; L. *dorsum* back) (*Mustela*).

strophiatus Mod. L. *strophiatus* breast-band bearing (L. *strophium* breast-band (>Gr. *strophion* breast-band); *-atus*, provided with (L.)) (J. Syn. Nepal Field Mouse, see *Mus cervicolor*).

subbadius Med. L. *subbadius* pale chestnut–coloured (L. *sub* beneath, somewhat; L. *badius* chestnut-coloured) (*Rhinolophus*).

subcanus Med. L. *subcanus* pale grey–coloured (L. *sub* beneath, somewhat; L. *canus* grey-coloured) (J. Syn. Northern Kelaart's Pipistrelle or Gujarat Pipistrelle, see *Pipistrellus ceylonicus*).

subcristata L. adj. *subcristata* slightly crested (L. *sub* slightly, somewhat; L. *cristatus* crested (L. *crista* crest)) (Subsp. *Hystrix brachyura* Chinese Porcupine).

subflaviventris Mod. L. *subflaviventris* somewhat yellow-bellied (L. *sub* beneath, somewhat; L. *flavus* yellow; L. *ventris* belly) (J. Syn. Nepal Squirrel, see *Dremomys lokriah*).

subfulva Mod. L. *subfulva* somewhat yellow (L. *sub* beneath, somewhat; L. *fulvus* tawny, fulvous); in allusion to its somewhat yellow throat (J. Syn. Yellow-throated Shrew, see *Suncus stoliczkanus*).

subgutturosa Mod. L. *subgutturosa* somewhat goitered (L. *sub* slightly, somewhat; L. *gutturosus* goitred (*guttur* throat, gullet)) (*Gazella*).

subhemachalana/subhemachalus Mod. L. *subhemachalana/subhemachalus* sub-Himalayan (L. *sub* beneath; Hin. *himachal* in the lap of the Himalayas) (Subsp. *Mustela sibirica* Himalayan Weasel/J. Syn. Gauri Bull, see *Bos gaurus*).

sublimis L. adj. *sublimis* lofty (J. Syn. Blanford's House Mouse, see *Mus musculus*).

sublineatus Mod. L. *sublineatus* faint-lined (L. *sub* slight; L. *-lineatus* -lined (*linea*, from *līnum* flax, for line or cord)) (*Funambulus*).

Submyotodon Mod. L. *submyotodon* near or having close affinity to myotodon (L. *sub* under, close to; Gr. *mŷs* mouse; Gr. *ous*, ear; Gr. *ódôn* tooth), in allusion to its nearly myotodont development of the postcristid in the lower molars (*S. caliginosus*).

subsolanus L. *subsolanus* eastern (J. Syn. Eastern Swarthy Gerbil, see *Gerbillus aquilus*).

sumatrensis L. adj. *sumatrensis* meaning from Sumatra, Indonesia (*-ensis*, geographic, occurrence in (location, toponym) (L.)) (*Dicerorhinus*).

Suncus Mod. L. *suncus* shrew, from Arabic name '*far sunki*' (*S. dayi*, *S. etruscus*, *S. fellowesgordoni*, *S. montanus*, *S. murinus*, *S. niger*, *S. stoliczkanus*, and *S. zeylanicus*).

surda L. *surdus* silent, still (Subsp. *Tupaia nicobarica* Little Nicobar Island Tree Shrew).

surkha Mod. L. *surkha* ruddy, from Urdu *surkh* ruddy (J. Syn. Grey Small Spiny Mouse, see *Mus phillipsi*).

Sus L. *sus* (Gr. *hûs*) pig (*S. scrofa*).

swinhoei Eponym after Lt.-Col. Charles Swinhoe (b. 1838–d. 1923), a British army officer and a naturalist, also a founding member of the Bombay Natural History Society (*-i*, commemorating (dedication, eponym) (L.)) (J. Syn. Swinhoe's Jird, see *Meriones crassus*).

Tt

taciturnus L. *taciturnus* silent, quiet (J. Syn. Miller's Long-footed Rat, see *Rattus stoicus*).

Tadarida Etymology not known, probably from *Tadaris* (Rafinesque, 1815) a *nomen nudum* (*T. aegyptiaca* and *T. teniotis*).

talpinus Mod. L. *talpinus* talpa-like (L. *talpa* European mole, *-inus* pertaining to (L.)) (*Ellobius*).

Tamiops Mod. L. *tamiops* bearing an aspect of *tamias* chipmunk (Gr. *Tamias* chipmunk – a treasurer or the one who stores, Gr. *óps* aspect) (*T. macclellandi*).

tanezumi Mod. L. *tanezumi* rat (>Japanese *ta-nezumi* the rat) (*Rattus*).

Taphozous Mod. L. *taphozous* tomb-dwelling (Gr. *táphos* grave, tomb; *zòós* living); in allusion to its preference to tombs to dwell (*T. perforatus*, *T. longimanus*, *T. nudiventris*, *T. melanopogon*, and *T. theobaldi*).

tarayensis L. adj. *tarayensis* meaning from Terai region of the Himalayas in Nepal (*-ensis*, geographic, occurrence in (location, toponym) (L.)) (J. Syn. Terai Mole Rat, see *Bandicota bengalensis*; Terai Otter, see *Lutrogale perspicillata*).

tardigradus L. *tardigradus* slow walker (L. adj. *tardus* slow or limping; L. *gradus* step) (*Loris*).

tatei Eponym after George Henry Hamilton Tate (b. 1894–d. 1953), an English-born American zoologist, and botanist who worked in Southeast and East Asia, subsequently curator of American Museum of Natural History (*-i*, commemorating (dedication, eponym) (L.)) (*Eptesicus*).

Tatera Etymology unknown (*T. indica*).

taxicolor Mod. L. *taxicolor* badger-coloured (Mod. L. *taxus* badger; L. *color* colour) (*Budorcas*).

temminckii Eponym after Dutch ornithologist Coenraad Jacob Temminck (b. 1778–d. 1858) (*-ii*, commemorating (dedication, eponym) (L.)) (*Pardofelis*/J. Syn. Lesser Yellow Bat, see *Scotophilus kuhlii*).

temon Mod. L. *temon* >Tib. *temon*, a local name for the weasel (Subsp. *Mustela altaica* Tibetan Mountain Weasel).

tenebricus L. *tenebricus* dark, gloomy (Subsp. *Trachypithecus pileatus* Tenebrous Capped Langur, also Dusky Capped Langur, Tenebrous Capped Leaf Monkey).

teniotis Mod. L. *teniotis* banded ears (L. *taenia* (Gr. *taina*) head-band; Gr. *otis*, *ōtos* ear) (*Tadarida*).

tenuis L. *tenuis*, *tenue* weak, slight; in allusion to its size (*Pipistrellus*).

terricolor L. *terricolor* earth-coloured (L. *terra* earth; L. *color* colour) (*Mus*).

Tetracerus Mod. L. *tetracerus* four-horned (Gr. *tetra* four, *kéras* horn) (*Tetracerus*).

tetracornis Mod. L. *tetracornis* four-horned (L. *tetra* four; L. *cornu* horn) (J. Syn. Chikara or Chouka, see *Tetracerus quadricornis*).

thar Mod. L. *thar* a local name for the species in Nepal (*Capricornis*).

theobaldi Eponym after William Theobald (b. 1829–d. 1908), an English geologist, naturalist, and malacologist who worked in Burma (*-i*, commemorating (dedication, eponym) (L.)) (*Taphozous*/J. Syn. Theobald's House Mouse, see *Mus musculus*; Theobald's Mouse-bat, see *Myotis longipes*).

thersites Gr. Myth. *Thersites*, a soldier in Greek army in Trojan war – a character in Homer's *Iliad* (Subsp. *Semnopithecus priam* Tufted Gray Langur, also Gray Langur, Grey Wanderoo).

thibetana/thibetanus L. adj. *thibetana*, *thibetanus* meaning from Tibet (*-ana*, *-anus*, belonging to, pertaining to (L.) geographic (location, toponym)) (*Ochotona*/*Ursus*/J. Syn. Tibetan Shrew, see *Sorex minutus*).

thomasi Eponym after Oldfield Thomas (b. 1858–d. 1929), an English zoologist who worked in British Museum (Natural History), who worked relentlessly on mammals, and described many new taxa based on specimens collected during the Mammal Survey of British India of the Bombay Natural History Society (*-i*, commemorating (dedication, eponym) (L.)) (Subsp. *Tadarida aegyptiaca* Thomas's Free-tailed Bat/J. Syn. Thomas's Jungle Palm Squirrel, see *Funambulus tristriatus*; Thomas's Pygmy Jerboa, see *Salpingotulus michaelis*).

thysanurus Mod. L. *thysanurus* tufted-tail (Gr. *thusanos* tassel, fringe, tuft; Gr. *-ouros* -tailed (Gr. *oúrá* tail)) (Subsp. *Urva smithii* Kashmir Ruddy Mongoose).

tibetanus L. adj. *tibetanus* meaning from or belonging to Tibet (*-anus*, belonging to, pertaining to (L.) geographic (location, toponym)) (*Lepus*/J. Syn. Tibet Marmot, see *Marmota himalayana*; Tibetan Dwarf Hamster, see *Cricetulus alticola*).

tickelli Eponym after Colonel Samuel Richard Tickell (b. 1811–d. 1875), an English ornithologist in India and Burma (*-i*, commemorating (dedication, eponym) (L.)) (*Hesperoptenus*).

tigris L. *tigris* (Gr. *tígris*) tiger (*Panthera*).

tolai Mod. L. *tolai* >Transbaikalia or Tungus *tolai*, *tulai*, or *tolui* names by which this species is known (*Lepus*).

topali Eponym after Dr. György Topal (b. 1931), a well-known bat taxonomist and biologist, former curator of mammals at the Hungarian Museum of Natural History (*-i*, commemorating (dedication, eponym) (L.)) (Subsp. *Rhinolophus macrotis* Topal's Big-eared Horseshoe Bat.

torquata L. *torquatus* collared (L. *torques* collar) (J. Syn. Spotted Wild Cat, see *Felis silvestris*).

toufoeus Mod. L. *toufoeus* >Tib. *toufee*, a local name for this marten in Tibet; toufee pelts are highly prized in Tibet and China (Subsp. *Martes foina* Lhasa Beech Marten).

Trachypithecus Mod. L. *Trachypithecus* 'Shaggy-ape' (Gr. *trachus* rough or shaggy, in allusion to longer hair; Gr. *pithecus* ape) (*T. geei*, *T. phayrei*, and *T. pileatus*).

tragatus Mod. L. *tragatus* in allusion to the well-developed antitragus (a projection of the auricle of the ear extending to the tragus) (L. *tragus* and *-atus*, provided with (L.)) (Subsp. *Rhinolophus ferrumequinum* Nepal Greater Horseshoe Bat and *Tadarida aegyptiaca* Dobson's Free-tailed Bat).

tragocamelus Mod. L. *tragocamelus* goat-camel (name combination of Mod. L. *tragos* (Gr. *trágos*) goat and L. *camelus* camel) (*Boselaphus*).

traubi Eponym after Robert Traub (b. 1916–d. 1996), an American entomologist (*-i*, commemorating (dedication, eponym) (L.)) (J. Syn. Swat Vole, see *Hyperacrius wynnei*).

travancorensis L. adj. *travancoensis* meaning from Travancore region of the peninsular India (=Kerala) (*-ensis*, geographic, occurrence in (location, toponym) (L.)) (J. Syn. Travancore Pygmy Shrew, see *Suncus etruscus*).

trevelyani Eponym after CoL. W.R.F. Trevelyan (dates not known), a British naturalist and sportsm an, who collected the type specimens (*-i*, commemorating (dedication, eponym) (L.)) (Subsp. *Prionailurus bengalensis* Kashmir Leopard Cat).

Triaenops Mod. L. *triaenops* trident-faced (Gr. *triaena* trident; Gr. *óps* aspect, face); in allusion to the posterior part of the noseleaf, which terminates in three pointed projections resembling the three prongs of a trident (*T. persicus*).

trichotis Mod. L. *trichotis* tufted-eared (Gr. *thríx* hair; Gr. *-otis* -eared (*ōtos* ear)) (J. Syn. Tufted-eared Flying Squirrel, see *Belomys pearsonii*).

tridens L. *tridens* three-pronged, trident (L. *tri-* three-; L. *dens* tooth); in allusion to its trident-shaped posterior noseleaf (*Asellia*).

trifoliatus L. *trifoliatus* three-leaved (L. *tri* (*tres*) >Sans. *tri* three; L. adj. *foliatus* leaved from *folium* leaf) (*Rhinolophus*).

trilineatus Mod. L. *trilineatus* three-lined (L. *tri* (*tres*) >Sans. *tri* three; L. adj. *lineatus* lined (*linea* line) (J. Syn. Three-lined Dusky-striped Squirrel).

tristriatus Mod. L. *tristriatus* three-striped (L. *tri* (*tres*) >Sans. *tri* three; L. adj. *striatus* striped (*stria* furrow or channel)) (*Funambulus*).

trivirgata Mod. L. *trivirgata* three-striped (L. *tri* (*tres*) >Sans. *tri* three; L. adj. *virgatus* striped (L. *virga* stripe) (*Arctogalidia*).

tropicalis Mod. L. *tropicalis* (>L. *tropicus*) tropical (Subsp. *Delphinus capensis* Arabian Common Dolphin).

truncatus L. *truncatus* mutilated, cut short (L. *truncare* to mutilate) (*Tursiops*).

tubinaris L. *tubinaris* tube-nosed (L. *tubis* tube; L. *naris* nose) (*Murina*).

Tupaia Mod. L. *tupaia* >Malay *tupai* (for various small squirrel-like creatures) (*T. belangeri* and *T. nicobarica*).

turnbulli Eponym of unclear origin; no explanation was given by J.E. Gray (1837), possibly either after Mrs. Turnbull or Major Turnbull, who are listed as collectors of mammals in British India (*-i*, commemorating (dedication, eponym) (L.)) (J. Syn. Turnbull's Flying Squirrel, see *Hylopetes alboniger*).

turkestanicus Mod. L. *turkestanicus* belonging to Turkestan, Kazakhstan (*-icus* belonging to, pertaining to (L./Gr.)) (J. Syn. Turkestan Rat, see *Rattus pyctoris*).

Tursiops Mod. L. *tursiops* dolphin-like (L. *tursio* a kind of fish resembling a dolphin; Gr. *óps* aspect) (*T. aduncus* and *T. truncatus*).

Tylonycteris Mod. L. *tylonycteris* padded bat (Gr. *týlos* knob, knot; Gr. *nikterís* bat); in allusion to the expanded fleshy pads on the under surface of the base of the thumb (*T. aurex*, *T. fulvida*, and *T. malayana*).

typus L. *typus* type (J. Syn. Luwack, see *Paradoxurus hermaphroditus*).

tytleri Eponym after Robert Christopher Tytler (b. 1818–d. 1872), a British soldier, naturalist, and photographer, worked as the superintendent of the convict settlement on Port Blair, Andaman (*-i*, commemorating (dedication, eponym) (L.)) (Subsp. *Paguma larvata* Tytler's Masked Palm Civet and Subsp. *Pteropus melanotus* Tytler's Black-eared Flying Fox/J. Syn. Long-haired Mouse, see *Mus musculus*; Tytler's Hare, see *Lepus nigricollis*; Dehra Shrew, see *Suncus murinus*).

Uu

umbrinus Mod. L. *umbrinus* dark, shady, umber-colour (>L. *umbra* shade, dark) (Subsp. *Episoriculus sacratus* Dark Brown Sichuan Brown-toothed Shrew).

umbrosus L. adj. *umbrosus* (>L. *umbra* shade, dark) shady, of shade, of twilight (Subsp. *Macaca fascicularis* Nicobar Long-tailed Macaque – also Nicobar Crab-eating Macaque).

uncia L. *uncia* >Fr. *once* (originally used for lynx, occasionally for other similar-sized cats) (*Panthera*).

unicolor L. *unicolor* plain, uniform (L. *uni-* single-; L. *color* colour) (*Rusa*).

unicornis L. *unicornis* one-horned (L. *uni* one; L. *cornu* horn) (*Rhinoceros*).
unitus Mod. L. *unitus* united (L. *uni*, *unus*, one) (J. Syn. Central India Leaf-nosed Bat, see *Hipposideros lankadiva*).
urbanus L. adj. *urbanus* urban (L> *urbs*, *urbis* town) (J. Syn. Common Indian Mouse, see *Mus musculus*).
ursinus L. *ursinus* resembling a bear (*Melursus*).
Ursus L. *ursus* a bear (*U. arctos* and *U. thibetanus*).
Urva/urva Mod. L. *urva* >Nep. *arva*, the name for the crab-eating mongoose (*U. auropunctatus*, *U. fuscus*, *U. edwardsii*, *U. palustris*, *U. smithii*, *U. urva*, and *U. vitticollis*/*Urva*).

Vv

vaginalis Mod. L. *vaginalis* sheathed (L. *vagina* sheath); in allusion to its short antler that is mostly enclosed by hair-covered pedicels (*Muntiacus*).
Vandeleuria Etymology not known; no explanation was given by J.E. Gray (1842) (*V. nilagirica*, *V. nolthenii*, and *V. oleracea*).
vellerosus Mod. L. *vellerosus* thick-furred (L. *vellera* fur, >*vellus* fleece, wool; *-osus*, abundance, fullness, quality of (L.)) (Subsp. *Paradoxurus hermaphroditus* Kashmir Palm Civet).
venaticus L. *venaticus* hunting-, of hunting (L. *venatus* hunting, the chase) (Subsp. *Acinonyx jubatus* Indian Cheetah, extinct).
Vespertilio L. *vespertilio* a bat, so-called due its flying habit in the evening – probably from the L. *vespertinus* meaning of the evening (*Vespertilio*).
vestita L. *vestitus* clothes (L. *vestis* garment, clothing) (J. Syn. Tibetan Rhesus Monkey, see *Macaca mulatta*).
vetulus L. *vetulus* old, a little old man (*Semnopithecus*/Subsp. *Semnopithecus vetulus* Southern Purple-faced Leaf-monkey – also Black Wanderoo).
vicerex L. *vicerex* viceroy (J. Syn. North Asian Rat, see *Rattus pyctoris*).
vilaliya Mod. L. *vilaliya* >Sinh. *vil* - *aliya*, the name for the large-footed population of elephants in Sri Lanka that show affinity to floodplain and marshy habitats (from Sinhalese *vil* lake, pit, and *aliya* cow-elephant or elephant) (J. Syn. Marsh Elephant, Ceylon Marsh Elephant, see *Elephas maximus*).
villosus L. *villosus* hairy (L. *villus* shaggy hair) (J. Syn. Kashmir Rhesus Monkey, see *Macaca mulatta*; Upper Assam Flying Squirrel, see *Belomys pearsonii*).
vignei Eponym after Godfrey Thomas Vigne (b. 1801–d. 1863), a British traveler and naturalist, who perhaps was the firt Englsihman to visit Afghanistan (*-i*, commemorating (dedication, eponym) (L.)) (Subsp. *Ovis orientalis* Ladhak Urial).
virgata L. *virgatus* striped, streaked (L. *virga* streak) (Subsp. *Panthera tigris* Caspian Tiger, extinct).
vittatus L. *vittatus* banded (L. *vitta* ribbon, band) (Subsp. *Sus scrofa* Asian Wild Pig).
vitticollis L. *vitticollis* striped-necked (L. *vitta* band, ribbon; Mod. L. *-collis* -throated (>L. *collum* neck) (*Urva*).

Viverra L. *viverra* civet (*V. civettina* and *V. zibetha*).
viverriceps Mod. *viverriceps* civet-headed (L. *viverra* civet; L. *-ceps* headed (L. *caput* head)) (J. Syn. Hodgson's Fishing Cat, see *Prionailurus viverrinus*).
Viverricula L. *viverricula* small civet, dim. of *viverra* civet (*-ula*, diminutive (comparison); somewhat (adj.) (L.)) (*V. indica*).
viverrinus Mod. L. *viverrinus* civet-like (L. *viverra* civet; *-inus* pertaining to (L.)) (*Prionailurus*).
vorax L. *vorax* voracious (*Crocidura*).
Vormela Mod. L. *vormela* >German *würmlein* meaning little worm, in allusion to its body shape (*V. peregusna*).
Vulpes/vulpes L. *vulpes* Fox (*V. bengalensis*, *V. cana*, *V. corsac*, *V. ferrilata*, *V. rueppelli*, and *V. vulpes*).
vulturna Rom. Myth. *Vulturnus*, a river god, god of the Tiber (Baluchistan Pika, see *Ochotona rufescens*).

Ww

wallichi Eponym after Dr. Nathaniel Wallich (b. 1786–d. 1854), a Danish surgeon, botanist, and naturalist who worked in India, the first curator of Indian Museum and force beyond the development of Calcutta Botanical Garden (*-i*, commemorating (dedication, eponym) (L.)) (Subsp. *Cervus elaphus* Sikkim Stag).
wardi Eponym after Colonel A. E. Ward (dates not known), an amateur zoologist, sportsman and collector, active in the Kashmir Himalayas (*-ii*, commemorating (dedication, eponym) (L.)) (*Plecotus*/Subsp. *Bandicota bengalensis* Wroughton's Lesser Bandicoot-rat/J. Syn. Baltistan Ibex, see *Capra sibirica*).
watsoni Eponym after H.E. Watson (dates not known), who worked for the Sindh Commission and collected in Sindh Province, Pakistan (*-i*, commemorating (dedication, eponym) (L.)) (J. Syn. Northern Indian Bush-rat, see *Golunda ellioti*).
wellsi Eponym after H.W. Wells (dates not known), collected in India for the Mammal Survey of British India (*-i*, commemorating (dedication, eponym) (L.)) (Subsp. *Viverricula indica* Well's Small Civet/J. Syn. Wells' Squirrel, see *Callosciurus erythraeus*; Wells' White-toothed Rat, see *Berylmys bowersi*).
whitei Eponym after John Claude White (b. 1853–d. 1918), British Commissioner in Sikkim, who presented two pairs of horns based on which R. Lydekker described this species (*-i*, commemorating (dedication, eponym) (L.)) (Subsp. *Budorcas taxicolor* Bhutan Takin).
williamsi Eponym after Major W.H. Williams (dates not known), H.M. Consul at Van (=Artemita); in Kurdistan (= Turkey), who collected the type (*-ii*, commemorating (dedication, eponym) (L.)) (*Allactaga*).
wollastoni Eponym after Alexander Frederick Richmond 'Sandy' Wollaston (b. 1875–d. 1930), a British medical doctor, ornithologist, and botanist (*-i*, commemorating (dedication, eponym) (L.)) (Subsp. *Ochotona macrotis* Wollaston's Large-eared Pika or Everest Pika).

wroughtoni Eponym after Robert Charles Wroughton (b. 1849–d. 1921), an English officer of Indian Forest Service in Bombay Presidency, member of the Bombay Natural History Society, described many new taxa resulting from Mammal Survey of British India (-*i*, commemorating (dedication, eponym) (L.)) (*Funambulus* and *Otomops*/Subsp. *Paguma larvata* Wroughton's Masked Palm Civet/J. Syn. Wroughton's Tree Shrew, see *Anathana ellioti*; Wroughton's Large-toothed Rat, see *Dacnomys millardi*; Wroughton's Black Rat, see *Rattus rattus*; Wroughton's Tree Mouse, see *Vandeleuria oleracea*; Wroughton's Bat, see *Scotophilus kuhlii*).

wynnei Eponym after Arthur Beavor Wynne (b. 1835–d. 1906), an officer with the Indian Geological Survey, worked in the Bombay and Punjab presidencies (-*i*, commemorating (dedication, eponym) (L.)) (*Hyperacrius*).

Yy

yarkandensis L. adj. *yarkandensis* meaning from Yarkand, Xinjiang Uyghur Autonomous Region, China (-*ensis*, geographic, occurrence in (location, toponym) (L.)) (Subsp. *Cervus elaphus* Yarkand Stag).

yunanensis L. adj. *yunanensis* meaning from Yunnan, China (-*ensis*, geographic, occurrence in (location, toponym) (L.)) (*Rhinolophus*).

Zz

zarudnyi Eponym after Dr. N.A. Zarudny (b. 1859–d. 1919), curator of Tashkent Museum, Uzbekistan, who collected in Persia (-*i*, commemorating (dedication, eponym) (L.)) (*Crocidura* and *Meriones*/Subsp. *Vulpes rueppelli* Zarudny's Fox).

zeylanicus L. adj. *zeylanicus* belonging to Ceylon (= Sri Lanka) (-*icus* belonging to, pertaining to (L./Gr.)) (*Suncus*/Subsp. *Urva smithii* Ceylon Ruddy Mongoose).

zeylanius Mod. L. *zeylanius* belonging to Ceylon (= Sri Lanka) (-*ius* belonging to, pertaining to (L. /Gr.)) (Subsp. *Urva smithii* Ceylon Ruddy Mongoose).

zeylonensis Mod. L. *zeylonensis* belonging to Ceylon (= Sri Lanka) (-*ensis* belonging to, pertaining to (L./Gr.)) (J. Syn. Ceylon Crested Porcupine, see *Hystrix indica*; Golden Palm Civet or Golden Palm Cat or Ceylon Palm Civet or Red Palm Civet, see *Paradoxurus hermaphroditus*/S. Syn. Golden Palm Civet, see *Paradoxurus aureus*).

zibetha Med. L. *zibethum* or ItaL. *zibetto*, both >Arab. *zabād*, civet perfume (*Viverra*).

Ziphius Mod. L. *ziphius* swordfish (Gr. *xiphos* sword) (*Z. cavirostris*).

zygomaticus Mod. L. *zygomaticus* cheek bone (Gr. *zygoma* yoke) (J. Syn. Phillips' Vole, see *Hyperacrius fertilis*).

Appendix I: List of Languages Used in Autochthonous and Vernacular Names

Abor A Tibeto-Burman language spoken by the ethnic Abor group of Abor and the adjoining Abor Hills, Arunachal Pradesh, India.

Adi A Tibeto-Burman language spoken by the ethnic Adi group of Abor Hills, Arunachal Pradesh, India.

Arab. Arabic, an Afro-Asiatic language primarily spoken in Arab League countries, and also in countries with an Islamic population.

Assam. Assamese or Asamiya, an Eastern Indo-Aryan language used mainly in the state of Assam; also spoken by some in adjoining regions in India and Bangladesh.

Balochi An Indo-European Northwestern Iranian language spoken by the Baloch people of Pakistan, Afghanistan, Iran, and Turkmenistan.

Balt. Balti, a Sino-Tibetan language predominantly spoken in the disputed Baltistan region of Kashmir, Jammu; Kashmir, India; and the adjoining districts in Pakistan.

Beng. Bengali or Bangla, an Eastern Indo-Aryan language spoken predominantly in the Indian states of West Bengal, Tripura, Assam (southern), and Bangladesh.

Bhut. Bhutia, a Sino-Tibetan language spoken by the Bhutia and Sherpa communities of Sikkim and adjacent Nepal; its dialect is akin to Sikkimese.

Bihar. Bihari, an Indo-Aryan language primarily spoken by people of the state of Bihar, India. It includes many sub-languages and dialects such as Angika, Bajjika, Bhojpuri, Kudmali, Maghai, Maithili, Mahji, Musasa, Panchapargania, Sadri, Khortha, and Surajpuri.

Bundel. Bundeli (also known as Bundelkhandi) is a western Hindi language, belonging to the Indo-Aryan language family, spoken by the people of the Bundelkhand region of Madhya Pradesh and southern parts of Uttar Pradesh, India.

Dakh. Dakhini (also known as Dakhani or Deccani) is a language spoken by the Hindi- and Urdu-speaking population of the erstwhile states of Hyderabad and Mysore. It is a language akin to Urdu with a strong influence of Arabic, Persian, Konkani, Marathi, Telugu, and Kannada, spoken in the states of Maharashtra, Telangana, Andhra Pradesh, and Karnataka, India.

Daphla A Tibeto-Burman language spoken by the ethnic Daphla group of Daphla Hills, Arunachal Pradesh, India.

Dima. Dimasa, a Sino-Tibetan language spoken by the Dimasa people in the state of Assam and Nagaland in North East India.

Garo A Sino-Burman language spoken by the ethnic people of Garo Hills, Meghalaya, India.

Gond. Gondi, a south-central Dravidian language spoken by the Gondi people of Telangana, Chhattisgarh, Madhya Pradesh, Maharashtra, Odisha.

Guj. Gujarati, an Indo-Aryan language spoken by the people of the state of Gujarat, India.

Hebrew An Afro-Asiatic language, primarily spoken in Israel at the present time.

Hind. Hindi, a standardised and Sanskritised register of the Hindustani language, spoken throughout, but mostly in northern India.

Ho An Austroasiatic language spoken by the Ho people of East Singhbhum district, Jharkhand, and Mayurbhanj and Keonjhar districts, Odisha, India.

Kachari A Tibeto-Burman language spoken by the Kachari-Bodo ethnic group of Assom, India.

Kan. Kannada or Kanarese, a Dravidian language spoken predominantly in the state of Karnataka.

Kash. Kashmiri (also known as Koshur and Kishtwari), an Indo-Aryan language primarily spoken in Kashmir Valley, Jammu and Kashmir, India. This language is also spoken in the Neelam district, Pakistan.

Kathi. Kathiawari, a dialect of Gujarati, an Indo-Aryan language spoken by the people of the state of Gujarat, India.

Khám. Khámti, a Tai-Kadai language spoken by the Khámti people in Assam and Arunachal Pradesh, India.

Khan. Khandeshi, an Indo-Aryan language primarily spoken by the people of the Khandesh region of Maharashtra, India.

Khas. Khasi, an Austroasiatic language spoken primarily in the Meghalaya state in India by the Khasi people, and also in adjoining districts in Assam, India; and Bangladesh.

Kin. Kinnauri (also known as Kunawar or Kanauri), a Sino-Tibetan dialect cluster primarily spoken by the ethnic people of the Kinnaur district, Himachal Pradesh, India.

Kiran. Kiranti, a Sino-Tibetan language spoken by the Kiranti people of Sikkim, Darjeeling; and adjacent Nepal. Its dialect is akin to Limbu.

Kirghiz A Turkic language, primarily spoken by people in Kyrgyzstan, and also by a few people in China, Afghanistan, Kazakhstan, Tajikistan, Turkey, Uzbekistan, Pakistan, and Russia.

Kol An Austroasiatic language spoken by the ethnic Oraon tribals of Bihar, Uttar Pradesh; Chhattisgarh, Madhya Pradesh; Assam and Tripura, India; and Bangladesh and Nepal.

Kon. Konda (also known as Konda-Dora), a south-central Dravidian language spoken by the Konda-Dora people from Andhra Pradesh and Odisha, India.

Korku An Austroasiatic language spoken by the Korku people of Madhya Pradesh and Maharashtra, India.

Kúki One of the Sino-Tibetan Kukish languages spoken by ethnic groups in North East India.

Kum. Kumaoni, an Indo-Aryan language primarily spoken by people of Kumaon region of Uttarakhand, India; it consists of many dialects.

Kurg. Kurgi (also known as Kodava), a Dravidian language spoken by the Kodavas of the Kodagu district, Karnataka, India.

Kurukh A Dravidian language spoken by the Uraon and Kisan tribals of the Odisha, Bihar, Jharkhand, Madhya Pradesh, Chhattisgarh, and West Bengal, India. This language is also spoken by a small number of people in Bangladesh, Bhutan, and Nepal.

Kut. Kutchi (also known as Kacchhi), an Indo-Aryan language spoken in the Kutch region in the state of Gujarat, India and also some areas in Sindh Province, Pakistan.

Lad. Ladakhi (also known as Bhoti), a Sino-Tibetan language predominantly spoken in the Leh district, Jammu and Kashmir, India.

Lah. Lahauli, a Sino-Tibetan language predominantly spoken in the Lahul and Spiti districts of Himachal Pradesh and adjoining districts in Jammu and Kashmir, India.

Lepch. Lepcha (or Róng), a Sino-Tibetan language spoken by the Lepcha people in Sikkim and parts of West Bengal, Nepal, and Bhutan.

Limbú Sino-Tibetan language spoken by ethnic communities of Nepal and adjoining districts of West Bengal and Sikkim, India.

Mal. Malayalam, a Dravidian language spoken predominantly in the state of Kerala.

Malto Also known as Paharia, a northern Dravidian language spoken in Eastern India (in Jharkhand, Bihar, Odisha, West Bengal) and Bangladesh.

Mani. Manipuri (also known as Meithei), a Sino-Tibetan language spoken in the state of Manipur.

Mar. Marathi, an Indo-Aryan language spoken predominantly in the state of Maharashtra.

Marw. Marwari, a western Indo-Aryan language spoken by the Marwari people living in the states of Rajasthan, Gujarat, and Haryana, India; and Sindh, Pakistan.

Mis. Mishmi, a Sino-Tibetan language spoken by the Idu-Mishmi people of Arunachal Pradesh, India.

Mizo One of the Sino-Tibetan Kukish languages spoken by ethnic groups in Mizoram, India.

Mon. Monpa languages, principally spoken by the Monpa people of Arunachal Pradesh; several dialects, such as Brokpa, Central Monpa, and Souther Monpa, are subsumed under this group of languages.

Mongol. Mongolian, a Mongolic language, spoken chiefly by the natives of Mongolia.

Naga One of the Sino-Tibetan Kukish languages spoken by ethnic groups in Nagaland, India.

Nep. Nepali, an Indo-Aryan language, principally spoken by people from Nepal; it also includes the dialects of Khas and Gorkhali.

Nic. Nicobarese, an Austroasiatic language, principally spoken by the Nicobarese people of Nicobar and Andaman Islands and Nicobar Archipelago, India.

Odi. Odiya, an eastern Indo-Aryan language spoken by the people of the state of Odisha, India.

Pahari W. Western Pahari (including the Dogri-Kangri languages, also known as Himachali), an Indo-Aryan family, is a range of languages and dialects spoken across the Himalayan range in Pakistan, India, and Nepal.

Pard. Pardhi, a member of the Western Indo-Aryan language family, is a Bhil language spoken by nomadic and hunting tribes of central India. Dialects such as Neelishikari, Pittala Bhasha, Takari, and Haran Shikari belong to this language.

Pash. Pashto, an Indo-Iranian language, principally spoken by Pashtun people from central Asia.

Pers. Persian, an Indo-European language, primarily spoken in Iran, Afghanistan, Tajikistan, Uzbekkistan, Iraq, Russia, and Azerbaijan.

Punj. Punjabi, an Indo-Aryan language spoken by the Punjabi people of the Punjab region of both India and Pakistan.

Russian An Indo-European language, primarily spoken in Russia, Belarus, Kazakhstan, Kyrgyzstan, and also in a few other countries under the former USSR.

Sans. Sanskrit is primarily a liturgical language of Vedic India belonging to the Indo-European and Indo-Aryan family of languages. It is still being spoken by some people in the Shimoga district, Karnataka, Rajgarh and Narsinghpur districts, Madhya Pradesh, Banswara and Bundi districts, Rajasthan, Kendujhar district, Odisha, and Bagpat district, Uttar Pradesh. It is the second official language of the state of Uttarakhand, India.

Sant. Santali (also known as Santhali), an Austroasiatic language spoken by Santhal people mostly from Jharkhand, Assom, Bihar, Odisha, Tripura, and West Bengal, India.

Sind. Sindhi, an Indo-Aryan language of the historical Sindh region, spoken by Sindhi people in both Pakistan and India.

Sinh. Sinhalese (also Sinhala), an Indo-Aryan language of the Sinhalese people of Sri Lanka.

Sylh. Sylheti, an eastern Indo-Aryan language spoken by around 10 million in the Sylhet region of Bangladesh and in parts of India.

Tam. Tamil, a Dravidian language spoken predominantly in Tamil Nadu, India.

Tel. Telugu, a Dravidian language spoken predominantly in the states of Andhra Pradesh and Telangana.

Tib. Tibeti, a cluster of mutually unintelligible Sino-Tibetan languages spoken primarily by Tibetan peoples.

Toda A Dravidian language, spoken by the Toda people of the Nilgiri Hills, Tamil Nadu, India.

Turkish A Turkic language, primarily spoken by Turkish people in South Eastern Europe and Western Asia.

Urdu An Indo-European language, primarily spoken by Muslims in the Indian subcontinent.

Wad. Waddari, a dialect of the Dravidian language of Telugu, spoken by the Waddari people of Telangana, Karnataka, Andhra Pradesh, and parts of Maharashtra, India.

Yan. Yanadi, a dialect of the Dravidian language of Telugu, spoken by the Yanadi people of the Rayalaseema region of Andhra Pradesh, India.

Bibliography

Abel, C. (1826). On the supposed unicorn of the Himalayas. *Philosophical Magazine and Journal*, 68: 232–234.

Agrawal, V.C. and D.K. Ghosal. (1969). A new field rat [Mammalia: Rodentia: Muridae] from Kerala, India. *Proceedings of the Zoological Society of Calcutta*, 22: 41–45.

Allen, G.M. (1908). Notes on Chiroptera. *Bulletin of the Museum of Comparative Zoology at Harvard College*, 52(3): 23–62.

Allen, G.M. (1923). New Chinese insectivores. *American Museum Novitates*, 100: 1–11.

Allen, G.M. (1936). Two new races of Indian bats. *Records of the Indian Museum*, 38: 343–346.

Amato, G., M.G. Egan and A. Rabinowitz. (1999). A new species of muntjac, *Muntiacus putaoensis* (Artiodactyla: Cervidae) from northern Myanmar. *Animal Conservation*, 2: 1–7.

Andersen, K. (1905a). On some bats of the genus *Rhinolophus*, with remarks on their mutual affinities, and descriptions of twenty-six new forms. *Proceedings of the Zoological Society of London*, 1905(2): 75–145.

Andersen, K. (1905b). On the bats of the *Rhinolophus philippensis* group, with description of five new species. *Annals and Magazine of Natural History*, Ser 7, 16: 243–257.

Andersen, K. (1906). On some new or little-known bats of the genus *Rhinolophus* in the collection of the Museo Civico, Genoa. *Annali del Museo civico di storia naturale di Genova*, Ser 3, 2: 173–189.

Andersen, K. (1907). Chiropteran notes. *Annali del Museo civico di storia naturale di Genova*, Ser 3, 3: 5–45.

Andersen, K. (1908). Twenty new forms of *Pteropus*. *Annals and Magazine of Natural History*, Ser 8, 2: 361–370.

Andersen, K. (1911). Six new fruit-bats of the genera *Macroglossus* and *Syconycteris*. *Annals and Magazine of Natural History*, Ser 8, 7: 641–643.

Andersen, K. (1918). Diagnoses of new bats of the families Rhinolophidae and Megadermatidae. *Annals and Magazine of Natural History*, Ser 9, 2: 374–384.

Anderson, J. ('1878' 1879a). *Anatomical and Zoological Researches: Comprising and Account of the Two Expedition to Western Yunnan in 1868 and 1875, and A Monograph of the Two Cetacean Genera, Platanista and Orcella. Volume 1—Text.* Bernard Quaritch, London. xxv + 985.

Anderson, J. ('1878' 1879b). *Anatomical and Zoological Researches: Comprising and Account of the Two Expedition to Western Yunnan in 1868 and 1875, and A Monograph of the Two Cetacean Genera, Platanista and Orcella. Volume 2—Plates.* Bernard Quaritch, London. 81 plates.

Anderson, J. (1875). Description of some new Asiatic Mammals and Chelonia. *Annals and Magazine of Natural History*, Ser 4, 16: 282–285.

Anderson, J. (1877). Description of some new and little known Asiatic shrews in the Indian Museum, Calcutta. *Journal of the Asiatic Society of Bengal*, 46(2): 261–283.

Anderson, J. (1878). On the Indian species of the genus *Erinaceus*. *Journal of the Asiatic Society of Bengal*, 47(2): 195–211.

Anderson, J. (1881). *Catalogue of Mammalia in the Indian Museum, Calcutta. Part I. Primates, Prosimiae, Chiroptera, and Insectivora.* Indian Museum, Calcutta. xvi + 223 p.

Anderson, J. and W.E. de Winton. (1902). *Zoology of Egypt: Mammalia*. Hugh Rees Ltd, London. ill + 377 p.

Barrett-Hamilton, G.E.H. (1900). On geographical and individual variation in *Mus sylvaticus* and its allies. *Proceedings of the Zoological Society of London*, 1900: 387–428.

Barrett-Hamilton, G.E.H. (1906). Descriptions of two new species of *Pterygistes*. *Annals and Magazine of Natural History*, Ser 7, 17: 98–100.

Bates, P., O. Tun, A.M. Moe, A. Lu, M.R. Lum and S.M. Mie. (2015). A review of *Hipposideros lankadiva* Kelaart, 1850 (Chiroptera: Hipposideridae) with a Description of a new subspecies from Myanmar. *Tropical Natural History*, 15(2): 191–204.

Bechstein, J.M. (1800). *Thomas Pennant's allgemeine Übersicht der Vierfüssigen Thiere. Volume 2*. Verlag des Industrie-Comptoirs, Weimar.

Bechthold, G. (1936). Einige neue Unterrten asiatischer Herpestiden. *Zeitschrift für Säugetierkunde: Im Auftrage der Deutschen Gesellschaft für Säugetierkunde e. V.*, 11: 149.

Bennett, E.T. (1832). Characters of two new species of the Genus *Mus*, L., collected by Col. Sykes in Dukhun. *Proceedings of the Zoological Society of London*, 1832: 121.

Bennett, E.T. (1833). Proceedings of the Society—*Felis viverrinus*. *Proceedings of the Zoological Society of London*, 1833: 68–69.

Bennett, E.T. (1835a). Proceedings of the Society—*Herpestes vitticollis*. *Proceedings of the Zoological Society of London*, 1835(3): 67.

Bennett, E.T. (1835b). Proceedings of the Society—*Paradoxurus grayi*. *Proceedings of the Zoological Society of London*, 1835(3): 118.

Berkenhout, J. (1769). *Outlines of the Natural History of Great Britain and Ireland: Containing a Systematic Arrangement and Concise Description of All the Animals, Vegetables and Fossiles Which Have Been Hitherto Been Discovered in These Kingdoms. Vol. I. Comprehending the animal Kingdom*. P. Elmsly, London. XIII + 233 p.

Bianchi, V. (1917). Predvaritel'nyja zametki o letucih' mysah' (Chiroptera) Rossii [Notes preliminaires sur les chauve-souries ou Chiropteres de la Russie]. *Ezegodnik Zoologiceskogo Muzeja Akademii Nauk*, 21: 73–82 (in Russian, French subtitle).

Birula, A.A. (1912). [Materials on systematics and geographical distribution of mammals. III. Carnivora collected by N.A. Zarudny in Persia in 1896, 1898, 1900–1901, and 1903–1904]. *Ezhegodnik Zoologicheskogo Muzeya Imperatorskoi Akademii Nauk*. 17: 219–280 [in Russian].

Blainville, H.-M. D. de. (1816). Sur plusiers espèce d'animaux mammiferes, de l'orde des ruminans. *Bulletin de la Société Philomathique de Paris*, Ser 3, 3: 73–82.

Blainville, H.-M. D. de. (1817). *Dauphins*, 146–179 pp. In: *Nouveau dictionnaire d'histoire naturelle, appliquée aux arts, à l'agriculture, à l'économie rurale et domestique, à la médecine, etc. Par une société de naturalistes et d'agriculteurs*. Volume 9. Deterville, Paris. 624 p.

Blainville, H.-M. D. de. (1838). Sur les Cachalots. *Annales françaises et étrangéres d'anatomie et de physiologie, appliquées a la medécine et l'histoire naturelle*, 2: 335–337.

Blanchard, R. (1881). *De la Nomenclature des Êtres Organisés*. Sociéte Zoologique de France, Paris. 16 p.

Blanchard, R. (1889). De la nomenclature des Êtres Organisés 333–424 pp. In: Blanchard, R. (Ed.), *Compte-Rendudes Séances du Congrès International de Zoologie*. Sociéte Zoologique de France, Paris.

Blanford, W.T. (1873). Notes on the Gazelles of India and Persia, with description of a new species. *Proceedings of the Zoological Society of London*, 1873: 313–318.

Blanford, W.T. (1874). On two species of *Herpestes* and a hare collected by Dr. F. Day in Sind. *Proceedings of the Zoological Society of London*, 1874: 661–664.

Blanford, W.T. (1875a). Descriptions of new Mammalia from Persia and Balúchistán. *Annals and Magazine of Natural History*, Ser 4, 16: 309–313.

Blanford, W.T. (1875b). List of Mammalia collected by the late Dr. Stoliczka when attached to the embassy under Sir D. Foryth in Kashmir, Ladák, Eastern Turkestan, and Wakhán, with descriptions of new species. *Journal of the Asiatic Society of Bengal*, 44(2): 105–112.

Blanford, W.T. (1875c). Note on a large hare inhabiting high elevations in Western Tibet. *Journal of the Asiatic Society of Bengal*, 44(2): 214–215.

Blanford, W.T. (1877a). Notes on the species of Asiatic Bears, the "Mamh" of Baluchistan and *Ursus pruinosus*, Blyth, Tibet, and on an apparently undescribed fox from Baluchistan. *Journal of the Asiatic Society of Bengal*, 46(2): 315–323.

Blanford, W.T. (1877b). On an apparently new hare, and some other mammalia from Gilgit. *Journal of the Asiatic Society of Bengal*, 46(2): 323–327.

Blanford, W.T. (1877c). On an apparently undescribed weasel from Yarkand. *Journal of the Asiatic Society of Bengal*, 46(2): 259–261.

Blanford, W.T. (1878). On some mammals of Tenasserim. *Journal of the Asiatic Society of Bengal*, 47(2): 231–238.

Blanford, W.T. (1881). Description of an *Arvicola* from the Punjab Himalayas. *Journal of the Asiatic Society of Bengal*, 49(2): 244–245.

Blanford, W.T. (1885a). A monograph of the genus *Paradoxurus*, F. Cuv. *Proceedings of the Zoological Society of London*, 1885: 780–808.

Blanford, W.T. (1885b). Proceedings of the Society—*Paradoxurus jerdoni. Proceedings of the Zoological Society of London*, 1885: 613.

Blanford, W.T. (1888). *The Fauna of British India including Ceylon and Burma. Mammalia. Vol. I.* Taylor and Francis, London. xii + 250 p.

Blanford, W.T. (1891). *The Fauna of British India including Ceylon and Burma. Mammalia. Vol. II.* Taylor and Francis, London. 366 p.

Blyth, E. ('1850' 1851). Description of a new Mole (*Talpa leucura,* Blyth). *Journal of the Asiatic Society of Bengal*, 19: 215–217.

Blyth, E. ('1851' 1852a). Notice of a collection of Mammalia, Birds, and Reptiles, procured at or near the station of Chérra Punji in the Khásia hills, north of Sylhet. *Journal of the Asiatic Society of Bengal*, 20: 517–524.

Blyth, E. ('1851' 1852b). Report on the Mammalia and the more remarkable species of birds inhabiting Ceylon. *Journal of the Asiatic Society of Bengal*, 20: 153–185.

Blyth, E. ('1855' 1856). Report for October Meeting, 1855. *Journal of the Asiatic Society of Bengal*, 24(7): 711–723.

Blyth, E. (1840 '1841'). An amended list of the species of the genus *Ovis*. *Proceedings of the Zoological Society of London,* 1840: 62–79.

Blyth, E. (1841). Report for the month of September, Animal Kingdom. *Journal of the Asiatic Society of Bengal*, V. 10, Pt. 2 (118): 836–842.

Blyth, E. (1842a). Proceedings of the Asiatic Society of Bengal—6th May, 1842. *Proceedings of the Asiatic Society of Bengal*, 11: 439–470.

Blyth, E. (1842b). The Curator's report. *Journal of the Asiatic Society of Bengal*, 10(2): 917–929.

Blyth, E. (1843). Proceedings of the Asiatic society. *Journal of the Asiatic Society of Bengal*, 12(1): 129–182.

Blyth, E. (1844a). Notices of various mammalia, with description of many new species. *Journal of the Asiatic Society of Bengal*, 13(1): 463–495.

Blyth, E. (1844b). Notices of various Mammalia, with descriptions of many new species, Part I—The Primates. *Journal of the Asiatic Society of Bengal*, 13(1): 463–494.

Blyth, E. (1845). In Hutton, T. Rough notes on Zoology of Candahar and the neighbouring districts. *Journal of the Asiatic Society of Bengal*, 14: 340–354.

Blyth, E. (1846). In Hutton, T. Rough notes on Zoology of Candahar and the neighbouring districts. *Journal of the Asiatic Society of Bengal*, 15: 135–170.

Blyth, E. (1847a). Report of Curator, Zoological Department, for September 1847. *Journal of the Asiatic Society of Bengal*, 16(2): 1176–1181.

Blyth, E. (1847b). Supplementary Report of the Curator of the Zoological Department. *Journal of the Asiatic Society of Bengal*, 16(2): 728–737.

Blyth, E. (1848). Report of Curator, Zoological Department. *Journal of the Asiatic Society of Bengal*, 17(1): 247–255.

Blyth, E. (1849). Note on the Sciuri inhabiting Ceylon and those of the Tenasserim provinces. *Journal of the Asiatic Society of Bengal*, 18(1): 600–603.

Blyth, E. (1851a). Notice of a collection of mammalia, birds and reptiles, procured at or near the station of Cherra Punji in the Khasia Hills, north of Sylhet. *Journal of the Asiatic Society of Bengal*, 20(6): 516–524.

Blyth, E. (1851b). Report on the mammalia and more remarkable species of birds inhabiting Ceylon. *Journal of the Asiatic Society of Bengal*, 20(2): 153–185.

Blyth, E. (1853). Report of Curator, Zoological Department. *Journal of the Asiatic Society of Bengal*, 22: 408–417.

Blyth, E. (1854). Report of Curator, Zoological Department, for September 1854. *Journal of the Asiatic Society of Bengal*, 23: 729–740.

Blyth, E. (1855). Proceedings of the Asiatic Society—Report for May Meeting, 1855. *Journal of the Asiatic Society of Bengal*, 24: 359–363.

Blyth, E. (1859a). Mammalia: Proceedings of the Asiatic Society for July, 1859. *Journal of the Asiatic Society of Bengal*, 28: 293–298.

Blyth, E. (1859b). On the great rorqual of the Indian Ocean, with notices of other cetals, and of the Syrenia or marine pachyderms. *Journal of the Asiatic Society of Bengal*, 28: 481–498.

Blyth, E. (1859c). Report of Curator, Zoological Department, for February to May Meetings, 1859. *Journal of the Asiatic Society of Bengal*, 28(3): 271–301.

Blyth, E. (1860). Report of Curator, Zoological Department. *Journal of the Asiatic Society of Bengal*, 29(1): 87–115.

Blyth, E. (1862). Report of Curator, Zoological Department, February 1862. *Journal of the Asiatic Society of Bengal*, 31: 331–345.

Blyth, E. (1863a). *Catalogue of the Mammalia in the Museum of the Asiatic Society*. Savielle and Cronenburgh, London. 187 p.

Blyth, E. (1863b). Report of Curator, Zoological Department. *Journal of the Asiatic Society of Bengal*, 32(1): 73–90.

Bobrinskii, N.A. (1926). Predvaritelnoje soobshtchnie o letutchikh myshach (Chiroptera) iz Centralnoj Azii (Note préliminaire sur les Chiropteres de l'Asie Centrale). *Comptes Rendus de l'Académie des Sciences URSS, A*, 1926: 95–98 (in Russian with a summary in French).

Boddaert, P. (1785). *Elenchus animalium, volumen 1: Sistens quadrupedia huc usque nota, erorumque varietates*. C. R. Hake, Rotterdam, 174 p.

Boie, H. (1828). Uittreksels uit Brieven van Heinrich Boie van Java aan H. Schlegel, Conservator Animalium Vertebratorum aan's Rijks Museum te Leyden. *Bijdragen tot de natuurkundige wetenschappen*, 3: 231–252.

Bonaparte, C.L. (1837–1841). *Iconografia della fauna italica: per le quattro classi degli animali vertebrati. Vol. 1*. Dalla Tipografia Salvuicci, Rome.

Bonhote, J.L. (1898a). Description of a new species of *Mus* from South India. *Journal of the Bombay Natural History Society*, 12: 99–100.

Bonhote, J.L. (1898b). On the species of genus *Viverricula*. *Annals and Magazine of Natural History*, Ser 7, 1: 119–122.

Borowski, G.H. (1781). *Balaena novae angliae*. *Gemeinnüzzige Naturgeschichte des Thierreichs*, 2(1): 21.

Brandt, J.F. ('1843' 1844). Observations sur les différentes espèces de sousliks de Russie, suivies de remarques sur l'arrangement et la distribution géographique du genre *Spermophilus*, ansé que sur la classification de la familie des ecureuils (Sciurina) en général. *Bulletin Scientifique l'Académie Impériale des Sciences de Saint-Pétersbourg*, 1844: 357–382.

Brandt, J.F. (1836). Melances. *Bulletin Scientifique l'Académie Impériale des Sciences de Saint-Pétersbourg, Tome*, 1(4): 32.

Brandt, J.F. (1855). Beitrage zur nahern Kenntniss der Säugethiere Russland's. *Kaiserlichen Akademie der Wissenschaften, Saint Petersburg, Mémoires Mathématiques, Physiques et Naturelles*, 7: 1–365.

Brünnich, M.T. (1782). *Dyrenes Historie og Dyre-Samlingen udi Universitetets Natur-Theater. Vol. 1.* Nicolaus Møller, Kiøbenhavn. xxxviii + 76 p.

Buchannan, F. (1800). Description of the *Vespertilio plicatus. Transactions of the Linnean Society of London*, 5: 261–263.

Büchner, E. (1891). Die Säugethiere der Ganssu-Expedition (1884–87). *Mélanges biologiques tirés du Bulletin de l'Académie impériale des sciences de St. Pétersbourg*, 13: 143–164.

Büchner, von E. (1892). Die säugethiere der Ganssu-Expedition (1884–87). *Bulletin de l'Académie impériale des sciences de St.-Pétersbourg*, 34: 97–118.

Buckland, F. (1872). A new rhinoceros at the Zoological Gardens. *Land and Water*, 14: 89.

Cabrera, A. (1908). Sobre los loris, y en especial sobre la forma Filipina. *Boletin de la Real Sociedad Española de Historia Natural, Madrid*, 8: 135–139.

Cantor, T. (1846). Catalogue of mammalia inhabiting the Malayan peninsula and islands. *Journal of the Asiatic Society of Bengal*, 15: 170–203.

Cesalpino, A. (1583). *De plantis libri XVI.* George Marrefcottum, Florence.

Chakraborty, S. (1978). A new species of the genus *Crocidura* Wagler (Insectivora: Soricidae) from Wright Myo, S. Andaman Island, India. *Bulletin of the Zoological Survey of India*, 1: 303–304.

Chetry, D., U. Borthakur and R.K. Das. (2015). A short note on a first distribution record of White-cheeked Macaque *Macaca leucogenys* from India. *Asian Primates Journal*, 5(1): 45–47.

Choudhury, A.U. (2007). A new flying squirrel of the genus *Petaurista* Link from Arunachal Pradesh in north-east India. *The Newsletter & Journal of the Rhino Foundation for Nature in NE India*, 7: 26–34.

Choudhury, A.U. (2009). One more new flying squirrel of the genus *Petaurista* Link, 1795 from Arunachal Pradesh in north-east India. *The Newsletter & Journal of the Rhino Foundation for Nature in NE India*, 8: 26–34.

Choudhury, A.U. (2013a). Description of a new species of giant flying squirrel of the genus *Petaurista* Link, 1795 from Siang Basin, Arunachal Pradesh in North East India. *The Newsletter & Journal of the Rhino Foundation for Nature in NE India*, 9: 30–38.

Choudhury, A.U. (2013b). Description of a new subspecies of Hoolock gibbon *Hoolock hoolock* from North East India. *The Newsletter & Journal of the Rhino Foundation for Nature in NE India*, 9: 49–59.

Clerck, C. (1757). *Svenska Spindlar uti sina hufvud-slågter indelte samt under några och sextio särskildte arter beskrefne och med illuminerade figurer uplyste / Aranei Svecici, descriptionibus et figuris æneis illustrati, ad genera subalterna redacti, speciebus ultra LX determinati.* Laurentius Salvius, Stockholm. pp. 1–8, 1–154.

Coues, E., J. Allen, R. Ridgway, W. Brewster and H. Henshaw. (1886). *The Code of Nomenclature and Check-List of North American Birds adopted by the American Ornithologists Union.* American Ornithologists Union, New York.

Cretzschmar, P.J. (1826). Säugethiere. 1–78 pp. In: Rüppell, E. (Ed.), *Atlas zu der Reise im nördlichen Afrika.* No. 13. Gedruckt und in Commission bei Heinrich Ludwig Brönner, Frankfurt am Main.

Csorba, G. and P.J.J. Bates. (1995). A new subspecies of the Horseshoe bat *Rhinolophus macrotis* from Pakistan (Chiroptera, Rhinolophidae). *Acta Zoologica Academiae Scientiarum Hungaricae*, 41(3): 285–293.

Cuvier, F.G. (1822 '1823"). Examen des especes formation des genres ou sous-genres *Acanthion, Eréthizon, Sinéthère* et *Sphiggure. Mémoires du Muséum d'Histoire Naturelle (Paris)*, 9(1822): 413–484.

Cuvier, F.G. (1822). Du genre Paradxure et de deux espèces nouvelles qui s'y rapportent. *Mémoires du Muséum d'Histoire Naturelle Paris*, 9: 41–48.

Cuvier, F.G. (1823a). *Zoologie, Mammalogie.* In *Dictionnaire des Sciences Naturelles, dans lequel on traite méthodiquement des différens êtres de la nature, considérés soit en eux-mêmes, d'aprés l'état actuel de nos connoissances, soit relativement à l'utilité quén peuvent retirer la médecine, l'agriculture, le commerce et les arts. Vol. 26.* F.G. Levrault, Paris. 1–553 pp.

Cuvier, F.G. (1823b). *Zoologie, Mammalogie.* In *Dictionnaire des Sciences Naturelles, dans lequel on traite méthodiquement des différens êtres de la nature, considérés soit en eux-mêmes, d'aprés l'état actuel de nos connoissances, soit relativement à l'utilité quén peuvent retirer la médecine, l'agriculture, le commerce et les arts. Vol. 27.* F.G. Levrault, Paris. 1–555 pp.

Cuvier, G. [Baron] (1812). Fait à la classe des Sciences mathématiques et physiques, sur divers Cétacés pris sur les côtes de France, principalment sur ceux qui sont échoués près de Paimpol, le 7 Janvier 1812. *Annales du Muséum d'Histoire Naturelle* [Paris], 19: 1–16.

Cuvier, G. [Baron] (1823a). *Recherches sur les ossemens fossile, ou l'on rétablit les caractères de plusieurs espècies d'animaux dont les révolutions du globe ont detruit les espèces. T. 4.* Chez G. Dufour et E. D'Ocagne, Paris et chez les mémes, Amsterdam. 514 p.

Cuvier, G. [Baron] (1823b). *Recherches sur les ossemens fossile, ou l'on rétablit les caractères de plusieurs espècies d'animaux dont les révolutions du globe ont detruit les espèces. T. 5, 1er partie, Contenant les rongeirs, les édentés, et les mammifères marins.* C.C. Van der Hoek, Leiden et E. d'Ocagne, Paris. xxvii + 405 p.

Cuvier, G. [Baron] (1828). *Delphinus bredanensis.* 206–207 pp. In: Lesson, R.P. (Ed.), *Histoire naturelle générale et particulière des mammifères et des oiseaux découverts depuis 1788 jusqu'a nos jours—Cetaces.* Volume 1. Chez Baudouin frères, Paris. vii + 442 p.

Cuvier, G. [Baron] (1829). *Le règne animal distribué d'après son organisation: pour servir de base à l'Histoire naturelle des animaux. Nouvelle edition, Volume 2.* Chez Deterville, Paris. xv + 406 p.

Dall, W.H. (1877). *Nomenclature in Zoology and Botany: A Report to the American Association for the Advancement of Science.* Salem Press, Salem.

de Beaux, O. (1935). *Mustela kathiah caporiaccoi Atti della Società ligustica di scienze naturali e geografiche*, 14: 65.

de Roguin, L. (1988). Notes sur quelques mammiferes du Baluchistan iranien. *Revue suisse de zoologie*, 95(2): 595–606.

Deraniyagala, P.E.P. (1956). The Ceylon leopard, a distinct subspecies. *Spolia Zeylanica*, 28: 115–116.

Desmarest, A.-G. (1820). *Mammalogie, ou, Description des espèces de mammifères. Volume 1.* Chez Mme. Veuve Agasse, imprimeur-libraire, Paris. viii + 276 p.

Desmarest, A.-G. (1822). *Mammalogie, ou, Description des espèces de mammifères. Volume 2.* Chez Mme. Veuve Agasse, imprimeur-libraire, Paris. viii + 279 p.

Desmoulins, A. (1822). Baleine: 155–160, In: Audouin, Isid. Bourdon, Ad. Brongniart, De Candolle, d'Audebard de Férussac, A. Desmoulins, Drapiez, Edwards, Flurens, Geoffroy de Saint-Hilaire, A. De Jussieu, Kunth, G. De La Fosse, Lamouroux, Latreille, Lucas fils, -Duplessis, C. Prévost, A. Richard, Theibaut de Berneaud et Bory de Saint-Vincent. *Dictionaire classique d'Histoire naturelle, Volume 2.* Rey et Gravier, Paris, 621 p.

Dobson, G.E. (1871a). Description of four new species of Malayan Bats, from the collection of Dr. Stoliczka. *Journal of the Asiatic Society of Bengal*, 40(2): 260–267.

Dobson, G.E. (1871b). Notes on nine new species of Indian and Indo-Chinese Vespertilionidae, with remarks on the synonymy and classification of some other species of the same family. *Proceedings of the Asiatic Society of Bengal*, 1871: 210–215.

Dobson, G.E. (1871c). On a new genus and species of Rhinolophidæ, with description of a new species of Vesperus, and notes on some other species of insectivorous bats from Persia. *Journal of the Asiatic Society of Bengal*, 40(2): 455–461.

Dobson, G.E. (1871d). On some new species of Malayan bats from the collection of Dr. Stoliczka. *Proceedings of the Asiatic Society of Bengal*, 1871: 105–108.

Dobson, G.E. (1872a). Brief descriptions of five new species of Rhinolophine bats. *Journal of the Asiatic Society of Bengal*, 41(2): 336–338.

Dobson, G.E. (1872b). Notes on the Asiatic species of the genus *Taphozous*. *Proceedings of the Asiatic Society of Bengal*, 1872: 151–154.

Dobson, G.E. (1872c). Notes on some bats collected by Captain W.G. Murray, in the North-Western Himalaya, with description of a new species. *Proceedings of the Asiatic Society of Bengal*, 1872: 208–210.

Dobson, G.E. (1872d). In: Stoliczka, F. Notice of the Mammals and Birds inhabiting Kachh. *Journal of the Asiatic Society of Bengal*, 41(2): 211–258.

Dobson, G.E. (1873). On the genera *Murina* and *Harpyiocephalus* of Gray. *Proceedings of the Asiatic Society of Bengal*, 1873: 107–110.

Dobson, G.E. (1874a). Description of new species of Chiroptera from India and Yunan. *Journal of the Asiatic Society of Bengal*, 43(2): 237–238.

Dobson, G.E. (1874b). List of Chiroptera inhabiting the Khasia Hills, with description of a new species. *Journal of the Asiatic Society of Bengal*, 43(2): 234–236.

Dobson, G.E. (1874c). On the Asiatic species of Molossi. *Journal of the Asiatic Society of Bengal*, 43(2): 142–144.

Dobson, G.E. (1875). On the genus *Scotophilus*, with description of a new genus and species allied thereto. *Proceedings of the Zoological Society of London*, 1875: 368–373.

Dobson, G.E. (1876). *Monograph of the Asiatic Chiroptera, and Catalogue of the species of bats in the collection of the Indian Museum, Calcutta*. Taylor and Francis, London. 251 p.

Dobson, G.E. (1877). Notes on the collection of bats from India and Burma, with descriptions of new species. *Journal of the Asiatic Society of Bengal*, 46(2): 310–313.

Dobson, G.E. (1878). *Catalogue of the Chiroptera in the collection of the British Museum*. British Museum (Natural History), London. xlii + 567 + 30 plates.

Dobson, G.E. (1888). Description of two new species of Indian Soricidae. *Annals and Magazine of Natural History*, Ser 6, 1: 427–429.

Dufresne, C. (1797). Sur une nouvelle espece de singe. *Bulletin des Sciences, Par la Societe Philomathique, Paris*, Ser 1, 7: 49.

Ehrenberg, C.G. ('1832' 1833). *Delphinus aduncus*. In: Friderici Guilelmi Hemprich et Christiani Godofredi Ehrenberg, *Symbolae physicae, seu, Icones et descriptiones corporum naturalium novorum aut minus cognitorum: quae ex itineribus per Libyam Aegyptum Nubiam Dongalam Syriam Arabiam et Habessiniam publico institutis sumptu*, decas II. Berlin, Germany. folio k, ftn. 1.

Ellerman, J.R. (1947). Further notes on two little-known Indian Murine genera, and preliminary diagnosis of a new species of *Rattus* (Subgenus *Cremnomys*) from the Eastern Ghats. *Annals and Magazine of Natural History*, Ser 11, 13: 204–208.

Ellerman, J.R. and T.C.S. Morrison-Scott. (1951). *Checklist of Palaearctic and Indian mammals 1758 to 1946*. British Museum (Natural History), London. 810 p.

Ellerman, J.R. and T.C.S. Morrison-Scott. (1966). *Checklist of Palaearctic and Indian mammals 1758 to 1946*, 2nd ed. British Museum (Natural History), London. 810 p.

Elliot, W. (1839). A catalogue of species of Mammalia found in the Southern Mahratta country; with their synonymies in the native languages in use there. *Madras Journal of Literature and Science*, 10(25): 207–233.

Erxleben, J.C.P. (1777). *Systema regni animalis per classes, ordines, genera, species, varietates: cvm synonymia et historia animalivm: Classis I. Mammalia.* Impensis Weygandianis, Lipsiae.

Felten, H. (1977). In Felten, H., F. Spitzenberger and G. Storch. (1977). Zur Kleinsäugerfauna West-Anatoliens. Teil IIIa. *Senckenbergiana Biologica*, 58: 1–44.

Feng, Z. (1973). [A new species of *Ochotona* (Ochotonidae, Mammalia) from the Mount Jolmo-lungma area]. *Acta Zoologica Sinica*, 19: 69–75.

Feng, Z. and Y. Kao. (1974). [Taxonomic notes on Tibetan pika and allied species—including a new subspecies]. *Acta Zoologica Sinica*, 20: 76–87.

Fischer, G. (1814). *Zoognosia tabulis synopticis illustrata: In usum praelectionum Academiae imperialis medico-chirugicae mosquensis edita.* Volume 3. Quadrupedum reliquorum, Cetorum et Monotrymatum descriptionem continens. Typis Nicolai Sergeidis Vsevolozsky, Mosquae. Xxiv + 732 p.

Fischer de Waldheim, G. (1817). Adversaria zoologica, fasciculus primus. *Mémoires de la Société des Naturalistes de Moscou*, 5: 357–446.

Fischer, J.B. (1797). *Synopsis Mammalium. Addenda, Emendanda et Index.* Sumithus, J.G. Cottae, Stuttgardtiae. XLII + 752 p.

FitzGibbon, J. (1966). On a new species of *Salpingotus* (Dipodidae, Rodentia) from Northwestern Baluchistan. *Mammalia*, 30(3): 431–440.

Fraser, F.C. (1956). A new Sarawak dolphin. *Sarawak Museum Journal*, Ser 8, 7: 478–503.

Gaisler, J. (1970). The bats (Chiroptera) collected in Afghanistan by the Czechoslovak Expeditions of 1965–1967. *Acta Societatis Scientiarum Naturalium Moravo-Silesiacae (Brno, Československo)*, 4(6): 1–56.

Geoffroy Saint-Hilaire, É. (1803). *Catalogue des mammifères du Muséum National d'Histoire Naturelle.* Muséum National d'Histoire Naturelle, Paris. 272 p.

Geoffroy Saint-Hilaire, É. (1806). Mémoire sur le genre et les espèces de Vespertilion, l'un des genres de la famille des chauve-souris. *Annales du Muséum d'Histoire Naturelle [Paris]*, 8: 187–205.

Geoffroy Saint-Hilaire, É. (1810a). Description des roussettes et des céphalotes, deux nouveaux genres de la famille des Chauve-souris. *Annales du Muséum d'Histoire Naturelle [Paris]*, 15: 86–108.

Geoffroy Saint-Hilaire, É. (1810b). Sur les Phyllostomes et les Megadermes, deux genres de la famille des Chauve-souris. *Annales du Muséum d'Histoire Naturelle [Paris]*, 15: 157–198.

Geoffroy Saint-Hilaire, É. (1812). Suite au Tableau des Quadrummanes, ou des Animaux composant le premier Ordrea de la Class de mammiferes. *Annales du Muséum National d'Histoire Naturelle*, 19: 85–122, 156–170.

Geoffroy Saint-Hilaire, É. (1813). Sur un genre de Chauve-souris, sous le nom de Rhinolophes. *Annales du Muséum d'Histoire Naturelle [Paris]*, 20: 254–266.

Geoffroy Saint-Hilaire, É. (1818). Description des mammiferes qui se trouvent en Égypte. *Description de l'Égypte, ou, Recueil des observations et des recherches qui ont été faites en Égypte pendant l'expédition de l'armée française.Histoire Naturelle*, 2: 99–144.

Geoffroy Saint-Hilaire, É. and Cuvier, F. ('1824' 1825). *Histoire naturelle des mammifères: avec des figures originales, coloriées, dessinées d'aprèsdes animaux vivans, pt. 1–5.* Chez A. Belin, Paris.

Geoffroy Saint-Hilaire, I. (1826). Lutra perpicillata. p. 519. In Bory de Saint-Vincent (Ed.), *Dictionnaire Classique d'Histoire Naturelle, Paris, Volume 9.* Ray et Gravier, Paris. 596 p.

Geoffroy Saint-Hilaire, I. (1831). Mammifères. In: Bélanger, M.C. (ed.), *Voyage aux Indes-Orientales, Zoologie*. Bertrand, Paris.

Geoffroy Saint-Hilaire, I. (1832). Sur le genre Sciurus, et description de six nouvelles especes. *Magasin de Zoologie Journal*, 2: 5, pl. 4–6.

Geoffroy Saint-Hilaire, I. ('1842' 1843). Sur les singes de l'ancien monde, specialment sur les genres Gibbon et Semnopitheque. *Comptes rendus hebdomadaires des séances de l'Académie des sciences*, 1842(15): 716–720.

Geoffroy Saint-Hilaire, I. (1844). Description de la Marmotte a longue queue, Arctomys caudatus. pp. 66–68. In Jacquemont, V. (Ed.), *Voyage dans l'Inde, pendant les années 1828 à 1832*. Vol. 4, Pt. 4, Zoologique. Publié sous les auspices de M. Guizot, ministre de l'instruction publique, Paris. p. 183.

Ghose, R.K. (1965). A new species of Mongoose (Mammalia: Carnivora: Viverridae) from West Bengal, India. *Proceedings of Zoological Society of Calcutta*, 18(2): 174–178.

Gmelin, S.G. (1774). *Reise durch Russland. Volume 3*. St. Petersburg. 508 p. + 57 illus.

Gmelin, S.G. (1769 '1770'). *Erinaceus auritus. Novi commentarii Academiae Scientiarum Imperialis Petropolitanae*, 14: 519–524.

Goodwin, G.G. ('1938' 1939). Five new rodents from the Eastern Elburz Mountains and a new race of hare from Teheran. *American Museum Novitates*, 1050: 1–5.

Gray, J.E. and T. Hardwicke. (1830–1832). *Illustrations of Indian Zoology; chiefly collected from the collection of Major-General Hardwicke. Volume 1*. Treutel, Wurtz, Treutel, Jun. and Ricther, and Parbury, Allen and Co., London; Rue Bourbon, Paris; and Grande Ruse Strasburg. 100 plates.

Gray, J.E. and T. Hardwicke. (1833–1834). *Illustrations of Indian Zoology; chiefly collected from the collection of Major-General Hardwicke. Volume 2*. Adolphus Ritcher and Co. and Parbury, Allen and Co., London. 102 plates.

Gray, J.E. (1821). On the natural arrangement of vertebrose animals. *London Medical Repository*, 25: 296–310.

Gray, J.E. (1827). Synopsis of the species of the class Mammalia. pp. 1–391. In Griffith, E. and others, *The Animal Kingdom Arranged in Conformity with Its Organization, by the Baron (G) Cuvier, with Additional Descriptions by Edward Griffith and others*, Volume 5. George B. Whittaker, London.

Gray, J.E. (1828). *Spicilegia Zoologica or Original Figures and Short Systematic Descriptions of New and Unfigured Animals*. Treüttel, London. 12 p.

Gray, J.E. (1831a). Descriptions of some new genera and species of bats. *The Zoological Miscellany*, 1: 37–38.

Gray, J.E. (1831b). Proceedings of the Society—*Helictis moschata. Proceedings of the Committee of Science and Correspondence of the Zoological Society of London*, 1830–31: 94.

Gray, J.E. (1832). Proceedings of the Society—Specimens and drawings of numerous animals referable to the genus *Paradoxurus. Proceedings of the Committee of Science and Correspondence of the Zoological Society of London*, 1832: 63–68.

Gray, J.E. (1837a). Description of some new or little known Mammalia principally in the British Museum Collection. *Charlesworth's Magazine of Natural History*, 1: 578.

Gray, J.E. (1837b). Description of some new, or little known Mammalia, principally in the British Museum collection. *Charlesworth's Magazine of Natural History and Journal of Zoology, Botany, Mineralogy, Geology and Meteorology*, 1: 577–587.

Gray, J.E. (1838). A revision of the genera of bats (Vespertilionidae), and the description of some new genera and species. *Magazine of Zoology and Botany*, 2(12): 483–505.

Gray, J.E. (1842). Description of some new Genera and fifty unrecorded species of Mammalia. *Annals and Magazine of Natural History*, 10(65): 255–267.

Gray, J.E. (1843a). Description of some new genera and species of Mammalia in the British Museum Collection. *Annals and Magazine of Natural History*, 11: 117–119.

Gray, J.E. (1843b). *List of the Specimens of Mammalia in the Collection of the British Museum*. The Trustees, British Museum, London. 257 p.
Gray, J.E. (1846a). Mammalia. 1–53 pp. In: Richardson, J. and Gray, J.E. (Eds.), *The Zoology of the Voyage of the H.M.S. Erebus & Terror, under the Command of Captain Sir James Clark Ross, during the Years 1839 to 1843. Volume 1 Mammalia, Birds*. E.W. Janson, London. xii + 92 + 95 plates.
Gray, J.E. (1846b). New species of Mammalia. *Annals and Magazine of Natural History*, 18: 211–212.
Gray, J.E. ('1846' 1847). *Catalogue of the Specimens and Drawings of Mammalia and Birds of Nepal and Thibet*. Presented by the B.H. Hodgson, Esq. to the British Museum. British Museum, London. London. 156 p.
Gray, J.E. (1847). On the porcupines of the older or Eastern Continent, with descriptions of some new species. *Proceedings of the Zoological Society of London*, 1847(15): 97–104.
Gray, J.E. (1851). Notice of two Viverridae from Ceylon, lately living in the gardens. *Proceedings of the Zoological Society of London*, 1851(19): 131.
Gray, J.E. (1853). Observations on some rare Indian animals. *Proceedings of the Zoological Society of London*, 1853(21): 190–192.
Gray, J.E. (1863). Notice on the Chanco or Golden wolf (*Canis chanco*) from Chinese Tartary. *Proceedings of the Zoological Society of London*, 1863: 94.
Gray, J.E. (1870). *Catalogue of Monkeys, Lemurs, and Fruit-Eating Bats in the Collection of the British Museum*. British Museum, London. 168 p.
Gray, J.E. (1871). *Catalogue of the Monkeys, Lemurs, and Fruit-Eating Bats in the Collection of the British Museum*. viii + 137 p. British Museum, London.
Gray, J.E. (1875). *Feresa attenuate. Journal des Museum Godeffroy (Hamburg)*, Ser 3, 8: 184.
Griffith, E. (1821). *General and Particular Description of Vertebrated Animals, Arranged Conformably to the Modern Discoveries and Omprovement in Zoology. Order Carnivora*. Baldwin, Cradock and Joy, London. 143 p.
Groves, C. (1981). *Ancestors for the Pigs: Taxonomy and Phylogeny of the Genus Sus*. Research School of Pacific Studies, Australian National University, Canberra.
Groves, C. (1982). Geographic variation in the Barasingha or Swamp Deer (*Cervus duvauceli*). *Journal of the Bombay Natural History Society*, 79: 620–629.
Groves, C. and Meijaard, E. (2005). Intraspecific variation in *Moschiola*, the Indian Chevrotain. *The Raffles Bulletin of Zoology*. Supplement 12: 413–421.
Groves, C.P. (1967). Geographic variation in the hoolock or white-browed gibbon (*Hylobates hoolock* Harlan, 1834). *Folia Primatologica*, 7: 276–283.
Groves, C.P. (2003). Taxonomy of ungulates of the Indian subcontinent. *Journal of the Bombay Natural History Society*, 100: 341–362.
Groves, C.P., C. Rajapaksha and K. Manemandra-Arachchi. (2009). The taxonomy of the endemic golden palm civet of Sri Lanka. *Zoological Journal of the Linnean Society*, 155: 238–251. With 13 figures.
Grubb, P. (1982). The systematics of Sino-Himalayan musk deer (*Moschus*), with particular reference to the species described by BH Hodgson. *Säugetierkundliche Mitteilungen*, 30: 127–135.
Gruber, U.F. (1969). Tiergeographische, ökologische und bionomische Untersuchungen an kleinen Säugetieren in Ost-Nepal. In: Hellmich, W. (Ed.), *Khumbu Himalaya Band 3, Lieferung 2*. Universitatsverlag Wagner, Innsbruck, pp. 197–312.
Guldenstaedt, A.I. (1770). *Peregusna, nova Mustela species. Novi commentarii Academiae Scientiarum Imperialis Petropolitanae*, 14: 441–455.
Guldenstaedt, A.L. (1778). *Antilope subgutturosa. Acta Academiae Scientiarum Imperialis Petropolitanae*, 1778(1): 251–274.
Günther, A.L. (1875). Descriptions of some leporine mammals from Central Asia. *Annals and Magazine of Natural History*, Ser 4, 16: 228–231.

Hardwicke, T. (1807). Description of a species of Jerboa, found in the upper Provinces of Hindustan, between Benares and Hurdwar. *Transactions of the Linnean Society of London*, 8: 279–281.

Hardwicke, T. (1825a). Description of a new species of tailed bat (*Taphosous* of Geoff.) found in Calcutta. *Transactions of the Linnean Society of London*, 14: 525–526.

Hardwicke, T. (1825b). Descriptions of two species of Antelope from India. *Transactions of the Linnean Society of London*, 14: 518–524.

Harlan, R. (1834). Description of a species of Orang from the northeastern province of British East India, lately the kingdom of Assam. *Transactions of the American Philosophical Society*, 4: 52–59.

Hemmer, H. (1974). [Studies on the systematics and biology of the sand cat.] *Zeitschrift des Kölner Zoo*, 17(1): 11–20 (in German).

Hemprich, W. (1820). *Grundriss der Naturgeschichte für höhere Lehranstalten Entworfen von Dr. W. Hemprich.* August Rucker, Berlin.

Heptner, V.G. (1937). Notes on Gerbollidae (Mammalia, Glires). IX. Remarks on a new species of *Meriones* from Turkestan and on the systematic position of the gerbils belonging to the *Meriones persicus*-Group. *Bulletin de la Société des naturalistes de Moscou, Section Biologique*, 46(4): 189–190 (in Russian), 191–193 (in English).

Hermann. (1780). In: Zimmermann, E.A.W. (Ed.), *Geographische Geschichte des Menschen. Vierf. Thiere, 2.* Weygandschen, Leipzig. 382 p.

Heude, P.-M. (1901). *Mémoires concernant l'histoire naturelle de l'empire chinois par des pères de la Compagnie de Jésus. Volume 5, Part 1.* Mision Catholique, Chang-Hai.

Hilzheimer, M. (1905). Über einige Tigerschädel aus der Straßburger zoologischen Sammlung. *Zoologischer Anzeiger*, 28: 594–599.

Hinton, M.A.C. (1918). Scientific results from the Mammal Survey, No. 18: Report on the House Rats of India, Burma and Ceylon. *Journal of the Bombay Natural History Society*, 26(1): 59–88.

Hinton, M.A.C. (1923). Scientific results from the Mammal Survey, No. 36. On the Capped Langurs (*Pithecus pileatus*, Blyth, and its allies). *Journal of the Bombay Natural History Society*, 29(1): 77–83.

Hodgson, B.H. (1831). On the bubaline Antilope, Nobis. *Gleanings in Science*, 3: 122–123, 324.

Hodgson, B.H. (1833a). Descriptions of the wild dog of the Himalaya. *Asiatic Researches, or, Transactions of the Society instituted in Bengal for Inquiring into the History and Antiquities, the Arts, Sciences and Literature of Asia*, 18(2): 221–237.

Hodgson, B.H. (1833b). The wild goat, and the wild sheep, of Nepal. *Asiatic Researches, or, Transactions of the Society Instituted in Bengal for Inquiring into the History and Antiquities, the Arts, Sciences and Literature of Asia*, 18(2): 129–138.

Hodgson, B.H. (1835a). Description of the little Musteline animal, denominated Kathiah Nyul in the Catalogue of the Nepalese Mammalia. *Journal of the Asiatic Society of Bengal*, 4: 702–704.

Hodgson, B.H. (1835b). Synopsis of the Vespertilionidae of Nipal. *Journal of the Asiatic Society of Bengal*, 4: 699–701.

Hodgson, B.H. (1836a). Indication of a new genus of the Carnivora, with a description of the species on which it is founded. *Asiatic Researches, or, Transactions of the Society Instituted in Bengal for Inquiring into the History and Antiquities, the Arts, Sciences and Literature of Asia*, 19: 61.

Hodgson, B.H. (1836b). Synoptical description of sundry new animals, enumerated in the Catalogue of Nipalese Mammals. *Journal of the Asiatic Society of Bengal*, 5(52): 231–238.

Hodgson, B.H. (1837). On a new genus of the Plantigrades. *Journal of the Asiatic Society of Bengal*, 6: 560–564.

Hodgson, B.H. (1839a). On three new species of Musk (*Moschus*) inhabiting the Hemalayan districts. *Journal of the Asiatic Society of Bengal*, 8: 202–203.
Hodgson, B.H. (1839b). Summary description of four new species of Otter. *Journal of the Asiatic Society of Bengal*, 8: 319–320.
Hodgson, B.H. (1840a). On the common hare of Gengetic Provinces, and of the Sub-Hemalaya; with a slight notice of a strictly Hemalayan species. *Journal of the Asiatic Society of Bengal*, 9: 1183–1186.
Hodgson, B.H. (1840b). Three new species of monkey; with remarks on the genera *Semnopithecus* et *Macacus*. *Journal of the Asiatic Society of Bengal*, 9(2): 1211–1213.
Hodgson, B.H. (1841a). Classified catalogue of mammals of Nepal, corrected to the end of 1840, first printed in 1832. *Calcutta Journal of Natural History*, 2: 212–221.
Hodgson, B.H. (1841b). New species of *Rhizomys* discovered in Nepal. *Calcutta Journal of Natural History*, 2: 60–62.
Hodgson, B.H. (1841c). Notice of the Marmot of the Himalaya and of Tibet. *Journal of the Asiatic Society of Bengal*, 10(2): 777–778.
Hodgson, B.H. (1841d). Of a new species of *Lagomys* inhabiting Nepal, (with plate), *Lagomys Nepalensis* nob. *Journal of the Asiatic Society of Bengal*, 10(2): 854–855.
Hodgson, B.H. (1841e). On a new species of *Prionodon*, *Prionodon pardicolor* nobis. *Calcutta Journal of Natural History*, 2: 57–60.
Hodgson, B.H. (1842). Notice of the Mammals of Tibet, with descriptions and plates of some new species. *Journal of the Asiatic Society of Bengal*, 11: 275–289.
Hodgson, B.H. (1843). Notices on two marmots inhabiting respectively the plains of Tibet and the Himalayan slopes near to the snows, and also of a *Rhinolophus* of the central region of Nepal. *Journal of the Asiatic Society of Bengal*, 12: 409–414.
Hodgson, B.H. (1844). Classified catalogue of mammals of Nepal, corrected to the end of 1841, first printed in 1832. *Calcutta Journal of Natural History*, 4: 284–294.
Hodgson, B.H. (1845). On the rats, mice and shrews of the Central region of Nepal. *Annals and Magazine of Natural History*, Ser 1, 15: 266–270.
Hodgson, B.H. (1847a). Description of the wild ass and wolf of Tibet, with illustrations. *Calcutta Journal of Natural History*, 7: 469–477.
Hodgson, B.H. (1847b). On a new form of the Hog kind or Suidae. *Journal of the Asiatic Society of Bengal*, 16(1): 423–428.
Hodgson, B.H. (1847c). On a new species of *Plecotus*. *Journal of the Asiatic Society of Bengal*, 16: 894–896.
Hodgson, B.H. (1847d). The slaty blue Megaderme, *Megaderma schistacea*, N.S. *Journal of the Asiatic Society of Bengal*, 16: 889–894.
Hodgson, B.H. ('1847' 1848). On the Four-horned Antelopes of India. *Calcutta Journal of Natural History*, 8: 87–94.
Hodgson, B.H. (1850). On the Takin of Eastern Himalaya: *Budorcas Taxicolor* mihi. *Journal of the Asiatic Society of Bengal*, 19: 65–74.
Hodgson, B.H. ('1857' 1858). On a new *Lagomys* and a new *Mustela* inhabiting north region of Sikim and proximate parts of Tibet. *Journal of the Asiatic Society of Bengal*, 26: 207–208.
Holden, M.E. (1996). Description of a new species of *Dryomys* (Rodentia, Myoxidae) from Balochistan, Pakistan, including morphological comparisons with *Dryomys laniger* Felten and Storch, 1968, and *D. nitedula* (Pallas, 1778). *Bonner Zoologische Beiträge*, 46(1–4): 111–131.
Horsfield, T. ('1823' 1824). *Zoological Researches in Java, and the Neighbouring Islands*. Kingsbury, Parbury and Allen, London. 47 p.
Horsfield, T. ('1826' 1827). Notice of a species of *Ursus* from Nepaul. *Transactions of the Linnean Society of London, Zoology,* 15: 332–334.

Horsfield, T. (1831). On two species of bats, accompanying a large collection of birds from Madras, presented by J.M. Heath Esq. *Proceedings of the Committee of Science and Correspondence of the Zoological Society of London*, 1831(1): 113–125.

Horsfield, T. (1839). Communication to the society of Mr. McClelland's list of mammalia and birds collected in Assam. *Proceedings of the Zoological Society of London*, Part VII: 146–167.

Horsfield, T. (1851). *A Catalogue of the Mammalia in the Museum of the Hon. East-India Company.* J & H Cox, London. vi + 212 p.

Horsfield, T. (1855). Brief notices of several new or little-known species of Mammalia, lately discovered and collected in Nepal, by Brian Houghton Hodgson, Esq. *Annals and Magazine of Natural History*, (2)16: 101–114.

Hutton, T. (1842). "Markhore" or the "Snake-Eater" of the Afghans—Wild goat of Afghanistan. *Calcutta Journal of Natural History*, 2: 535–542.

Ichihara, T. (1966). The pygmy blue whale, *Balaenoptera musculus brevicauda*, a new subspecies from the Antarctic. 79–113 pp. In: Norris, K.S. (Ed.), *Whales, Dolphins and Porpoises*, University of California Press, Los Angeles. ill + 789 pp.

ICZN (International Commission on Zoological Nomenclature). (1905). *Règles Internationales de la Nomenclature Zoologique; International Rules of Zoological Nomenclature; Internationale Regeln der Zoologischen Nomenklatur.* F. R. de Rudeval, Paris.

ICZN (International Commission on Zoological Nomenclature). (1999). *International Code of Zoological Nomenclature. 4th ed.* The International Trust for Zoological Nomenclature, London.

Illiger, C. (1811). *Prodromus systematis mammalium et avium additis terminis zoographicis utriusque classis, eorumque versione germanica.* Sumptibus C. Salfeld, Berolini [Berlin]: [I]–XVIII, [1]–301, *Errata et Omissa.*

Illiger, J.K.W. (1815). Überblick der Säugthiere nach ihrer Vertheilung über die Welttheile. *Abhandlungen der Königlichen Akademie der Wissenschaften in Berlin*, 1804–1811: 39 –159.

Jefferson, T.A. and K.V. Waerebeek. (2002). The taxonomic status of the nominal dolphin species *Delphinus tropicalis* Van Bree, 1971. *Marine Mammal Science*, 18: 787–818.

Jerdon, T.C. (1847). In Blyth, E. Report of the Curator of the Zoological Department. *Journal of the Asiatic Society of Bengal*, V. 16, pt. 2, (July, 1847): 863–880.

Jerdon, T.C. (1867). *The Mammals of India.* Roorkee, India.

Johnsingh, A.J.T. and N. Manjrekar (Eds.) (2013). *Mammals of South Asia, Volume I.* University Press, Hyderabad, India. lxviii + 614 p.

Kastchenko, N.F. and M.P. Akimov. (1917). *Rhinolophus bocharicus* sp. n. *Annuaire du Musee Zoologique de l'Academie d. Sciences de St. Petersbourg*, 22: 221–223.

Kelaart, E.F. (1850a). Description of additional mammals. *Journal of the Ceylon Branch of the Royal Asiatic Society*, 2(5): 329–330 (1887 Reprint).

Kelaart, E.F. (1850b). Description of new species and varieties of mammals found in Ceylon. *Journal of the Ceylon Branch of the Royal Asiatic Society*, 2(5): 321–328.

Kelaart, E.F. (1852). *Prodromus Faunæ Zeylanicæ: Being Contributions to the Zoology of Ceylon. Part 1, Mammals of Ceylon.* 1–90 pp. Printed for the Author, Ceylon. xxxiii + 392 p.

Kerr, R. (1792). *The Animal Kingdom or Zoological System, of the celebrated Sir Charles Linnaeus. Class I. Mammalia: Containing a Complete Systematic Description, Arrangement, and Nomenclature, of All the Known Species and Varieties of the Mammalia or Animals Which Give Suck to Their Young, Being a Translation of that Part of the Systema Naturae as Lately Published with Great Improvements by Professor Gmelin of Gottingen. Together with Numerous More Additions from More Recent Zoological Writers and Illustrated with Copperplates.* A. Strahan and T. Cadell, London, and W. Creech, Edinburgh. xii + 644 p.

Khajuria, H. (1956). A new langur (Primates: Colobidae) from Goalpara district, Assam. *Annals and Magazine of Natural History*, 12(9): 86–88.

Khajuria, H. (1970). A new leaf-nosed bat from central India. *Mammalia*, 34(4): 622–627.

Kloss, C.B. (1915). Notes on some hares in the Indian Museum with descriptions of two new forms. *Records of the Indian Museum*, 15(2): 89–96.

Kock, D. and H.R. Bhat. (1994). *Hipposideros hypophyllus* n. sp. of the *H. bicolor*-group from peninsular India (Mammalia: Chiroptera: Hipposideridae). *Senckenbergiana Biologica*, 73: 25–31.

Kuhl, H. (1817). *Die Deutschen Fledermause*. Universitätsbibliothek Johann Christian Senckenberg, Hanau. 67 p.

Kuzyakin, A.P. (1950). *Letuchieye myschi* [Bats]. Sovetskaya nauka, Moscow. 444 p. [in Russian].

Lacépède, B.-G.-É. de La. (1804). *Histoire naturelle des cétacées*. Plassan, Paris. xliv + 329 p.

Lacepede, M. le comte de. (1800). *Séances des écoles normales: recueillies par des sténographes, et revues par les professeurs. [Leçons], Volume 8*. École normale supérieure, France.

Leach, W.E. (1822). The characters of three new genera of bats with foliaceous appendages to the nose. *Transactions of the Linnean Society of London*, 13: 69–72.

Lesson, R.-P. (1827). *Manuel De Mammalogie, ou Histoire naturelle des mammifères*. Roret, Paris. 440 p.

Lesson, R.P. (1838). *Complément de Buffon. Deuxieme edition. Races humaines et mammifeŕes*. P. Pourrat Freŕes, E'diteus, Paris, France.

Li, C. (1981). On a new species of Musk Deer from China. *Zoological Research*, 2: 157–161.

Li, C., C. Zhao and P.-F. Fan. (2015). White-cheeked macaque (*Macaca leucogenys*): A new macaque species from Medog, southeastern Tibet. *American Journal of Primatology*, 77: 753–766.

Lichtenstein, H. (1823a). *Verzeichniss der Doubletten des zoologischen Museums der Königl. Universität zu Berlin nebst Beschreibung vieler bisher unbekannter Arten von Säugethieren, Vögeln, Amphibien und Fischen*. T. Trautwein, Berlin. X + 118 p.

Lichtenstein, H. (1823b). Wortverzeichnifs aus der Afghanischen Sprache begleitet von einem naturhistorichen Anhange und einer Vorrede, Säugethiere. 118–125. In: Eversmann, E. (Ed.), *Reise von Orenburg nach Buchara*. Im Verlage vo E. H. G. Christiani, Berlin.

Lichtenstein, H. (1828). Über de Springmäuse order de arten der gattung *Dipus*. *Abhandlungen der Königlichen Akademie der Wissenschaften zu Berlin*, 1825: 133–161, X Tab.

Linnaeus, C. (1758). *Systema naturae per regna tria naturae, secundum classes, ordines, genera, species, cum characteribus, differentiis, synonymis, locis, Vol. 1: Regnum animale. Editio decima, reformata*. Laurentii Salvii, Stockholm, Sweden.

Linnaeus, C. (1766). *Systema naturæ per regna tria naturæ, secundum classes, ordines, genera, species, cum characteribus, differentiis, synonymis, locis. Tomus I. Editio duodecima, reformata*. 1–532. Holmiæ (Salvius).

Linnaeus, C. (1768). *Systema naturæ per regna tria naturæ, secundum classes, ordines, genera, species, cum characteribus & differentiis. Tomus III*. 1–236, [1–20] pp + Tab. I–III. Holmiæ (Salvius).

Linnaeus, C. (1771). Regni animalis. *In*: *Mantissa plantarum altera generum editionis VI & specierum editionis II*. Holmiae, Laurentius Salvius.

Loche, G.G. (1858). Description d'une nouvelle espece de Chat. *Revue et Magasin de Zoologie Pure et Appliquée, Paris*, (2)10: 49–50.

Longman, H.A. (1926). New records of Cetacea, with a list of Queensland species. *Memoirs of the Queensland Museum*, 8(3):266–278. pl. 43.

Lydekker, R. (1905). The Gorals of India and Burma. *The Zoologist*, Ser. 4, 9(765): 81–84.

Lydekker, R. (1907). The Bhutan takin. *Field, London*, 110: 887.

Lydekker, R. (1913). *Catalogue of the ungulate mammals in the British Museum (Natural History). Volume 1, Artiodactyla, Family Bovidae, Subfamilies Bovinae to Ovibovinae.* The Trustees BMNH, London. 249 p.

Lydekker, R. (1914). *Catalogue of the ungulate mammals in the British Museum (Natural History). Volume 2, Artiodactyla, Family Bovidae, Subfamilies Bubalinae to Reduncinae.* The Trustees BMNH, London. 295 p.

Lydekker, R. (1915). *Catalogue of the ungulate mammals in the British Museum (Natural History). Volume 4, Artiodactyla, Families Cervidae, Tragulidae, Camelidae, Suidae and Hippopotamidae.* The Trustees BMNH, London. 438 p.

Lydekker, R. (1916). *Catalogue of the ungulate mammals in the British Museum (Natural History). Volume 5, Perissodactyla, Hyracoidea and Proboscidae.* The Trustees BMNH, London. 207 p.

Lydekker, R. and G. Blaine. (1914). *Catalogue of the ungulate mammals in the British Museum (Natural History). Volume 3, Artiodactyla, Family Bovidae, Subfamilies Aepycerotinae to Tragelaphinae, Antilocapridae and Giraffidae.* The Trustees BMNH, London. 283 p.

Maeda, K. (1982). Studies on the classification of *Miniopterus* in Eurasia, Australia and Melanesia. *Honyurui Kagaku (Mamm. Sci.), Supple., No. 1*: 1–176.

Martin, W. (1837). Description of the new species of the genus *Felis*. *Proceedings of the Zoological Society of London*, 1836: 107–108.

Mason, G.E. (1908). On the fruit bats of the genus *Pteropus* inhabiting the Andaman and Nicobar Archipelagos, with a description of a new species. *Records of the Indian Museum*, 2(2): 159–166.

Matschie, P. (1901). Ueber rumänische Säugethiere. *Sitzungsberichte der Gesellschaft Naturforschender Freunde zu Berlin*, 1901: 220–238.

Matschie, P. (1912). Über einige Rassen des Steppenluchses *Felis* (*Caracal*) *caracal* (st. Müll.). *Sitzungsberichte der Gesellschaft naturforschender Freunde zu Berlin*, 1912: 55–67.

Mayen, F.J.F. (1833). Beiträge zue Zoologie, gesammelt auf einer reise um die erde. *Nova acta physico-medica Academiae Caesareae Leopoldino-Carolinae Naturae Curiosum*, 16(2): 549–610.

Mazák, V. (1968). Nouvelle sous-espèce de tigre provenant de l'Asie du sud-est. *Mammalia*, 32(1): 104.

McClelland, J. (1839). In: Horsfield, T. Communication to the society of Mr. McClelland's list of mammalia and birds collected in Assam. *Proceedings of the Zoological Society of London*, 1839(7): 146–167.

McClelland, J. (1842). Footnote on "Further notice of a Nondescript Deer" by Percy Eld. *Calcutta Journal of Natural History*, 2: 417.

Meegaskumbura, S.H., M. Meegaskumbura, K. Manamendra-Arachchi, R. Pethiyagoda and C.J. Schneider. (2007). *Crocidura hikmiya*, a new shrew (Mammalia: Soricomorpha: Soricidae) from Sri Lanka. *Zootaxa*, 1665: 19–30.

Meyer, F.A.A. (1794). Über de la Metheries schwarzen Panther. 394–396 pp. *Zoologische Annalen, Band I.* Im Verlage des Industrie-Comptoirs, Weimar.

Meyer, J.N. (1826). *Dissertatio inauguralis anatomico-medica de genere felium.* Doctoral thesis, University of Vienna, Austria.

Miller, G.S. Jr. (1898). List of bats collected by Dr. W.L. Abbott in Siam. *Proceedings of the Academy of Natural Sciences of Philadelphia*, 50: 316–325.

Miller, G.S. Jr. (1902). The mammals of the Andaman and Nicobar Islands. *Proceedings of the United States National Museum*, 24: 751–795.

Miller, G.S. Jr. (1903). Seventy new Malayan mammals. *Smithsonian Miscellaneous Collections*, 45: 1–73.

Miller, G.S. Jr. (1911). Two new shrews from Kashmir. *Proceedings of the Biological Society of Washington*, 24: 241–242.

Miller, G.S. Jr. (1913a). A new shrew from Baltistan. *Proceedings of the Biological Society of Washington*, 26: 113–114.

Miller, G.S. Jr. (1913b). Some overlooked names of Sicilian Mammals. *Proceedings of the Biological Society of Washington*, 26: 80–81.

Milne-Edwards, A. (1867). Description de quelques especes nouvelles d'Ecureuils de l'ancien continent. *Revue et magasin de zoologie pure et appliquée*, Ser. 2(19): 225–234.

Milne-Edwards, A. (1868–1874). Des observations sur L'Hippopotame de Liberia et Des etudes sur la faunae de la Chine et du Tibet oriental. In: Milne-Edwards, M.H. (Ed.), *Recherches pour servir à l'histoire naturelle des mammifères: comprenant des considérations sur la classification de ces Animaux, Volume 1 Text and Volume 2 Atlas.* G. Masson, Editor, Libriare de L'Academie de Medecine, Paris. 394 p and 105 plates.

Milne-Edwards, A. (1870). Note sur quelques Mammiferes du Thibet oriental. *Comptes rendus hebdomadaires des séances de l'Académie des sciences*, 70: 341–342.

Milne-Edwards, A. ('1871' 1872). *Arvicola melanogaster* 93 p (footnote), In: David, M.A. (Ed.), *des oiseaux de Chine observes dans la partie spetentrionale de l'empire* (*Au nord du fleuve-bleu*). De 1862 A 1870. *Bulletin des Nouvelles archives du Muséum*, vol. 7: 1–110.

Mishra, A.C. and V. Dhanda. (1975). Review of the Genus *Millardia* (Rodentia: Muridae), with description of a new species. *Journal of Mammalogy*, 56(1): 76–80.

Montagu, G. (1821). Description of a species of *Delphinus*, which appears to be new. *Memoirs of the Wernerian Natural History Society, Edinburgh*, 3: 75–82.

Moorcroft, W. (1841). *Travels in the Himalayan provinces of Hindustan and the Panjab; in Ladakh and Kashmir; in Peshawar, Kabul, Kunduz, and Bokhara. Volume 1 and 2.* lvi + 459 and viii + 508 + 12 pp. J. Murray, London.

Moore, J.C. (1959). Relationships among the living squirrels of the Sciurinae. *Bulletin of the American Museum of Natural History*, 118(4): 157–206.

Müller, O.F. (1776). *Zoologiæ Danicæ prodromus, seu animalium Daniæ et Norvegiæ indigenarum characteres, nomina, et synonyma imprimis popularium.* Havniæ, Hallager. XXXII + 282 p.

Müller, S. (1838). Over eenige nieuwe zoogdieren van Borneo. *Tijdschrift voor natuurlijke geschiedenis en physiologie*, 5(1): 134–150.

Muller, S. (1839). In Temminck, C.J. (Ed.), *Verhandelingen over de Natuurlijke Geschiedenis der Nederlandsche Overzeesche Bezittingen. Vol. 1 Zoologie.* S. and J. Luchtmans en C.C. van der Hoek, Leiden.

Murray, J.A. (1884). Additions to the present knowledge of the vertebrate zoology of Persia. *Annals and Magazine of Natural History*, Ser 5, 14: 97–106.

Murray, J.A. (1885). Description of a new species of Mus from Sind. *Proceedings of the Zoological Society of London*, 1885: 809–810.

Murray, J.A. (1886). Description of a new *Gerbillus* from Sind. *Annals and Magazine of Natural History*, Ser 5, 17: 246–248.

Nishiwaki, M. and T. Kamiya. (1958). A beaked whale *Mesoplodon* stranded at Oiso Beach, Japan. *The Scientific Reports of the Whales Research institute*, 13: 53–83.

Oebeck, H.P. (1765). *Reise nach Ostindien und China.* ins Deutsche übersetzt von J. G. Georgi. Koppe, Rostock.

Ogilby, W.M. (1838). Proceedings of the Asiatic Society: Report for August, 1837 Meeting. *Proceedings of the Zoological Society of London*, 1837: 81.

Ogilby, W.M. (1839). Memoir on the Mammalogy of the Himalayas. pp. lvi–lxxiv. In: Royle, J.F. (Ed.), *Illustrations of the Botany and Other Branches of the Natural History of the Himalayan Mountains and Flora of Cashmere. Volume 1 Text and Volume 2 Plates.* Wm. H. Allen and Co., London. Vol. 1 lxxvii + 472 p. and Vol. 2 97 plates.

Ognev, S.I. (1928). *Mammals of Eastern Europe and Northern Asia, Vol. I. Insectivora and Chiroptera*. Izdatel'stvo Akademi Nauk SSSR, Moscow-Leningrad.

Owen, R. (1846). *A history of British Fossil Mammals, and Birds: Illustrated by 237 Woodcuts*. J. Van Voorst, London. xlvi + 560 p.

Owen, R. (1853). *Descriptive Catalogue of the Osteological Series Contained in the Museum of the Royal College of Surgeons of England, Volume 2, Mammalia Placentalia*. Taylor & Francis, London. 563 p.

Owen, R. (1866). On some Indian Cetacea collected by Walter Elliot, Esq. *Transactions of the Zoological Society of London*, 6(1): 17–47.

Pallas, P.S. (1766). *Miscellanea zoologica: quibus novae imprimis atque obscurae animalium species describuntur et observationibus iconibusque illustrantur.* Apud Petrum van Cleef, Hagae Comitum. 224 p.

Pallas, P.S. (1767). *Spicilegia zoologica quibus novae imprimis et obscurae animalium species iconibus, descriptionibus atque commentariis illustrantur*, 3: 35 p.

Pallas, P.S. (1769 '1770'). *Mus talpinus. Novi commentarii Academiae Scientiarum Imperialis Petropolitanae*, 14(1): 568–573.

Pallas, P.S. (1771). *Reise durch verschiedene Provinzen des rußischen Reichs, Volume 1.* Kayserliche Akad. der Wiss., St. Petersburg.

Pallas, P.S. (1773). *Reise durch verschiedene Provinzen des rußischen Reichs, Volume 2.* Kayserliche Akad. der Wiss., St. Petersburg.

Pallas, P.S. (1774 '1775'). *Equus hemionus. Novi commentarii Academiae Scientiarum Imperialis Petropolitanae*, 19: 394–417.

Pallas, P.S. (1776a). Descriptio Ibicis sibirici. *Spicilegia zoologica: quibus novae imprimis et obscurae animalium species iconibus, descriptionibus atque commentariis illustrantur*, 11: 52–57.

Pallas, P.S. (1776b). *Miscellanea zoologica: quibus novae imprimis atque obscurae animalium species describuntur et observationibus iconibusque illustrantur.* Apud Petrum van Cleef, Hagae Comitum. 284 pp.

Pallas, P.S. (1776c). *Reise durch verschiedene Provinzen des rußischen Reichs, Volume 3.* Kayserliche Akad. der Wiss., St. Petersburg.

Pallas, P.S. (1777). In: Schreber, J.C.D. (1777). *Die Säugethiere in Abbildungen nach der Natur, mit Beschreibungen. 1776–1778. Wolfgang Walther, Erlangen.*

Pallas, P.S. (1778). *Novae species qvadrvpedvn e Glirivm ordine, cvm illvstrationibvs variis complvrivm ex hoc ordine Animalium.* Erlange, SVMTV, Wolfgangi Waltheri. Viii, 388, XXVII Tab.

Pallas, P.S. (1779). *Novae species quadrupedum e glirum ordine cum illustrationibus variis complurium ex hoc ordine animalium. Fasciculi I, II.* Academia Petropolitana, Erlangae [= Erlangen], 388 p.

Pallas, P.S. (1811) [1831]. *Zoographia Rosso-Asiatica: sistens omnium animalium in extenso Imperio Rossico, et adjacentibus maribus observatorum recensionem, domicilia, mores et descriptiones, anatomen atque icones plurimorum. 3 Vols.* In officina Caes, Acadamiae Scientiarum Impress, Petropoli. (Vol. 1 & 2, 1811; Vol. 3, 1814).

Peale, T.R. (1848). Mammalia and Ornithology, p. 47. In: Wilkes, C. (Ed.), *United States Exploring Expedition during the years 1838, 1839, 1840, 1841, 1842, under the command of Charles Wilkes, U.S.N. Vol. 8.* C. Sherman, Philadelphia. xxv + (17)–338 p.

Pearson, J.T. (1832). Proceedings of Societies: A stuffed specimen of a species of *Felis*. *Journal of the Asiatic Society of Bengal*, 1: 75.

Pearson, J.T. (1836). On the *Canis vulpes montana* or Hill Fox. *Journal of the Asiatic Society of Bengal*, 5; 313–314.

Pearson, J.T. (1839). List of mammalia and birds collected in Assam by John McClelland. *Proceedings of the Zoological Society of London*, 1839(7): 146–167.

Peters, H.W. ('1859' 1860). Neue beiträge zur kenntniss der chiropteren. *Monatsberichte der Königlichen Preussische Akademie des Wissenschaften zu Berlin*, 1859: 222–229.

Peters, H.W. ('1872' 1873). Über neue Flederthiere (*Phyllorhina micropus, Harpyiocephalus Huttonii, Murina grisea, Vesperugo micropus, Vesperus* (*Marsipolaemus*) *albigularis, Vesperus propinquus, tenuipinnis*). *Monatsberichte der Königlichen Preussische Akademie des Wissenschaften zu Berlin*, 1872: 256–264.

Peters, W. ('1866' 1867). Über einige neue oder weniger bekannte Flederthiere. *Monatsberichte der Königlichen Preussische Akademie des Wissenschaften zu Berlin*, 1866: 16–25.

Peters, W. ('1869' 1870). Las bemerkungen über neue oder weniger bekannte Flederthiere, besonders des Pariser Museums. *Monatsberichte der Königlichen Preussische Akademie des Wissenschaften zu Berlin*, 1869: 391–406.

Peters, W. (1871). In: Swinhoe, R. (Ed.), Catalogue of mammals of China (South of river Yangtsze) and of the island of Formosa. *Proceedings of the Zoological Society of London*, 1870: 615–652.

Petter, F. (1963). Un nouvel insectivore du nord de l'Assam: *Anourosorex squamipes schmidi* nov. sbsp. *Mammalia*, 27: 444–445.

Phillips, W.W.A. (1928). Ceylon shrews. *Spolia Zeylanica*, 14(2): 295–331.

Philips, W.W.A. (1929a). Two new rodents from the highlands of Ceylon. *Spolia Zeylanica*, 15: 165–168.

Phillips, W.W.A. (1929b). New and rare Ceylon shrews. *Spolia Zeylanica*, 15: 113–118.

Phillips, W.W.A. (1932a). Additions to the fauna of Ceylon. No. 1. Two new rodents from the hills of Central Ceylon. *Spolia Zeylanica*, 16(3): 323–327.

Phillips, W.W.A. (1932b). Additions to the fauna of Ceylon. No. 3. A new pigmy shrew from the mountains of Ceylon. *Spolia Zeylanica*, 17(2): 123–126.

Phillips, W.W.A. (1932c). Additions to the fauna of Ceylon. No.2. Some new and interesting bats, from the hills of the Central Province. *Spolia Zeylanica*, 16(3): 329–335, pl. 66.

Phillips, W.W.A. (1932d). Survey of the distribution of mammals in Ceylon. *Spolia Zeylanica*, 16(3): 337–351.

Pocock, R.I. ('1932' 1933a). The Black and Brown Bears of Europe and Asia. Part I – European and Asiatic representatives of the Brown Bear (*Ursus arctos*). *Journal of the Bombay Natural History Society*, 35: 771–823.

Pocock, R.I. ('1932' 1933b). The Black and Brown Bears of Europe and Asia. Part II – The Sloth Bear (*Melursus*), the Himalayan Black Bear (*Selenarctos*) and the Malayan Bear (*Helarctos*). *Journal of the Bombay Natural History Society*, 36(1): 101–138.

Pocock, R.I. ('1933' 1934a). The Civet cats of Asia. *Journal of the Bombay Natural History Society*, 36(2): 423–449.

Pocock, R.I. ('1933' 1934b). The Civet cats of Asia. Part II. *Journal of the Bombay Natural History Society*, 36(3): 629–656.

Pocock, R.I. ('1933' 1934c). The Palm Civets or 'Toddy Cats' of the genera *Paradoxurus* and *Paguma* inhabiting British India. *Journal of the Bombay Natural History Society*, 36(4): 855–877.

Pocock, R.I. ('1934' 1935a). The Palm Civets or 'Toddy Cats' of the genera *Paradoxurus* and *Paguma* inhabiting British India. *Journal of the Bombay Natural History Society*, 37(1): 172–192.

Pocock, R.I. ('1934' 1935b). The Palm Civets or 'Toddy Cats' of the genera *Paradoxurus* and *Paguma* inhabiting British India. *Journal of the Bombay Natural History Society*, 37(2): 314–346.

Pocock, R.I. ('1940' 1941a). Notes on some British Indian Otters, with description of two new subspecies. *Journal of the Bombay Natural History Society*, 41(3): 514–517.

Pocock, R.I. ('1940' 1941b). The Hog-Badgers (*Arctonyx*) of British India. *Journal of the Bombay Natural History Society*, 41(3): 461–469.

Pocock, R.I. (1914). Description of a new species of Goral (*Nemorhaedus*) shot by Captain F. M. Bailey. *Journal of the Bombay Natural History Society*, 23: 32–33.

Pocock, R.I. (1923). The classification of the Sciuridae. *Proceedings of the Zoological Society of London*, 1923: 209–246.

Pocock, R.I. (1928). The langurs, or leaf monkeys, of British India. *Journal of the Bombay Natural History Society*, 32(3): 472–504.

Pocock, R.I. (1929). "Tigers". *Journal of the Bombay Natural History Society*, 33: 505–541.

Pocock, R.I. (1936). The Asiatic wild dog or Dhole (*Cuon javanicus*). *Proceedings of the Zoological Society of London*, 106(1): 33–55.

Pocock, R.I. (1937). The mongooses of British India, including Ceylon and Burma. *Journal of the Bombay Natural History Society*, 39(2): 211–245.

Pocock, R.I. (1939). *Fauna of British India Mammalia Volume 1, 2nd ed.* Taylor and Francis Ltd., London. 463 p.

Pocock, R.I. (1941). *Fauna of British India Mammalia Volume 2, 2nd ed.* Taylor and Francis Ltd., London. 503 p.

Pocock, R.I. (1947a). Comments on some races of the Asiatic Wild Asses (*Microhippus*), with a description of the skull of the Chigetai. *Proceedings of the Zoological Society of London*, 1947: 764–767.

Pocock, R.I. (1947b). Two new local races of the Asiatic Wild Ass. *Journal of the Bombay Natural History Society*, 47: 143–144.

Pocock, R.I. (1948a). The larger deer of British India. Part I. *Journal of the Bombay Natural History Society*, 43(3): 298–317.

Pocock, R.I. (1948b). The larger deer of British India. Part II. *Journal of the Bombay Natural History Society*, 43(4): 553–572.

Przewalski, N.M. (1883). *Third Journey in Central Asia. From Zaisan through Khami into Tibet and to the sources of the Yellow River.* p. 191. [In Russian].

Pucheran, J. (1855). Melanges et nouvelles, Notes mammalogiques. *Revue et magasin de zoologie pure et appliquée,* Ser 2, 7: 392–394.

Raffles, T.S. (1821). Descriptive catalogue of a zoological collection, made on account of the honorable East Indian Company, in the island of Sumatra and its vicinity, under the direction of Sir Thomas Stamford Raffles, Lieutenant-Governor of Fort Marlborough, with additional notices illustrative of the natural history. *Transactions of the Linnaean Society of London*, 13: 239–274.

Rafinesque, C.S. (1814). *Cephalotes teniotis Précis des découvertes et travaux somiologiques de m.r C. S. Rafinesque-Schmaltz entre 1800 et 1814 ou Choix raisonné de ses principales découvertes en zoologie et en botanique, pour servir d'introduction à ses ouvrages futurs.* Royale typographie militaire, aux dépens de l'auteur, Palerme. 55 p.

Rafinesque, C.S. (1832). *Atlantic Journal, and Friend of Knowledge in Eight Numbers: Containing about 160 Original Articles and Tracts on Natural and Historical Sciences, the Description of about 150 New Plants, and 100 New Animals or Fossils; Many Vocabularies of Languages, Historical and Geological Facts, &c. &c. &c. /.* Philadelphia, PA. 212 p.

Robinson, H.C. and C.B. Kloss. (1922). New mammals from French Indo-China and Siam. *Annals and Magazine of Natural History*, Ser 9, 9: 87–99.

Roxburgh, W. (1801). An account of a new species of *Delphinus*, an inhabitant of the Ganges. *Asiatic researches, or, Transactions of the Society Instituted in Bengal for Inquiring into the History and Antiquities, the Arts, Sciences and Literature of Asia*, 7: 170–174, pl. V.

Ryley, K.V. (1914). A new field-mouse from Burma. F. *Mus cookii*, sp. n. *Journal of the Bombay Natural History Society*, 22(4): 663–664.

Saha, S.S. 1981. A new genus and a new species of flying squirrel (Mammalia: Rodentia: Sciuridae) from northwestern India. *Bulletin of the Zoological Survey of India*, 4(3): 331–336.

Sanborn, C.C. (1931). Bats from Polynesia, Melanesia, and Malaysia. *Field Museum of Natural History Publication, Zoological series*, 18: 7–29.

Satunin, K. (1901). Zwei neue Säugthiere aus Transkaukasien. *Zoologischer Anzeiger*, 24: 461–464.

Savi, P. (1822). Osservazioni sopra il Mustietto, o Mustiolo, nouva specie di Topo ragno Toscano. *Nuovo Giorn. De Litterati, Pisa*, 1: 60–71.

Scalter, W.L. (1891). Catalogue of Mammalia in the Indian Museum, Calcutta. Part II. Rodentia, Ungulata, Proboscidea, Hyracoidea, Carnivora, Cetacea, Sirenia, Marsupialia, Monotremata. Indian Museum, Calcutta. xxx + 375 p.

Schinz, H.R. (1825). *Das Tierreich eingeheilt nach dem Bau der Thiere als Grindlage ihrer Naturegeschichte und der Vergleichenden Anatomie von dem Herrn Ritter von Cuvier staatsgrah von Fransreich und bestandiger Secretar der Academie der Wissenchaften U.S.W. Aus dem Franzossichen frey ubersetzt und mit vielen Zustaten versehen. Saugthiere und Vogel.* J.G. Cotta'schen, Stuttgart. Volume 4.

Schinz, H.R. (1844). *Systematisches vertzeichniss aller bis jetzi bekannten Saugethiere oder Synopsis Mammalium nach dem Cuvier'schen system. Volume 1.* Jent und Gassmann, Solothurn.

Schlitter, D.A. and H.W. Stezer. (1972). New rodents (Mammalia: Cricetidae, Muridae) from Iran and Pakistan. *Proceedings of the Biological Society of Washington*, 86: 163–174.

Schneider, (1800). Description and illustration of a new bat from the East Indies. In: Schreber, J.C.D. von (1775–1855). *Die Saeugethiere in Abbildungen nach der Natur, mit Beschreibungen. Part 1.* Wolfgang Walther, Erlangen, Germany.

Schreber, J.C.D. (1774). *Die Säugethiere in Abbildungen nach der Natur, mit Beschreibungen. Vol. 1.* Siegfried Leberecht Crusius, Leipzig.

Schreber, J.C.D. (1777). *Die Säugethiere in Abbildungen nach der Natur, mit Beschreibungen. 1776–1778.* Wolfgang Walther, Erlangen.

Schwarz, E. (1912). Notes on the Malay tigers with description of a new form from Bali. *Annals and Magazine of Natural History*, Ser 8, 10: 324–326.

Schwarz, E. (1913). On a Kangaroo and a new Palm-Civet in the British Museum. *Annals and Magazine of Natural History,* Ser 8, 12: 288–289.

Scully, J. (1881). On the Mammals of Gilgit. *Proceedings of the Zoological Society of London*, 1881: 197–209.

Severcov, N.A. (1873). Vertikal'noye i gorizontal'noye rasprye—deleniye turkestanskikh zhivotnykh [Vertical and horizontal distribution of the animals of Turkestan]. *Izvestiya Imperator skago Obshestva Lyubiteley Yestestvoznaniya, Antropologii i Etno grafii* (*Moskva*), 8(2): 1–157. [in Russian].

Severtzov, N. (1873). The mammals of Turkestan. *Annals and Magazine of Natural History*, Ser 4, 18: 40–57.

Severtzov, N.A. (1879). Vertikal'noe i gorizontal'noe raspredelenie Turkestanskikh zhivotnykh [Vertical and horizontal distribution of Turkestan animals] (in Russian). *Izvestiya imperatorskago obshch. lyubiteley, estestvoznaniya, antropologii i etnografii [sostoyashchego pri Moskovskom universitete], Moscow*, 8(2): i–ii, 1–157.

Shaw, G. (1791). *The Naturalist's Miscellany, or Coloured Figures of Natural Objects. Vol. 2.* Nodder & Co., London.

Shaw, G. (1800a). *General Zoology, or Systematic Natural History with Plates from the First Authorities and Most Selected Specimens, Engraved Principally by Mr. Heath. Volume 1, Part 1 Mammalia.* G. Kearsley, London. 248 p.

Shaw, G. (1800b). *General Zoology, or Systematic Natural History with Plates from the First Authorities and Most Selected Specimens, Engraved Principally by Mr. Heath. Volume 1, Part 2 Mammalia.* G. Kearsley, London. 552 p.

Shaw, G. (1801a). *General Zoology, or Systematic Natural History with Plates from the First Authorities and Most Selected Specimens, Engraved Principally by Mr. Heath. Volume 2, Part 1 Mammalia.* G. Kearsley, London. 226 p.

Shaw, G. (1801b). *General Zoology, or Systematic Natural History with Plates from the First Authorities and Most Selected Specimens, Engraved Principally by Mr. Heath. Volume 2, Part 2 Mammalia.* G. Kearsley, London. 560 p.

Sinha, A., Datta, A., Madhusudan, M.D. and Mishra, C. (2005). *Macaca munzala*: A new species from western Arunachal Pradesh, northeastern India. *International Journal of Primatology*, 26(4): 977–989.

Sinha, Y.P. (1969). A new pipistrelle bat (Mammalia: Chiroptera: Vespertilionidae) from Burma. *Proceedings of the Zoological Society of Calcutta*, 22: 83–86.

Smith, H. ('1826' 1827). Ruminatia. pp. 1–498. In Griffith, E. and others, *The Animal Kingdom Arranged in Conformity with Its Organization, by the Baron (G) Cuvier, with Additional Descriptions by Edward Griffith and others*, Volume 4. George B. Whittaker, London.

Sparrman, A. (1778). Beskrifning på Sciurus Bicolor, et nytt Species Ikorn, från Java, insänd. Götheborgska Wetenskaps och Witterhets Samhällets Handlingar, *Wetenskaps Afdelningen*, 1: 70–71.

Spitzenberger, F. (2006). *Plecotus strelkovi* sp. nov. 207–209. In: Spitzenberger, F., P.P. Strelkov, H. Winkler and E. Haring. (Eds.), *A Preliminary Revision of the Genus* Plecotus *(Chiroptera, Vespertilionidae) Based on Genetic and Morphological Results: Zoological Scripta*, 35: 187–230.

Srinivasulu, C. and B. Srinivasulu. (2006). First record of *Hipposideros ater* Templeton, 1848 from Andhra Pradesh, India, with a description of a new subspecies. *Zoos' Print Journal*, 21(5): 2241–2244.

Srinivasulu, C. and B. Srinivasulu. (2012). *South Asian Mammals: Their Diversity, Distribution and Status.* Springer, New York. xii + 468 p.

Strickland, H.E. (1841). *Proposed Plan for Rendering the Nomenclature of Zoology Uniform and Permanent (Draft, September 1841),* vol. [Strickland Correspondence, N-089]. Richard & John E. Taylor, London.

Strickland, H.E. (1842). *Proposed Report of the Committee on Zoological Nomenclature: For the Use of the Members of the Committee*, vol. [Strickland Correspondence, N-119]. Richard & John E. Taylor, London.

Sundevall, C.J. (1842). Om Professor J. Hedenborgs insamlingar af Daggdjur i Nordostra Africa och Arabien. Kongl. *Svenska vetenskaps-akademiens handlingar,* Ser 3(30–31): 189–244.

Swinehoe, R. (1870). Catalogue of the mammals of China (South of river Yangtsze and of the Island of Formosa. *Proceedings of the Zoological Society of London*, 1870: 615–652.

Sykes, W.H. (1831). Catalogue of the Mammalia of Dukhun (Deccan); with observations on their habits; &c; and characters of new species. *Proceedings of the Committee of Science and Correspondence of the Zoological Society of London*, 1830–1831: 99–105.

Taylor, J. (1891). Description of a new species of Palm-civet (*Paradoxurus*) found in Orissa. *Journal of the Bombay Natural History Society*, 6(4): 429–432.

Temminck, C.J. (1834). Over een geslacht der vleugelhandige zoogdieren, *Bladneus* Genaamd. (*Rhinolophus* Geoff., Cuv., Illig., Desm.; *Vespertilio* Linn., Erxleb.; *Noctilio* Kuhl). *Tijdschrift voor natuurlijke geschiedenis en physiologie*, 1: 1–30.

Temminck, C.J. (1835). *Monographies de mammalogie, ou Description de quelques genres de mammifères, dont les espèces ont été observées dans les différens musées de l'Europe; par C. J. Temminck. Ouvrage accompagné de planches d'ostéologie, pouvant servir de suite et de complment aux notices sur les animaux vivans, publiées par M. le baron G. Cuvier, dans ses Recherches sur les ossemens fossiles.* C.C. Van der Hoek, Leiden et E. d'Ocagne, Paris.

Temminck, C.J. (1838). Over de geslachten *Taphozous, Emballonura, Urocryptus* en *Disclidurus. Tijdschrift voor natuurlijke geschiedenis en physiologie*, 5: 1–34.

Temminck, C.J. (1840). Treizieme monographies sur les cheiropteres vespertilionides formant les genres Nyctice, Vespertilion et Furie. 141–272. In *Monographies de mammalogie, ou Description de quelques genres de mammifères, dont les espèces ont été observées dans les différens musées de l'Europe* (1835–41). C.C. van der Hock, Leiden; E. d'Ocagne et A. Bertrand, Paris. 2: 1–392 pp, pls. 28–70.

Temminck, C.J. (1842–1844). Aperçu general et spécifique sur les mammifères qui habitant le Japon et les îles que en dependent: 159, pls. 120. In: P.F. de Siebold, C.J. Temminck and H. Schlegel Fauna Japonica. Arnz et Socii, Lugduni Batavorum.

Temminck, C.J. (1853). Les Mammiferes. *Esquisses zoologiques sur la Côte de Guine. Vol. 1.* Brill, Leiden. 256 p.

Temminck, C.J. (publ.) (1839–1847). Verhandelingen over de Natuurlijke Geschiedenis der Nederlandsche overzeesche bezittingen, door de leden der Natuurkundige commissie in Indië en andere schrijvers. Uitgegeven op last van den Koning door C. J. Temminck. Geredigeerd door J. A. Suzanna. La Lau, Leiden. 3 volumes.

Templeton, R. (1848). In: Blyth, E. Report of Curator Zoological Department. *Journal of the Asiatic Society of Bengal*, 17(1): 247–255.

Thomas, O. (1881). Description of a new species of Mus from Southern India. *Annals and Magazine of Natural History*, Ser 5, 7: 24.

Thomas, O. (1882). Description of a new species of rat from China. *Proceedings of the Zoological Society of London*, 1882: 587–588.

Thomas, O. (1886). On the mammals presented by Allan O Hume, Esq. C.B., to the Nautral History Museum. *Proceedings of the Zoological Society of London*, 1886: 54–79.

Thomas, O. (1887). Description of a new rat from north Borneo. *Annals and Magazine of Natural History*, Ser 5, 20: 269–270.

Thomas, O. (1888). On *Eupetaurus*, a new form of Flying Squirrel from Kashmir. *Journal of the Asiatic Society of Bengal*, V. 57, Pt. 2 (3): 256–260.

Thomas, O. (1891a). Diagnoses of three new Mammals collected by Signor L. Fea in the Carin Hills, Burma. *Annali del Museo civico di storia naturale di Genova*, Ser 2, 10: 884.

Thomas, O. (1891b). On the Mammalia collected by Signor Leonardo Fea in Burma and Tenasserim. *Annali del Museo civico di storia naturale di Genova*, Ser 2, 10: 913–949.

Thomas, O. (1895). On the representatives of *Putorius ermineus* in Algeria and Ferghana. *Annals and Magazine of Natural History,* Ser 6, 15: 451–454.

Thomas, O. (1897). On two new rodents from V, Kurdistan. *Annals and Magazine of Natural History*, Ser. 6, 20: 308–310.

Thomas, O. (1902). On two new mammals from China. *Annals and Magazine of Natural History*, Ser 7, 10: 163–166.

Thomas, O. (1903). On the species of the genus *Rhinopoma. Annals and Magazine of Natural History,* Ser 7, 11: 496–499.

Thomas, O. (1905). On a collection of mammals from Persia and Armenia presented to the British Museum by Col. A. C. Bailward. *Proceedings of the Zoological Society of London 1905*, 2: 519–527.

Thomas, O. (1907). On mammals from Northern Persia, presented to the National Museum by Col. A. C. Bailward. *Annals and Magazine of Natural History,* Ser 7, 20: 196–202.

Thomas, O. (1911a). *Abstract of the Proceedings of the Zoological Society of London*. October 24th, 1911. No. 100: 48–50.
Thomas, O. (1911b). Mammals collected in the provinces of Kan-su and Sze-chwan, Western China by Mr. Malcom Anderson, for the Duke of Bedford's Exploration of Eastern Asia. *Abstract of the Proceedings of the Zoological Society of London*, 1911: 3–5.
Thomas, O. (1911c). New Asiatic Muridae. *Annals and Magazine of Natural History*, Ser 8, 7: 205–209.
Thomas, O. (1911d). The Duke of Bedford's Exploration of Eastern Asia—XIII. On mammals from the provinces of Kan-su and Sze-chwan, Western China. *Proceedings of the Zoological Society of London*, 1911: 158–180.
Thomas, O. (1912a). The Duke of Bedford's Zoological exploration of Eastern Asia. XV. On mammals from the provinces of Szechwan and Yunnan, Western China. *Proceedings of the Zoological Society of London*, 1912: 127 É.141.
Thomas, O. (1912b). Two new Asiatic voles. *Annals and Magazine of Natural History*, Ser 8, 9: 348–350.
Thomas, O. (1913a). A new shrew from the Andaman Islands. *Annals and Magazine of Natural History,* Ser 8, 11: 468–469.
Thomas, O. (1913b). On a remarkable new Free-tailed Bat from Southern Bombay. *Journal of the Bombay Natural History Society*, 22(1): 87–91.
Thomas, O. (1913c). Some new Ferœ from Asia and Africa. *Annals and Magazine of Natural History*, (8)12: 88–92.
Thomas, O. (1914a). A new *Soriculus* from the Mishmi Hills. *Journal of the Bombay Natural History Society*, 22: 683.
Thomas, O. (1914b). On small mammals collected in Tibet and the Mishmi Hills by Capt. F.M. Bailey. *Journal of the Bombay Natural History Society*, 23(2): 230–233.
Thomas, O. (1915a). On bats of the genera *Nyctalus, Tylonycteris* and *Pipistrellus. Annals and Magazine of Natural History*, Ser 8, 15: 225–232.
Thomas, O. (1915b). Scientific results from the Mammal Survey X—A. The Indian bats assigned to the genus *Myotis*. *Journal of the Bombay Natural History Society*, 23: 607–612.
Thomas, O. (1915c). Scientific results from the Mammal Survey XI—A. On pipistrels of the genera *Pipistrellus* and *Scotozous*. *Journal of the Bombay Natural History Society*, 24(1): 29–34.
Thomas, O. (1915d). Scientific results from the Mammal Survey XI—G. A second species *Coelomys* from Ceylon. *Journal of the Bombay Natural History Society,* 24(1): 49–50.
Thomas, O. (1915e). Scientific results from the Mammal Survey XI—I. On some specimens of *Vandeleuria* from Bengal, Bihar and Orissa. *Journal of the Bombay Natural History Society*, 24(1): 54–55.
Thomas, O. (1915f). Scientific results from the Mammal Survey XI—K. Notes on *Taphozous* and *Saccolaimus*. *Journal of the Bombay Natural History Society*, 24(1): 57–63.
Thomas, O. (1915g). Scientific results of the Mammal Survey: A new murine Genus and Species from Ceylon. *Journal of the Bombay Natural History Society*, 23(3): 414–416.
Thomas, O. (1916a). List of Microchiroptera, other than leaf-nosed bats, in the collection of the Federated Malay State Museum. *Journal of the Federated Malay States Museum*, 7: 1–6.
Thomas, O. (1916b). Scientific results from the Mammal Survey XIII—A. On Muridae from Darjiling and the Chin Hills. *Journal of the Bombay Natural History Society*, 24: 404–415.
Thomas, O. (1916c). Scientific results from the Mammal Survey XIII—B. Two new Indian bats. *Journal of the Bombay Natural History Society*, 24: 415–417.
Thomas, O. (1916d). Scientific results from the Mammal Survey XIII—C. On squirrels of the genus *Dremomys*. *Journal of the Bombay Natural History Society*, 24: 417–418.

Thomas, O. (1916e). Scientific results from the Mammal Survey XIII—D. The squirrels of the *Tomeutes lokroides* and *mearsi* groups. *Journal of the Bombay Natural History Society*, 24: 419–422.

Thomas, O. (1916f). Scientific results from the Mammal Survey XIV—A. A new bat of the genus *Murina* from Darjiling. *Journal of the Bombay Natural History Society*, 24: 639–680.

Thomas, O. (1916g). Scientific results from the Mammal Survey. No. XIII. On Muridae from Darjiling and Chin Hills. *Journal of the Bombay Natural History Society*, 24(3): 404–407.

Thomas, O. (1916h). Scientific results from the Mammal Survey. No. XIII. A new rat allied to *Epimys sabanus*, from Darjiling. *Journal of the Bombay Natural History Society*, 24(3): 407–409.

Thomas, O. (1916i). Scientific results from the Mammal Survey. No. XIII. On the large rats allied to *Epimys bowersi*. *Journal of the Bombay Natural History Society*, 24(3): 409–412.

Thomas, O. (1916j). Scientific results from the Mammal Survey. No. XIII. The rats of the *Epimys berdmorei* group. *Journal of the Bombay Natural History Society*, 24(3): 412–414.

Thomas, O. (1916k). Scientific results from the Mammal Survey. No. XIII. A new mouse from Sikkim. *Journal of the Bombay Natural History Society*, 24(3): 414–415.

Thomas, O. (1917a). On the small hamsters that have been referred to *Cricetulus phaeus* and *campbelli*. *Annals and Magazine of Natural History*, Ser 8, 19: 452–457.

Thomas, O. (1917b). Scientific results from the Mammal Survey. No. XVI. A new genus of Muridae. *Journal of the Bombay Natural History Society*, 25(2): 203–205.

Thomas, O. (1919). Scientific results from the Mammal Survey XX—A. Notes on the genus *Cheliones*. *Journal of the Bombay Natural History Society*, 26(3): 726–727.

Thomas, O. (1920). Scientific Results from the Mammal Survey, XXI. A. Some new mammals from Baluchistan and Northwest India. *Journal of the Bombay Natural History Society*, 26(4): 933–940.

Thomas, O. (1921). Scientific results from the Mammal Survey XXX. The mongooses of the *Herpestes smithii* group. *Journal of the Bombay Natural History Society*, 28(1): 23–26.

Thomas, O. (1922a). On some new forms of *Ochotona*. *Annals and Magazine of Natural History*, Ser 9, 9: 187–193.

Thomas, O. (1922b). Scientific results from the Mammal Survey XXXII—C. A new Ferret Badger (*Helictis*) from the Naga Hills. *Journal of the Bombay Natural History Society*, 28: 432.

Thomas, O. (1923a). On mammals from the Li-Kiang Range, Yunnan being a further collection obtained by Mr. George Forrest. *Annals and Magazine of Natural History*, Ser 9, 11: 655–663.

Thomas, O. (1923b). Scientific results from the Mammal Survey XL. A new mouse from Madura, S. India. *Journal of the Bombay Natural History Society*, 29(1): 87.

Thomas, O. (1923c). Scientific results from the Mammal Survey XLI. On the forms contained in the genus *Harpiocephalus*. *Journal of the Bombay Natural History Society*, 29(1): 88–89.

Thomas, O. (1924a). A new genus and species of shrew from Ceylon. *Spolia Zeylanica*, 13(1): 93–95.

Thomas, O. (1924b). On some Ceylon mammals. *Annals and Magazine of Natural History*, Ser 9, 13: 239–242.

Thomas, O. (1924c). Scientific results from the Mammal Survey. No. XLIV. On a new Field-Mouse from Nepal, with a note on the classification of the Genus *Apodemus*. *Journal of the Bombay Natural History Society*, 29(4): 888–889.

Thomas, O. and M.A.C. Hinton. (1922). The mammals of the 1921 Mount Everest Expedition. *Annals and Magazine of Natural History*, 9(9): 178–186.

Thomas, O. and R.C. Wroughton. (1915a). Scientific results from the Mammal Survey XI—B. The Giant Squirrels of Ceylon. *Journal of the Bombay Natural History Society*, 24(1): 34–37.

Thomas, O. and R.C. Wroughton. (1915b). Scientific results from the Mammal Survey XI—C. The Singhalese species of *Funambulus. Journal of the Bombay Natural History Society*, 24(1): 37–41.

Thonglongya, K. (1972). A new genus and species of fruit bat from South India (Chiroptera: Pteropodidae). *Journal of the Bombay Natural History Society*, 69(1): 151.

Tomes, R.F. (1856). In: Blyth, E. Memoir on Indian species of shrews. *Annals and Magazine of Natural History*, Ser 2, 17: 11–28.

Tomes, R.F. (1857). Descriptions of four undescribed species of bats. *Proceedings of the Zoological Society of London*, Pt 25: 50–54.

Tomes, R.F. (1859). Descriptions of six hitherto undescribed species of bats. *Proceedings of the Zoological Society of London*, Pt 27: 68–79.

Topál, G. (1997). A new mouse-eared bat species, from Nepal, with statistical analyses of some other species of subgenus *Leuconoe* (Chiroptera, Vespertilionidae). *Acta Zoologica Academiae Scientiarum Hungaricae*, 43(4): 375–402.

True, F.W. (1894). Notes on mammals of Baltistan and the vale of Kashmir, presented to the National Museum by Dr. W.L. Abbott. *Proceedings of the United States National Museum*, 17: 1–16.

Tytler, R.C. (1864). Description of a new species of *Pardoxurus* from Andaman Islands. *Journal of the Asiatic Society of Bengal*, 33: 188.

Vahl, M. (1797). Beskrivelse paa nye Arter Flagermuse. *Skrivter af Naturhistorie-Selskabet, Copenhagen*, 4(1): 121–138.

Vigors, N.A. and T. Horsfield. (1827). Descriptions of two species of the genus *Felis*, in the collection of the Zoological Society. *Zoological Journal*, 3: 449–451.

von Kiesenwetter, E. (1858). Gesetze der entomologischen nomenclatur. *Berliner entomologische Zeitschrift*, 2: 11–22.

Wagner, A. (1839a). Beschreibung einiger neuer oder wenig bekannten Säugetiere welche von herrn Baron Von Hugel in Indien gesammelt wurden. *Gelehrte Anzeiger der Bayerische Akademie der Wiessenschaften, Munchen*, 9(183): 429–432.

Wagner, A. (1839b). Beschreibung einiger neuer oder wenig bekannten Säugethiere. *Gelehrte Anzeiger der Bayerische Akademie der Wiessenschaften, Munchen*, 9(184): 433–440.

Wagner, A. (1841). Das peguanische Spitzhörnchen. pp. 42–43. In: *Die Säugethiere in Abbildungen nach der Natur mit Beschreibungen. Supplementband 2. Erlangen: Expedition des Schreber'schen Säugethier- und des Esper'schen Schmetterlingswerkes.*

Waterhouse, G. (1837). Observations on Palm Squirrel (*Sciùrus palmàrum of Authors*). *Charlseworth's The Magazine of Natural History and Journal of Zoology, Botany, Mineralogy, Geology and Meteorology, New Series*, 1: 496–499.

Waterhouse, G. (1841). Lepus tibetanus. *Proceedings of the Zoological Society of London*, Pt. 9: 7–8.

Waterhouse, G. (1850). Description of a new species of *Tupaia* discovered in continental India by Walter Elliot Esq. *Proceedings of the Zoological Society of London*, 1849: 106–108.

Waterhouse, G.R. (1838). On two new species of Mammalia, from Society's collection belonging to the genera *Gerbillus* and *Herpestes. Proceedings of the Zoological Society of London*, 1838(6): 55–56.

Wroughton, R.C. (1899). On some Konkan bats. *Journal of the Bombay Natural History Society*, 12(4): 716–725.

Wroughton, R.C. (1905). "The" Common striped palm squirrel. *Journal of the Bombay Natural History Society*, 16(3): 406–413.

Wroughton, R.C. (1912a). Some new Indian mammals. *Journal of the Bombay Natural History Society*, 21(3): 767–773.

Wroughton, R.C. (1912b). Some new Indian rodents. *Journal of the Bombay Natural History Society*, 21(1): 338–342.

Wroughton, R.C. (1912c). Some new Indian rodents. *Journal of the Bombay Natural History Society*, 21(3): 767–773.

Wroughton, R.C. (1915a). Scientific results from the Mammal Survey XI—D. The Ceylon Hare. *Journal of the Bombay Natural History Society*, 24(1): 41–42.

Wroughton, R.C. (1915b). Scientific results from the Mammal Survey XI—E. The Indian Ribbed-faced Deer or Muntjac. *Journal of the Bombay Natural History Society*, 24(1): 42–46.

Wroughton, R.C. (1915c). Scientific results from the Mammal Survey XI—F. The genus *Epimys* in Ceylon. *Journal of the Bombay Natural History Society*, 24(1): 46–49.

Wroughton, R.C. (1915d). Scientific results from the Mammal Survey XI—H. The common Indian mongoose. *Journal of the Bombay Natural History Society*, 24(1): 50–54.

Wroughton, R.C. (1916a). Scientific Results from the Mammal Survey, XIII. A. On Muridae from Darjiling and Chin Hills. *Journal of the Bombay Natural History Society*, 24(3): 404–415.

Wroughton, R.C. (1916b). Scientific results from the Mammal Survey XIII—G. New rodents of Sikkim. *Journal of the Bombay Natural History Society*, 24: 424–430.

Wroughton, R.C. (1916c). Scientific results from the Mammal Survey XIV—D. The squirrels of the *Funambulus palmarum–trisriatus* group in the peninsula. *Journal of the Bombay Natural History Society*, 24: 644–649.

Wroughton, R.C. (1916d). Scientific results from the Mammal Survey XIV—E. The Indian Jackals. *Journal of the Bombay Natural History Society*, 24: 649–653.

Wroughton, R.C. (1916e). Scientific results from the Mammal Survey XIV—F. The langurs of Assam. *Journal of the Bombay Natural History Society*, 24: 653–655.

Wroughton, R.C. (1919). Scientific results from the Mammal Survey XX—D. On the genus *Tadarida* (Wrinkle-lip Bats). *Journal of the Bombay Natural History Society*, 26(3): 731–733.

Wroughton, R.C. (1921). Scientific results from the Mammal Survey XXVI—B. A new Palm-civet from Assam. *Journal of the Bombay Natural History Society*, 27: 600–601.

Wroughton, R.C. and K.V. Ryley. (1913a). Scientific results from the Mammal Survey III—A. A new species of *Myotis* from Kanara. *Journal of the Bombay Natural History Society*, 22(1): 13–14.

Wroughton, R.C. and K.V. Ryley. (1913b). Scientific results from the Mammal Survey III—B. A new species of *Kerivoula* from N. W. Mysore. *Journal of the Bombay Natural History Society*, 22(1): 14–15.

Wroughton, R.C. and W.M. Davidson. (1919c). Scientific results from the Mammal Survey XX—C. Two new forms of the "*Funambulus tristriatus*" group. *Journal of the Bombay Natural History Society*, 26(3): 728–730.

Yenbutra, S. and H. Felten. (1983). A new species of the fruit bat genus *Megaerops* from SE Asia [Mammalia: Chiroptera: Pteropodidae]. *Senckenbergiana Biologica*, 64: 1–11.

Zalkin, V. (1945). On the taxonomic position of *Capra falconeri* Wag. In the USSR. *Comptes Rendus de l'Académie des Sciences URSS, A*, 46(5): 231.

Zelebor, J. (1869). *Säugethiere, Vol. 1. Wirbelthiere*. 1–42 pp mit 3 Tafeln. In: Scherzer K. Ritter von (Ed.), *Reise der oesterreichischen Fregatte Novara um die Erde: in den Jahren 1857, 1858, 1859, unter den Befehlen des Commodore B. von Wüllerstorf-Urbair.* Kaiserlich-Königlichen Hof- und Staatsdruckerei, Wien. 819 p.

Ziegler, R. (2003). Bats (Chiroptera, Mammalia) from Middle Miocene karstic fissure fillings of Petersbuch near Eichstätt, Southern Franconian Alb (Bavaria). *Geobios*, 36: 447–490.

Zimmermann, E.A.W., von. (1780). Geographische Geschichte des Menschen, und der allgemein verbreiteten vierfussigen Thiere, nebst einer Hieher gehorigen zoologischen Weltcharte. Vol. 2. Geographische Geschichte des Menschen, und der vierfussigen Thiere. Zweiter Band. Enthalt ein vollstandiges. Verzeichniss aller bekannten Quadrupeden. Weygandschen Buchhandlung, Leipzig.

Zukowsky, L. (1914). Drei neue Kleinkatzenrassen aus Westasien. *Archiv für Naturgeschichte, Berlin*, 80(10): 124–142.

Index

Note: Page numbers in italic refer to figures.